BIOMEDICAL
TEXTURE ANALYSIS

THE ELSEVIER AND MICCAI SOCIETY BOOK SERIES

Advisory Board

Titles:

Balocco, A., et al., Computing and Visualization for Intravascular Imaging and Computer Assisted Stenting, 9780128110188

Dalca, A.V., et al., Imaging Genetics, 9780128139684

Depeursinge, A., et al., Biomedical Texture Analysis, 9780128121337

Munsell, B., et al., Connectomics, 9780128138380

Pennec, X., et al., Riemannian Geometric Statistics in Medical Image Analysis, 9780128147252

Wu, G., and Sabuncu, M., Machine Learning and Medical Imaging, 9780128040768

Zhou, S.K., Medical Image Recognition, Segmentation and Parsing, 9780128025819

Zhou, S.K., et al., Deep Learning for Medical Image Analysis, 9780128104088

Zhou, S.K., et al., Handbook of Medical Image Computing and Computer Assisted Intervention, 9780128161760

MICCAI

BIOMEDICAL TEXTURE ANALYSIS

Fundamentals, Tools and Challenges

Edited by

ADRIEN DEPEURSINGE
University of Applied Sciences Western Switzerland
(HES-SO)

OMAR S. AL-KADI
King Abdullah II School for Information Technology,
University of Jordan

J. ROSS MITCHELL
Mayo Clinic College of Medicine,
and Department of Biomedical Informatics
at Arizona State University, USA

ELSEVIER

ACADEMIC PRESS
An imprint of Elsevier

Academic Press is an imprint of Elsevier
125 London Wall, London EC2Y 5AS, United Kingdom
525 B Street, Suite 1800, San Diego, CA 92101-4495, United States
50 Hampshire Street, 5th Floor, Cambridge, MA 02139, United States
The Boulevard, Langford Lane, Kidlington, Oxford OX5 1GB, United Kingdom

Notices

Knowledge and best practice in this field are constantly changing. As new research and experience broaden our understanding, changes in research methods, professional practices, or medical treatment may become necessary.

Practitioners and researchers must always rely on their own experience and knowledge in evaluating and using any information, methods, compounds, or experiments described herein. In using such information or methods they should be mindful of their own safety and the safety of others, including parties for whom they have a professional responsibility.

To the fullest extent of the law, neither the Publisher nor the authors, contributors, or editors, assume any liability for any injury and/or damage to persons or property as a matter of products liability, negligence or otherwise, or from any use or operation of any methods, products, instructions, or ideas contained in the material herein.

Library of Congress Cataloging-in-Publication Data
A catalog record for this book is available from the Library of Congress

British Library Cataloguing-in-Publication Data
A catalogue record for this book is available from the British Library

ISBN: 978-0-12-812133-7

For information on all Academic Press publications
visit our website at https://www.elsevier.com/books-and-journals

Working together
to grow libraries in
developing countries

www.elsevier.com • www.bookaid.org

Publisher: Mara Conner
Acquisition Editor: Tim Pitts
Editorial Project Manager: Andrae Akeh
Production Project Manager: Paul Prasad Chandramohan
Designer: Victoria Pearson Esser

Typeset by VTeX

CONTENTS

4. Deep Learning in Texture Analysis and Its Application to Tissue Image Classification   **95**
Vincent Andrearczyk, Paul F. Whelan

5. Fractals for Biomedical Texture Analysis   **131**
Omar S. Al-Kadi

PREFACE

Computerized recognition and quantification of texture information has been an active research domain for the past 50 years, with some of the pioneering work still widely used today. Recently, the increasing ubiquity of imaging data has driven the need for powerful image analysis approaches to convert this data into knowledge. One of the most promising application domains is biomedical imaging, which is a key enabling technology for precision medicine (e.g., radiomics and digital histopathology) and biomedical discovery (e.g., microscopy). The colossal research efforts and progress made in the general domain of computer vision have led to extremely powerful data analysis systems. Biomedical imaging relies upon well-defined acquisition protocols to produce images. This is quite different from general photography. Consequently, the analysis of biomedical images requires a paradigm change to account for the quantitative nature of the imaging process. Texture analysis is a broadly applicable, powerful technology for quantitative analysis of biomedical images.

The aims of this book are:

- Define biomedical texture precisely and describe how it is different from general texture information considered in computer vision;
- Define the general problem to translate 2D and 3D texture patterns from biomedical images to visually and biologically relevant measurements;
- Describe with intuitive concepts how the most popular biomedical texture analysis approaches (e.g., gray-level matrices, fractals, wavelets, deep convolutional neural networks) work, what they have in common, and how they are different;
- Identify the strengths, weaknesses, and current challenges of existing methods including both handcrafted and learned representations, as well as deep learning. The goal is to establish foundations for building the next generation of biomedical texture operators;
- Showcase applications where biomedical texture analysis has succeeded and failed;
- Provide details on existing, freely available texture analysis software. This will help experts in medicine or biology develop and test precise research hypothesis.

This book provides a thorough background on texture analysis for graduate students, and biomedical engineers from both industry and academia who have basic image processing knowledge. Medical doctors and biologists with no background in image processing will also find available methods and software tools for analyzing textures in medical images.

By bringing together experts in data science, medicine, and biology, we hope that this book will actively promote the translation of incredibly powerful data analysis methods into several breakthroughs in biomedical discovery and noninvasive precision medicine.

Adrien Depeursinge, Omar S. Al-Kadi, J. Ross Mitchell

CHAPTER 1

Fundamentals of Texture Processing for Biomedical Image Analysis
A General Definition and Problem Formulation

Adrien Depeursinge*, Julien Fageot[†], Omar S. Al-Kadi[‡]

*École Polytechnique Fédérale de Lausanne (EPFL), Biomedical Imaging Group, Lausanne, Switzerland
[†]University of Applied Sciences Western Switzerland (HES-SO), Institute of Information Systems, Sierre, Switzerland
[‡]University of Jordan, King Abdullah II School for Information Technology, Amman, Jordan

Abstract

This chapter aims to provide an overview of the foundations of texture processing for biomedical image analysis. Its purpose is to define precisely what biomedical texture is, how is it different from general texture information considered in computer vision, and what is the general problem formulation to translate 2D and 3D textured patterns from biomedical images to visually and biologically relevant measurements. First, a formal definition of biomedical texture information is proposed from both perceptual and mathematical point of views. Second, a general problem formulation for biomedical texture analysis is introduced, considering that any approach can be characterized as a set of local texture operators and regional aggregation functions. The operators allow locally isolating desired texture information in terms of spatial scales and directions of a texture image. The type of desirable operator invariances are discussed, and are found to be different from photographic image analysis. Scalar-valued texture measurements are obtained by aggregating operator's response maps over regions of interest.

Keywords

Quantitative image analysis, Spatial stochastic process, Texton, Heterogeneity, Texture analysis

1.1 INTRODUCTION

Everybody agrees that nobody agrees on the definition of *texture* information. According to the Oxford Dictionaries[1] texture is defined as "the feel, appearance, or consistency of a surface or a substance." The context in which the word *texture* is used is fundamental to attach unambiguous semantics to its meaning. It has been widely used in extremely diverse domains to qualify the properties of images, food, materials, and even music. In the context of food and sometimes material sciences, characterizing texture information often involves measuring the response of the matter subject to forces such as shearing, cutting, compressing, and chewing [1,2]. Starting from the early

---

[1] https://en.oxforddictionaries.com/definition/texture, as of October 10, 2016.

Biomedical Texture Analysis
DOI: 10.1016/B978-0-12-812133-7.00001-6

developmental months of newborn babies, tactile perception of textured surfaces is an important stage of the human brain development [3]. It is an essential step to be successfully acquainted with the physical properties of the surrounding environment [4]. To some extent, estimating the properties and diversity of the latter without having to touch every surface can be efficiently carried out through vision. Human vision learns to recognize texture patterns through extensive experimentation confronting visual and tactile perception of textured surfaces [5]. This provides hints on why human visual texture recognition performs much beyond the use of low level descriptive terms such as *coarse*, *edgy*, *directional*, *repetitive*, and *random*.

In the context of biomedical imaging, texture information relates to the micro- and macro-structural properties of biomedical tissue. Radiologists, pathologists, and biologists are trained to establish links between visual image patterns and underlying cellular and molecular content of tissue samples [6]. Unfortunately, very large variations of this complex mapping occur, resulting in image interpretation errors with potentially undesirable consequences [7–9]. These variations are partly due to the diversity of human biology and anatomy as well as image acquisition protocols and reconstruction, compounded by observer training. Important efforts were initiated by medical imaging associations to construct unified terminologies and grading scores in the context of radiology and histopathology, aiming to limit variations in image interpretation and reporting [10–13]. However, in the particular context of biomedical texture information, the terms used (*e.g.*, *heterogeneous enhancement*, *hypervascular* [12]) are often as inadequate as low level descriptive terms of general textures (*e.g.*, *coarse*, *edgy*) while the perception of human observers is much richer (see Sections 9.4.1 and 9.4.2 of Chapter 9). When considering three-dimensional architectures of biomedical tissue, human observers have limited intuition of these 3D solid textures, because they cannot be fully visualized [14]. Only virtual navigation in Multi-Planar Rendering (MPR) and semitransparent visualizations are made available by computer graphics and allow observing 2D projections.

Computer-based quantitative image texture analysis has a tremendous potential to reduce image interpretation errors and can make better use of the image content by yielding exhaustive, comprehensive, and reproducible analysis of imaging features in two and three dimensions [15–17]. Nevertheless, besides the lack of a clear definition of biomedical texture information, several challenges remain, such as: the lack of an appropriate framework for multiscale, multispectral analysis in 2D and 3D; validation; and, translation to routine clinical applications. The goal of this book is to illustrate the importance of these aspects and to propose concrete solutions for optimal biomedical texture analysis. This chapter will first propose a definition of texture in the particular context of biomedical imaging (Section 1.2). Second, a general theoretic framework for Biomedical Texture Analysis (BTA) will be proposed in Section 1.3. The latter is designed to best leverage the specific properties of biomedical textures. Differences with the classical texture analysis paradigm in computer vision will be highlighted.

Important aspects of texture operator and aggregation function design will be further discussed and illustrated through several concrete examples in Chapter 2. It will also recapitulate key aspects of biomedical texture processes and analysis, with an aim of raising awareness of limitations of popular texture operators used in the biomedical literature while providing directions to design the next generation of BTA approaches. Chapter 3 will use the comparison axes established in Chapters 1 and 2 to compare most popular modern biomedical texture analysis approaches. With the purpose of guiding neophyte or experienced users, a simple checklist is proposed to assess the relevance of the BTA approach in a particular medical or biological applicative context.

1.2 BIOMEDICAL TEXTURE PROCESSES

This section proposes an extensive definition of biomedical texture information under biological, medical, physical, statistical, and mathematical viewpoints.

1.2.1 Image intensity versus image texture

Low-level quantitative image analysis (*i.e.*, pixel-level[2]) can be separated into two main categories: intensity and texture. Image intensity relates to the statistical distribution of the pixel values inside a defined Region Of Interest (ROI). The pixel values can be either normalized across images (*e.g.*, Hounsfield Units (HU) in X-ray Computed Tomography (CT), Standardized Uptake Values (SUV) in Positron Emission Tomography (PET)), or unnormalized (*e.g.*, Hematoxylin and Eosin (H&E) stains in histopathology, Magnetic Resonance Imaging (MRI)). Classic quantitative measures of image intensity are the four statistical moments of the pixel values' distribution (mean, variance, skewness, and kurtosis). Other measures are specific to the considered imaging modality (*e.g.*, SUV max or Total Lesion Glycolysis (TLG) in PET) [18]. The latter are extremely useful to characterize the image content, but cannot measure the spatial relationships between pixel values (see Fig. 1.1). A qualitative keyword such as *tumor heterogeneity* is ambiguous because it is unclear if the heterogeneity concerns pixel values (intensity) or their spatial organization (texture). It is though commonly used to describe the visual aspect of tumors in radiological images with ambivalent meaning [19–21].

The spatial relationships (*i.e.*, the transitions) between pixel values are precisely what texture information is encoding. Haidekker defined texture as "a systematic local variation of image values" [22]. Petrou stated that "the most important characteristic of texture is that it is scale dependent" and that "different types of texture are visible at different scales" [23]. This highlights the importance of the variation *speed* or *slope* or *oscillation* between pixel values, which will be different in *smooth* versus *rough* textures (see Fig. 1.1 left and right). This first notion of the texture scale relates to the spatial

---

[2] The word *pixel* is used to design both 2D and 3D (*voxels*) image samples.

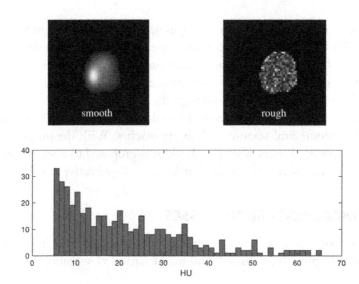

Figure 1.1 The two simulated tumors have identical distribution of the pixel's values and cannot be differentiated using intensity image measures only. They differ in the spatial relationships between the pixels, which is specifically captured by image texture analysis.

frequencies in the image. The higher the spatial frequency, the finer the scale of the transition between proximate pixel values. A second important notion is the direction of the transition. These two notions of spatial scale and direction are fundamental for visual texture discrimination (see Fig. 1.2) [24].

Blakemore et al. provided initial evidence that the human visual system possesses neurons that are selectively sensitive to directional spatial frequencies [25], which has been widely confirmed later on [26]. Most approaches proposed for computerized texture analysis are leveraging these two properties either explicitly (*e.g.*, Gray-Level Cooccurrence Matrices (GLCM) [27], Gray-Level Run Length Matrices (GLRLM) [28], Gray-Level Size Zone Matrices (GLSZM) [29], directional filterbanks and wavelets [30], Histogram of Oriented Gradients (HOG) [31], Local Binary Patterns (LBP) [32], Scattering Transforms (ST) [33,34]) or implicitly (*e.g.*, Convolutional Neural Networks (CNN) [35,36], Dictionary Learning (DL) [37–39]).

A natural mathematical tool to study directional spatial frequency components in D-dimensional signals and images is the Fourier transform and is defined in Eq. (1.1). It is straightforward to see that the Fourier transform for $\boldsymbol{\omega} = \boldsymbol{0}$ computes the mean of the function, which is not considered as texture information since it relates the mean intensity of the pixels in the image. For $\|\boldsymbol{\omega}\| > 0$ the modulus of the Fourier transform quantifies the magnitude of the transitions, where $\|\boldsymbol{\omega}\|$ is inversely proportional to the scale and the orientation of the vector $\boldsymbol{\omega}$ defines the direction of the spatial frequencies.

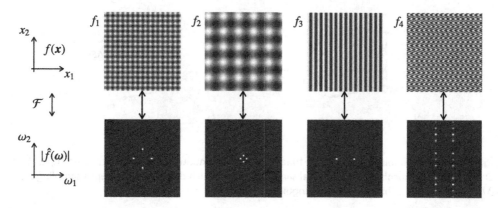

Figure 1.2 Importance of image scales and directions to unequivocally describe texture functions $f(x)$. The latter are efficiently represented by the modulus of their dual Fourier representation $\hat{f}(\omega)$ in terms of their directional frequency components. $\omega = 0$ corresponds to the centers of the images of the bottom row. f_1 and f_2 have identical directional information but differ in image scales. f_1, f_3, and f_4 have identical spatial frequencies along x_1, but differ in their frequency components along x_2.

The correspondence between texture images $f_i(x)$ and their Fourier transforms $\hat{f}_i(\omega)$ is illustrated in Fig. 1.2.

Another important notion of texture information relates to the order of the transition between the pixel values (*i.e.*, the order of directional image derivatives). For instance, first-order (gradient) transitions will characterize step-like transitions, whereas second-order transitions (Hessian) describe local image curvature. For instance, modeling these two types can identify various types of transitions either at the margin or inside a tumor region [12,40].

1.2.2 Notation and sampling

Throughout this section, images are defined on an unbounded continuum. This is motivated by the fact that many physical phenomena are intrinsically continuous and results in analog signals. Also, important mathematical operations arising in texture analysis, including geometric transformations (*e.g.*, shifts, rotations, scalings) and random processes, are better defined in the continuous domain without boundaries. For instance, a discrete rotation is defined as the approximation of its continuous domain counterpart. Moreover, stationary random processes (see Section 1.2.3.1) are implicitly defined over an unbounded domain, since any shifted process should be well-defined. The design of algorithms then calls for the discretization of continuous images, together with their restrictions on bounded domains.

An image is modeled as a D-dimensional function of the variable $x = (x_1, \ldots, x_D) \in \mathbb{R}^D$, taking values $f(x) \in \mathbb{R}$. The Fourier transform of an integrable function $f(x)$ is

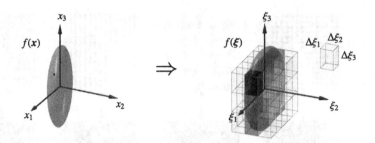

Figure 1.3 Image sampling of 3D texture functions (left: continuous $f(x)$, right: discretized $f(\xi)$). Most 3D bioimaging protocols have equal axial sampling steps, which differ from the depth sampling step ($\Delta\xi_1 = \Delta\xi_2 \neq \Delta\xi_3$). This results in rectangular voxels.

noted $\hat{f}(\boldsymbol{\omega}) \in \mathbb{C}$ and is defined as

$$f(\boldsymbol{x}) \xleftrightarrow{\mathcal{F}} \hat{f}(\boldsymbol{\omega}) = \int_{\mathbb{R}^D} f(\boldsymbol{x})\, e^{-j\langle \boldsymbol{\omega}, \boldsymbol{x}\rangle}\, d\boldsymbol{x}, \tag{1.1}$$

where $\langle \cdot, \cdot \rangle$ denotes the scalar product.

As we said, in practice, images and their Fourier transforms have to be sampled and restricted to bounded domains. We introduce now the notation that goes together with these operations. In what follows, however, we rely mostly on the continuous domain formulation. A D-dimensional discretized image function f of restricted domains is noted as $f(\boldsymbol{\xi}) \in \mathbb{R}$. Pixel positions $\boldsymbol{\xi} = (\xi_1, \ldots, \xi_D)$ are indexed by the vector $\boldsymbol{k} = (k_1, \ldots, k_D)$, the sampling steps $\Delta\xi_1, \ldots, \Delta\xi_D$, and the dimension-wise offsets $c_1, \ldots, c_D$ as

$$\begin{pmatrix} \xi_1 \\ \vdots \\ \xi_D \end{pmatrix} = \begin{pmatrix} \Delta\xi_1 \cdot k_1 \\ \vdots \\ \Delta\xi_D \cdot k_D \end{pmatrix} - \begin{pmatrix} c_1 \\ \vdots \\ c_D \end{pmatrix}, \tag{1.2}$$

with $k_i \in \{1, \ldots, K_i\}$. The domain $\boldsymbol{F} \subset \mathbb{R}^D$ of f is $F_1 \times \cdots \times F_D$ with $F_i = \Delta\xi_i \cdot \{1, \ldots, K_i\} - c_i$. Image sampling of 3D texture functions is illustrated in Fig. 1.3. The sampling steps are most often varying between patients and dimensions. This needs to be taken into account when designing texture operators and is further discussed in Section 1.3.2.

The discrete Fourier transform of $f(\boldsymbol{\xi})$ is noted $\hat{f}(\boldsymbol{v}) \in \mathbb{C}$ and is defined as

$$f(\boldsymbol{\xi}) \xleftrightarrow{\mathcal{F}} \hat{f}(\boldsymbol{v}) = \sum_{\boldsymbol{\xi} \in \boldsymbol{F}} f(\boldsymbol{\xi})\, e^{-j\langle \boldsymbol{v}, \boldsymbol{\xi}\rangle}, \tag{1.3}$$

with $\boldsymbol{v} = (v_1, \ldots, v_D) \in \boldsymbol{\Omega}$ in the discrete Fourier domain. In the context of discrete texture functions, the realization of the texture process is defined on the sampling grid

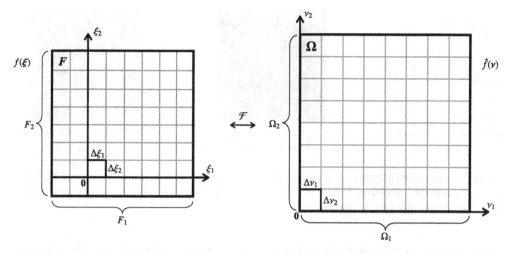

Figure 1.4 Sampled texture function $f(\xi)$, $\xi \in F$, and its corresponding Fourier plane $\hat{f}(v)$, $v \in \Omega$. The number of elements in Ω is the same as in F: $|\Omega| = |F|$. The sampling steps $\{\Delta\xi_1, \Delta\xi_2\}$ of the spatial domain are inversely proportional to the sampling steps $\{\Delta v_1, \Delta v_2\}$ of the Fourier domain: $\Delta v_i = 1/\Delta\xi_i$. Note that 0 need not lie on the grid for the spatial image domain F.

determined by the sampling step Δx_i. Therefore the sharpest possible transition can only be measured on a minimal physical distance corresponding to Δx_i. As a consequence the maximum spatial frequency measurable along the image axes is $\Delta v_i = 1/\Delta x_i$. The domain $\Omega \subset \mathbb{R}^D$ of $\hat{f}$ is $\Omega_1 \times \cdots \times \Omega_D$ with $\Omega_i = \{1, \ldots, K_i\}/\Delta x_i$ (see Fig. 1.4). The number of elements in Ω is the same as in F: $|\Omega| = |F|$.

The discretization implies that sampled images are approximations of continuous images. Moreover, for the sake of simplicity, although the domains F of image functions are defined as bounded, the potential issues raising at the frontiers are not discussed here. The hypothesis that all processing operations are happening far away from the domain boundaries will be made. This will be particularly important when discussing the stationarity of texture processes in Section 1.2.3.1, as well as texture operators and their application to image functions in Section 1.3.1.

1.2.3 Texture functions as realizations of texture processes

Biomedical texture patterns are produced by stochastic biological processes, where randomness results from the diversity of human biology and tissue development. Therefore, it is important to clarify that observed texture functions $f(\boldsymbol{x})$ are realizations of spatial stochastic texture processes of various kind [41]. We define stochastic texture processes of $\mathbb{R}^D$ as

$$\{X(\boldsymbol{x}), \boldsymbol{x} \in \mathbb{R}^D\},$$

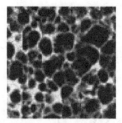

moving average Gaussian
$|F| = 128 \times 128$

pointwise Poisson
$|F| = 32 \times 32$

biomedical: lung fibrosis in CT
$|F| = 84 \times 84$

Figure 1.5 Discretized texture processes $X(\xi)$ of $F \subset \mathbb{R}^2$ (see Section 1.2.2 for a definition of F). PDFs of moving average Gaussian and pointwise Poisson are known, which is never the case for biomedical texture processes (*e.g.*, lung fibrosis in CT).

where $X(x)$ is the value at the position indexed by $x \in \mathbb{R}^D$. The values of $X(x)$ follow a Probability Density Function (PDF) $p_{X(x)}(q)$, where q is the value of the pixel at the position x. Classic examples of such processes are moving average Gaussian or pointwise Poisson (see Fig. 1.5), for which the probability laws are well controlled. However, biomedical texture processes (*e.g.*, lung fibrosis in CT, see Fig. 1.5 right) are much less studied and described in the literature, because we can only fit a PDF from sets of observations containing large variations caused by differences in anatomy of individuals and imaging protocols.

1.2.3.1 Texture stationarity

A spatial stochastic process $X(x)$ is stationary in the strict sense if it has the same probability law than all its shifted versions $X(x - x_0)$ for any shift $x_0 \in \mathbb{R}^D$. As a consequence, the PDF $p_{X(x)}(q) = p_X(q)$ does not depend on the position x anymore.

A discretized stochastic process inherits the stationary property by restricting to discretized shifts. The stationarity is not well-defined at the boundary of the finite domain F of discrete bounded image, but is still meaningful when we are far from the boundaries. For instance, let us consider a patch of dimension 3×3 sliding over the domain F while being fully included in F (no boundary conditions required). For any position of the patch and for each of its nine elements, the statistics of its pixels values will follow the same probability law. In images, texture processes may be piecewise-stationary in a well-defined ROI. Realizations of homo and heteroscedastic (therefore nonstationary) moving average Gaussian processes are shown in Fig. 1.6.

Since the PDFs are not known in the context of biomedical texture processes, strict stationarity must be relaxed to rather assess the similarity between local PDFs. Even more generally, the notion of spatial stationarity of natural texture processes must be redefined in terms of visual perception and/or tissue biology. It is important to note that the notions of spatial stationarity and texture classes are not tightly bound, since natural

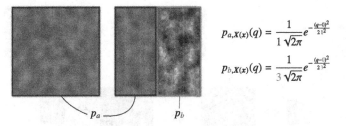

$$p_{a,X(x)}(q) = \frac{1}{1\sqrt{2\pi}}e^{-\frac{(q-0)^2}{2 \cdot 1^2}}$$

$$p_{b,X(x)}(q) = \frac{1}{3\sqrt{2\pi}}e^{-\frac{(q-0)^2}{2 \cdot 3^2}}$$

Figure 1.6 Texture stationarity illustrated with homo (left) and heteroscedastic (center) moving average Gaussian processes. The central image contains the realizations of two texture processes that are piecewise-stationary over the blue or the red region. (For interpretation of the references to color in this figure legend, the reader is referred to the web version of this chapter.)

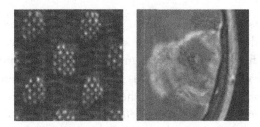

Figure 1.7 Relaxed spatial stationarity of natural texture processes defined in terms of visual perception and/or tissue biology. Left: the Outex texture class "canvas039" [42] consists of two distinct visual patterns. Right: brain glioblastoma multiforme in an axial slice of gadolinium-injected T1-MRI includes both an enhancing region and a necrotic core defining distinct subregions. In these two examples the textures are containing two clear distinct patterns that can be considered homogeneous although not stationary in the strict sense.

texture classes can be constituted by distinct texture processes (see Fig. 1.7). Likewise, biomedical texture classes may be defined based on anatomical or disease levels and can include several cell and tissue types. The image analysis task when the textures processes of interest are stationary relate to texture classification. Texture segmentation of ROIs is considered when the processes are piecewise-stationary.

1.2.4 Primitives and textons

As opposed to stochastic approaches, more deterministic methods for texture recognition consider that homogeneous classes are constituted by sets of fundamental elementary units (*i.e.*, building blocks) called *texture primitives* [43]. Examples of such texture classes are shown in Fig. 1.8. The bottom row of Fig. 1.8 highlights the importance of the geometric relationships between primitives, where f_1 and f_2 differ only in terms of the density and geometric transformations (*e.g.*, local rotations) of the same primitive. Occlusions are observed in the upper left texture of Fig. 1.8.

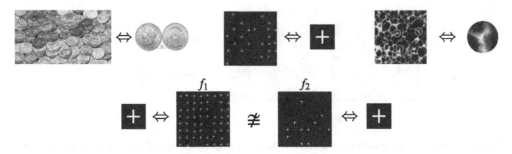

Figure 1.8 Top: various textures and their corresponding primitives or textons. Bottom: importance of the spatial organization (*e.g.*, density, local geometric transformations, occlusions) of the primitives.

Most biomedical texture patterns are of stochastic nature and are constituted by primitives that are neither well-defined, nor known in advance (see Fig. 1.10). A more general definition of fundamental texture microstructures was proposed by Julesz in 1981, who coined the word *texton* [44]. This was one of the earliest texture perception studies of human vision, which laid down the fundamentals of preattentive vision [45, 44,46–48]. The concept of texture being "effortlessly or preattentively discriminable," *i.e.*, the spontaneous perception of texture variations without focused attention, is illustrated in Fig. 1.9A, where the left side area is considered preattentively discriminable while the right side is not and can only be distinguished after careful visual analysis. Similarly, for Fig. 1.9B, the inner square is easily discerned from the outer square. Julesz attributed this to the order of image statistics of texture patterns. Textures having equal second-order statistics in the form of identical joint distribution of gray level pairs, tend to not be preattentively discriminable. An example of this focused attention is the case between the **L**-shaped background and the **T**-shaped right region of Fig. 1.9A. This is known as Julesz conjecture, which he and his colleagues refuted afterward using carefully constructed textures having equal second order and different third and higher order statistics [47]. Nevertheless, this conjecture gave a better understanding of texture perception which led to the proposal of the texton theory. The theory states that texture discrimination is achieved by its primitive or fundamental elements, named textons [44,46], or sometimes called texels (*i.e.*, texture elements) [27]. Several approaches were proposed to derive collections of textons from texture images using texture operators [49,50] (*e.g.*, Gabor, Laplacians of Gaussians, see Section 3.2.2 of Chapter 3), or directly from pixel values spatial organization in local neighborhoods [37,51,38] (*e.g.*, dictionary learning, see Section 3.2.3 of Chapter 3).

Understanding the types of primitives and textons, as well as their (local) spatial organization in biomedical texture classes is of primary importance to efficiently design BTA approaches and in particular texture operators. This is further discussed in Section 1.3 as well as in Chapters 2 and 3.

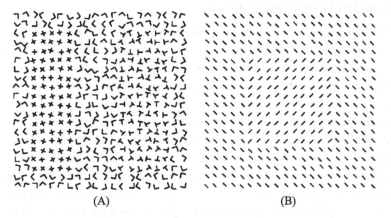

(A)   (B)

Figure 1.9 An example of preattentive discrimination [52]. (A) texture composed of two regions: left +-shaped textons are preattentively (*i.e.*, effortlessly) distinguishable against **L**-shaped textons, while in the right the **T**-shaped textons needs focused attention (*i.e.*, using long-term memory). (B) texture composed of line segments where the difference in orientations segregates preattentively the middle region from the outer region.

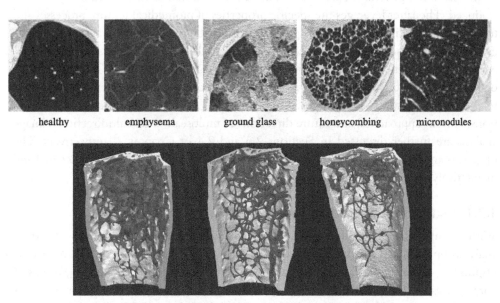

healthy   emphysema   ground glass   honeycombing   micronodules

Isodensity visualization of 3D trabeculae bone microarchitecture.

Figure 1.10 2D and 3D biomedical textures are mainly stochastic and do neither have known nor well-defined primitives. Top: normal and altered 2D lung parenchyma from interstitial lung diseases in axial high-resolution CT [54]. Bottom: 3D bone microarchitecture of normal and osteoporotic trabeculae (blue) in micro CT [55]. (For interpretation of the references to color in this figure legend, the reader is referred to the web version of this chapter.)

1.2.5 Biomedical image modalities

When compared to the broad research field of computer vision, analyzing texture in biomedical images can rely on the highly specific properties of the latter. In particular, most biomedical imaging modalities rely on controlled acquisition protocols allowing the standardization of fundamental image properties such as pixel/voxel size and intensity, as well as image orientation and sample/patient position [56]. The setting is therefore fundamentally different from general computer vision based on photographic imagery resulting from scene captures obtained with varying viewpoints. The properties of most common biomedical imaging protocols are listed in Table 1.1. Imaging protocols with unknown or inaccurate physical size of pixels are rare but exist. Examples of the latter are scanned film-based X-ray radiographs (obsolete), digital photography for dermatology and ophthalmology, and endoscopic videos (see Section 11.3 of Chapter 11).

1.3 BIOMEDICAL TEXTURE ANALYSIS (BTA)

In this section a general theoretic framework for D-dimensional texture analysis is introduced. The latter leverages the specific properties of biomedical texture processes and functions defined in Section 1.2. A general formulation of texture operators and aggregation functions is introduced in Section 1.3.1. Popular texture analysis approaches are exemplified as particular cases of operators and aggregation functions. The importance of intersubject and interdimension scale normalization, as well as operator robustness to rigid transformations is highlighted in Sections 1.3.2 and 1.3.3. Fundamental limitations, current approaches, and future directions of multiscale and multidirectional image analysis are further discussed in Sections 2.3 and 2.4 of Chapter 2, respectively. The operators and aggregation functions of most popular BTA approaches are described and qualitatively compared in Chapter 3.

1.3.1 Texture operators and aggregation functions

Without any loss of generality, we propose to consider that every D-dimensional texture analysis approach can be characterized by a set of N local operators $\mathcal{G}_n$ and their corresponding spatial supports $\mathbf{G}_n \subset \mathbb{R}^D$, assumed to be bounded. The value $\mathcal{G}_n\{f\}(\boldsymbol{x}_0) \in \mathbb{R}$ corresponds to the application of the operator $\mathcal{G}_n$ to the image f at location $\boldsymbol{x}_0$. $\mathcal{G}_n$ is local in the sense that $\mathcal{G}_n\{f\}(\boldsymbol{x}_0)$ only depends on the values of f on the shifted spatial support $\mathbf{G}_n + \boldsymbol{x}_0$ (see Fig. 1.11 for an illustration in 2D).

The operator $\mathcal{G}_n$ is typically applied to the input texture function $f(\boldsymbol{x})$ by *sliding* the spatial support of its function over all positions $\boldsymbol{x}_0$ in $\mathbb{R}^D$. This yields response maps[3]

---

[3] They are commonly called *feature maps* in CNNs.

Table 1.1 Typical specifications of common biomedical imaging modalities

Modality	Signal measured	Dimensionality	Image dimensions	Physical dimensions of pixels/voxels	Image type
40× digital histopathology	light absorption	2D color	$\approx 50{,}000^2$	$\Delta x_1, \Delta x_2$: ≈ 0.275 μm.	structural
Digital radiography	X-ray absorption	2D grayscale	$\approx 2048^2$	$\Delta x_1, \Delta x_2$: ≈ 0.1 mm.	structural
MRI	nuclear magnetic resonance	3D grayscale	$\approx 512^3$	$\Delta x_1, \Delta x_2$: 0.5–1 mm, Δx_3: 0.5–5 mm.	structural
CT	X-ray absorption	3D grayscale	$\approx 512^3$	$\Delta x_1, \Delta x_2$: 0.5–1 mm, Δx_3: 0.5–5 mm.	structural
3-D US	acoustic reflection	3D grayscale	$\approx 256^3$	$\Delta x_1, \Delta x_2, \Delta x_3$: 0.5–1 mm.	structural
μCT	X-ray absorption	3D grayscale	$\approx 512^3$	$\Delta x_1, \Delta x_2, \Delta x_3$: ≈ 1 μm.	structural
PET	γ-rays emitted indirectly by a radiotracer	3D grayscale	$\approx 256^3$	$\Delta x_1, \Delta x_2, \Delta x_3$: 2–5 mm.	functional
OCT	optical scattering	3D grayscale	$\approx 1024^2 \times 256$	$\Delta x_1, \Delta x_2, \Delta x_3$: <10 μm.	structural

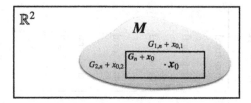

Figure 1.11 At a fixed position x_0, texture functions $f(x)$ are analyzed by local texture operators $\mathcal{G}_n$ with spatial supports $\mathbf{G}_n = G_{1,n} \times \cdots \times G_{D,n}$, where $\mathbf{G}_n \subset \mathbb{R}^D$. When applied to all positions, $x_0 \in \mathbb{R}^D$, $\mathcal{G}_n$ yields response maps $h_n(x_0) = \mathcal{G}_n\{f\}(x_0)$, which can be aggregated over a ROI $\mathbf{M} \subset \mathbb{R}^D$ to obtain a scalar-valued texture feature η_n.

$h_n(\boldsymbol{x}_0)$ as

$$h_n(\boldsymbol{x}_0) = \mathcal{G}_n\{f\}(\boldsymbol{x}_0). \tag{1.4}$$

The structure of the output space of the operator will depend on the desired properties and invariances of operators [57]. The operator $\mathcal{G}_n$ can be linear (*e.g.*, linear filter) or nonlinear (*e.g.*, max, linear filter combined with a rectifier in CNNs, GLCMs, LBPs). Accordingly, the response maps h_n can highlight desired properties of the input texture function (*e.g.*, spatial scales corresponding to a well–defined frequency band, image gradients along x_1, cooccurrences, local binary patterns, or circular frequencies). Examples of response maps are shown in Figs. 1.12 and 1.13. The properties of popular texture operators are discussed and compared in Chapter 3. In the particular case of linear operators, the application of the operator $\mathcal{G}_n$ to the image function $f(\boldsymbol{x})$ at a given position $\boldsymbol{x}_0$ is a *scalar product* between $f(\boldsymbol{x})$ and a function $g_n(\boldsymbol{x}_0 - \boldsymbol{x})$, where $g_n(\boldsymbol{x})$ has support $\mathbf{G}_n$ (see Eq. (3.2) of Chapter 3). Applying a linear operator $\mathcal{G}_n$ to the input texture function $f(\boldsymbol{x})$ by *sliding* the spatial support of its function over all positions $\boldsymbol{x}_0$ is called *convolution* (see Eq. (3.1) of Chapter 3). Convolutional texture operators are discussed in Section 3.2 of Chapter 3.

In order to extract collections of scalar measurements $\boldsymbol{\eta} = (\eta_1, \ldots, \eta_N)$ from N response maps $h_n(\boldsymbol{x})$ an aggregation function is required to gather and summarize the operators' responses over a defined ROI domain $\mathbf{M} \subset \mathbb{R}^D$ (see Fig. 1.11). The values of the vector $\boldsymbol{\eta}$ define coordinates of a texture instance in the feature space $\mathbb{R}^N$. Integrative aggregation functions are commonly used to extract estimations of features statistics (*e.g.*, counts, means, covariances). For instance, the mean can estimate the average responses

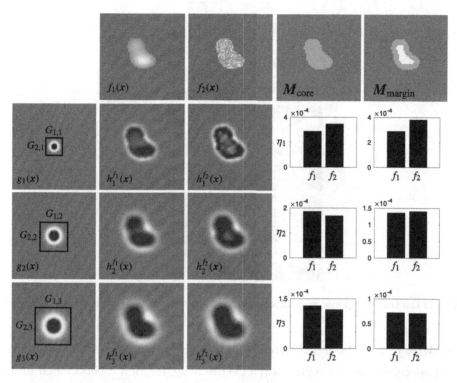

Figure 1.12 2D texture functions f_1 and f_2 (synthetic tumors) are analyzed with a collection of convolutional circularly symmetric texture operators $\{\mathcal{G}_1, \mathcal{G}_2, \mathcal{G}_3\}$ with functions $\{g_1, g_2, g_3\}$. The latter are 2D Laplacians of Gaussians (LoG, see Section 3.2.1 of Chapter 3) at different scales and have different spatial supports $G_{1,n} \times G_{2,n}$. The resulting response maps $h_n^{f_j}$ are revealing distinct properties of both the core and the margin of the tumors. In particular, $h_1^{f_2}(x)$ highlights the core texture of f_2. This is verified when averaging the response maps[4] over the core region M_{core} to obtain scalar measurements η_n, where $\eta_1^{f_1} < \eta_1^{f_2}$. Likewise, $h_1^{f_1}(x)$ highlights the margin of the tumor in f_2, where $\eta_1^{f_1} < \eta_1^{f_2}$. The texture scales captured by g_2 and g_3 are too large and do not discriminate well between f_1 and f_2, neither for the core nor for the margin of the tumors.

of a given operator over M as

$$\eta = \begin{pmatrix} \eta_1 \\ \vdots \\ \eta_N \end{pmatrix} = \frac{1}{|M|} \int_M \Big(h_n(x) \Big)_{n=1,\dots,N} \, \mathrm{d}x, \tag{1.5}$$

---

[4] The average of the absolute values of $h_n^{f_j}(x)$ is computed since LoGs are band-pass functions in the Fourier domain and have zero mean.

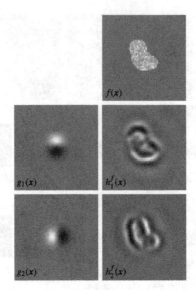

Figure 1.13 Directional response maps of scaled gradient operators along horizontal and vertical axes.

where $|M| = \int_M d\boldsymbol{x}$ is the area[5] covered by M. Aggregation functions are not limited to integral operations. For example, $\max_{\boldsymbol{x} \in M}\Big(h_n(\boldsymbol{x})\Big)$ is an aggregation function used in deep CNNs for the max-pooling of feature maps (see Section 4.2.4.1 of Chapter 4). It is worth noting that the aggregation operation can itself be seen as an operator applied to the feature maps $h_n(\boldsymbol{x})$. However, it differs from the definition introduced in Eq. (1.4) as it relies on irregular spatial supports defined by M, which are not slid over the entire image domain.

Examples of simple convolutional texture operators, response maps, and aggregation functions are shown in Fig. 1.12 (multiscale circularly symmetric operators) and Fig. 1.13 (directional operators). Both figures show that specific operator design allow highlighting desirable or discriminatory texture properties (*e.g.*, scale, directionality). However, it appears clearly in Fig. 1.10 that the properties of biomedical textures are very subtle and have big intraclass variations. To some extent, the latter type of variation is due to local rigid transformations of the structures (*e.g.*, a local rotation of a vessel or bronchus following the anatomy of lung lobes). In order to limit the impact of these variations on the performance of BTA algorithms, we will define more precisely the kind of texture transformations that can be expected, as well as strategies to make BTA approaches more robust to them. In particular, the characteristics of texture normalization and operator invariances are introduced in Sections 1.3.2 and 1.3.3, respectively.

---

[5] Or volume, hypervolume when $D > 2$.

Operator invariances are further discussed in Chapter 7. The management of intra and interclass complexity observed in the hyperspaces spanned by aggregated responses of the operators (texture features η) is developed in Chapter 6.

Designing sets of texture operators that are able to accurately identify tissue type, function, genomics, or response to treatment while accounting for intraclass variation is a very challenging task. How to design such operators? In the rare case where class-specific texture primitives are known the ideal set of operators would be detectors of those primitives (*e.g.*, detectors of crosses, coins, or collagen junctions for the textures shown in Fig. 1.8), where aggregation functions would count the occurrences of the primitives to quantify their density. Knowing texture primitives is rare in practice and more general approaches for operator design are required. The most general strategy for operator design relies on understanding and/or characterizing the whole span of possible texture functions and, within this span, identifying the most important combinations of image scales and directions with discriminative capabilities for each considered tissue class in the specific BTA task in hand. This can be done by exhaustively parceling the Fourier domain by groups of spatial frequencies and their directions (see Section 1.2.1 and Section 3.2.2 of Chapter 3). However, the exhaustive analysis of spatial scales and directions is computationally expensive for discrete texture functions $f(\xi)$ with densely sampled domains $F_1 \times \cdots \times F_D$ and choices are required (see Chapter 2). Making such choices has been studied in the literature by following essentially two opposing strategies: feature *handcrafting* versus feature *learning*. Feature handcrafting requires defining strong assumptions on expected discriminative types of image scales, directions, and transitions. Classic examples of handcrafted texture operators are circularly/spherically symmetric or directional Gabor wavelets [58], GLCMs [27], HOG [31], LBPs [32], and the ST [33]. Examples of such choices are depicted in Fig. 1.14. Making these choices (*i.e.*, feature handcrafting) raise two issues: How much information is missed or mixed in between two sampled directions or frequency bands? And, among the chosen image scales and directions, which ones are useful for texture discrimination? To address both issues, feature learning approaches relying on either supervised or unsupervised machine learning have recently been proposed to identify context-specific intended texture properties from data. Unsupervised approaches do not require labeled data to design image features and characterize image textons. Notable examples are Principal Component Analysis (PCA) [59], Bags of Visual Words (BoVW) [51], and unsupervised DL [37]. However, although the learned texture operators are data–driven, nothing guarantees that the latter are useful for discriminating a desired texture class from the background or from other types. To overcome this limitation, supervised feature learning approaches can design image features based on a set of labeled data and learning rules. Noteworthy examples are supervised DL [38], learned wavelets [54,53,60], and deep CNNs [61,35]. They are presented in Section 3.2.3 of Chapter 3. Deep CNNs for texture analysis are thoroughly discussed in Chapters 4, 9, and 10.

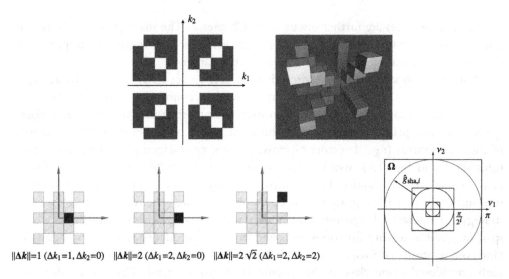

Figure 1.14 Popular choices of image scales and directions for handcrafted texture operator design. Top: popular 2D and 3D regular sampling of image directions along image axes and diagonals used in GLCMs, RLE, and HOG [14]. Bottom left: 2D sampling of image scales and directions for computing GLCMs. Bottom right: systematic sampling of the Fourier domain Ω into dyadic bands i with 2D Shannon circularly symmetric wavelets $\hat{g}_{\text{sha},i}\left(||\mathbf{v}||\right)$ [62,63].

The importance of image normalization and operator invariances has already been mentioned in the previous sections and warrants further clarification. They are discussed in Sections 1.3.2 and 1.3.3.

1.3.2 Normalization

Image normalization ensures optimal comparisons across data acquisition methods and texture instances. The normalization of pixel values (intensity) is recommended for imaging modalities that do not correspond to absolute physical quantities. Various advanced strategies have been proposed to normalize values and are often modality-specific (*e.g.*, MRI [64], histopathology [65]). Examples of pixel value normalization can be found in Section 10.2.1 of Chapter 10, and Section 11.2.4 of Chapter 11. The normalization of image sampling steps $\Delta\xi_d$ ($d = 1,\dots,D$) across subjects and dimensions is crucial to ensure accurate comparisons between scales and directions, which is illustrated in Fig. 1.15. Image resampling[6] with identical sampling steps both across image series and image dimensions can be used to normalize image scales and directions. Resampling can be carried out either on the image itself, or on the texture operator. In both cases, care must be taken on the transfer function of the resampling strategy, which can

---

[6] Classic resampling strategies can be used (*e.g.*, nearest neighbor, multilinear, multicubic).

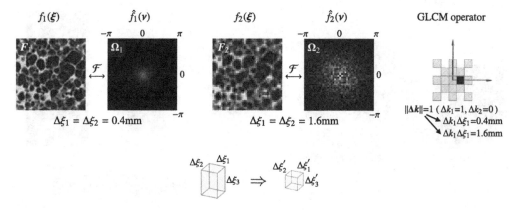

Figure 1.15 Importance of image sampling normalization across image series and image dimensions. Top row: sampling steps $\{\Delta\xi_1, \Delta\xi_2\}$ are four times smaller in the left image than in the center one. This results in a dilated distribution of spatial frequencies in $[-\pi, \pi]$ in the Fourier domain (the moduli of $\hat{f}_i(\nu)$ are displayed), because the normalized Nyquist frequencies π are not obtained with the same normalization. It is worth noting that $|F_1| = 4^2 \cdot |F_2|$. Likewise, a GLCM operator parameterized by a horizontal displacement of one pixel $(||\Delta k||=1)$ corresponds to a physical displacement of either 0.4 mm or 1.6 mm. Image resampling with a fixed step defined in physical dimensions (*e.g.*, $\Delta\xi_1 = \Delta\xi_2 = 0.4$ mm) is required to normalize scales across image series. Bottom row: image resampling across dimensions $\{\Delta\xi_1, \Delta\xi_2, \Delta\xi_3\} \Rightarrow \{\Delta\xi'_1, \Delta\xi'_2, \Delta\xi'_3\}$ to obtain cubic voxels and to ensure an isotropic description of image directions.

have an important influence on texture properties [66]. Examples of image resampling for spatial normalization can be found in Section 12.2.4.1 of Chapter 12, as well as in Sections 4.4.2.2 and 4.5.2 of Chapter 4.

1.3.3 Invariances

An operator or a measurement is *invariant* if its response is insensitive to the transformation of the input image. When the output of the operator is affected by the transformation in the same way the input is, we say that it is *equivariant*.

Invariances of the final texture measures η are required to provide robust recognition of all intraclass variants, while preserving interclass variations (see also Chapter 6 for a machine learning perspective of the problem). Whereas the sources of intraclass variations can be extremely diverse in the context of biomedical tissue (*e.g.*, imaging protocol, subject age, genetics, history) an important subcategory is related to geometric transformations of the tissue architecture. Examples of the effect of such transformations on texture functions are depicted in Fig. 1.16. Assumptions on types of geometric transformations expected in biomedical images are different from those expected in photographic images, because the well-controlled acquisition protocols have little viewpoint variations and allow to control the physical size of pixels (see Section 1.2.5). As a consequence, while robustness of the texture measures to *affine* transformations (*e.g.*,

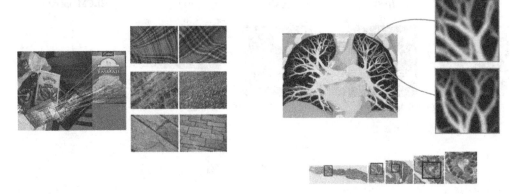

Figure 1.16 Typical geometric transformations of textures encountered in photographic imagery [67] (left) versus biomedical images [68] (right). In most cases, biomedical textures are observed in images with known pixel sizes. As a consequence, texture measures that are robust to translations and rotations (rigid transformations) but sensitive to changes in image scale will yield optimal descriptors. As opposed to photographic image analysis, it is not desirable to enforce any form of scale invariance which truly entails the risk of regrouping patterns of different nature.

translation, rotation, scaling) are desired for analyzing photographic images, measures that are robust to a smaller group called *Euclidean* or *rigid* transformations (combinations of translations and rotations) are recommended for biomedical image analysis (see Chapter 7). Scale, itself, is a powerful discriminative property in biomedical image analysis.

To better understand the types of geometric invariance that are desirable in the context of a particular image analysis task at hand, we must disentangle invariances of the operator's output (*i.e.*, at the level of the response maps $h_n(\boldsymbol{x}) = \mathcal{G}_n\{f\}(\boldsymbol{x})$) and the ones of the final texture measures η.

1.3.3.1 Invariance and equivariance of operators

We will first discuss the desired invariances of texture operators to design BTA approaches that are robust to rigid transformations. Let us consider a general group of geometric transformations **A** which can be used, *e.g.*, translation, rotation, scaling, rigid, or affine. In a strict sense, a texture operator $\mathcal{G}_n$ that is invariant to the group **A** observes

$$\mathcal{G}_n\{f\}(\cdot) = \mathcal{G}_n\{\mathcal{A}\{f\}\}(\cdot) \quad \forall f, \; \forall \mathcal{A} \in \mathbf{A}, \tag{1.6}$$

where $\mathcal{A}$ is a geometric transformation operator implementing a concrete transformation of the group **A** (*e.g.*, a 2D translation by the vector $(1, 1)$). Eq. (1.6) imposes an extremely strong requirement that would severely limit the ability of operators to extract any useful texture information. For instance if $\mathcal{A}$ is a translation operator as $\mathcal{T}\{f\}(\cdot) = f(\cdot - \boldsymbol{x}_0)$, enforcing $\mathcal{G}_n$ to be invariant to $\mathcal{T}$ means that the response map

$h_n(x)$ (see Eq. (1.4)) will not change under the effect of $\mathcal{T}$. Therefore the spatial positions of the texture "events" will be lost when such a translation-invariant operator is used. This operator will be useless for texture segmentation and difficult to use for texture classification. To preserve the ability of operators to characterize local texture properties, the operators must be *equivariant* to rigid transformations [69], which implies that if the input image $f(x)$ is translated and rotated, and the response map $h_n(x)$ must undergo the same rigid transformation. This means that $\mathcal{G}_n$ must commute with $\mathcal{A}$ as

$$\mathcal{A}\{\mathcal{G}_n\{f\}\}(\cdot) = \mathcal{G}_n\{\mathcal{A}\{f\}\}(\cdot) \quad \forall f, \ \forall \mathcal{A} \in \mathbf{A}. \tag{1.7}$$

For some geometric transformations, the origin plays a crucial role. This is the case for rotations and scalings, for which the origin is a fixed point ($\mathcal{A}\{f\}(0) = f(0)$ for any function f). On images the position of the origin is arbitrary. For instance, we are not only interested to know the effect of rotations around the origin, but also on rotations centered around any location x_0. For an operator $\mathcal{A}$, we denote by $\mathcal{A}_{x_0}$ its shifted version that is centered around x_0 instead of 0. Mathematically, we have $\mathcal{A}_{x_0} = \mathcal{T}_{x_0}\mathcal{A}\mathcal{T}_{-x_0}$, where $\mathcal{T}_{x_0}$ is the translation operator with shift x_0. We therefore refine the notion of equivariance as follows. We say that a texture operator is *locally equivariant* to the group of transformation $\mathbf{A}$ if it commutes with any transformation $\mathcal{A}_{x_0}$ for any $\mathcal{A} \in \mathbf{A}$ and any $x_0 \in \mathbb{R}^D$; *i.e.*,

$$\mathcal{A}_{x_0}\{\mathcal{G}_n\{f\}\}(\cdot) = \mathcal{G}_n\{\mathcal{A}_{x_0}\{f\}\}(\cdot) \quad \forall f, \ \forall \mathcal{A} \in \mathbf{A}, \ \forall x_0 \in \mathbb{R}^D. \tag{1.8}$$

We remark that the texture operators introduced in Section 1.3.1 are precisely the ones that are equivariant to translations, due to the sliding property. This has an important consequence: if the texture operator is equivariant to another group of transformations for which the origin plays a central role (such as scalings or rotation), then it is also locally equivariant in the sense of Eq. (1.8).

Strict equivariance is a difficult requirement to achieve in the practical design of operators. Therefore the equivariance constraint can be approximated to allow robustness of texture measurements to (local) rigid transformations. The desirable approximated operator invariances/equivariances to global/local geometric transformations are also depending on the image analysis task at hand. This is illustrated in Fig. 1.17 where the requirements substantially differ for detecting cars in photographic imagery and for detecting collagen junctions in lung CT. The former requires operator local equivariance to scaling transformations (all cars must be detected regardless of their distance to the viewpoint) and equivariance to translations (important for localizing the positions of the cars), while no equivariance to rotations (important to rule out objects that look like cars but are, *e.g.*, upside down). Detecting collagen junctions in lung CT requires operator local equivariance to rotations (all junctions must be detected in spite of their local orientation), equivariance to translations (important for localizing the junctions)

 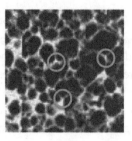

Figure 1.17 The requirements in terms of geometric invariance/equivariance of image analysis methods substantially differ depending on the image analysis task at hand. Left: detecting cars in photographic imagery[7] versus right: detecting collagen junctions in lung CT [70].

but no equivariance to scaling (important to rule out objects that look like junctions but are physically too big or small to be a junction). Since the texture primitives (or textons, see Section 1.2.4) can have arbitrary local orientations in most biomedical images (see Figs. 1.8 and 1.9), it is mostly interesting to design operators that are equivariant to local rotations [57] (see Section 2.4.1 of Chapter 2).

1.3.3.2 Invariances of texture measurements

Invariance of texture measurements to geometric transformations are further obtained through the aggregation function. Most of aggregation functions (*e.g.*, integral/summation, max) are adding invariance of the texture measurements over the ROI M in the sense of Eq. (1.6). This is reasonable under the condition that the texture processes are considered stationary (in the relaxed sense, see Section 1.2.3.1) over M. For instance, however, special care must be taken when choosing M to avoid undesirable and destructive side effects of aggregation. This is discussed in Sections 2.3.2 and 2.4 of Chapter 2.

1.3.3.3 Nongeometric invariances

Similar to the general categories of feature design mentioned in Section 1.3.1, designing texture operators that are invariant or equivariant to certain geometric transformations can be considered as *handcrafted* because it involves prior knowledge of biomedical texture variants. However, more subtle intraclass variations are caused by the diversity of, *e.g.*, biology, anatomy, subject age [71]. In this context, texture operators can be trained using machine learning to respond invariantly to subtle intraclass variation of the data. Deep CNNs have recently shown to perform very well on learning cascades of nonlinear operators and aggregation functions[8] minimizing intraclass variations and maximizing

---

[7] https://www.youtube.com/watch?v=xVwsr9p3irA, as of March 7 2017.

[8] The forward function of image operators results from the composition of linear and slightly nonlinear operations (*e.g.*, rectified linear unit, ReLU).

interclass separability [33,61]. However, they solely rely on class representations available in the training set, *i.e.*, the collection of texture instances (realizations) with ground truth (labels) available. The latter are difficult to acquire in practice for biomedical images, and data augmentation has been used to include geometric handcrafted invariances to deep CNNs [36] (see Section 3.2.3.4 of Chapter 3). Transfer learning has also been proposed to reuse deep networks trained on very large and publicly available databases (*e.g.*, ImageNet [72]) for other image recognition tasks [73,74] (see Section 4.4.3.3 of Chapter 4). In this particular context, using networks that were trained with images acquired with digital cameras (ImageNet) on biomedical images carries the risk of introducing undesired operator robustness to image scale for instance, discarding a strong discriminative property.

1.4 CONCLUSIONS

In this chapter, we present the foundations of texture processing for biomedical image analysis. We begin with a generic definition for the type of textures encompassing those observed in biomedical imaging (Section 1.2). We clarify the difference between intensity and texture in ROIs, where the former relies on the statistical distribution of pixel values, and the latter is characterized by the spatial transitions between the pixel values (see Fig. 1.1). The direction, scale, and order of these spatial transitions were found to be fundamental properties of biomedical texture and are naturally described in the Fourier domain. From a mathematical point of view, we defined biomedical texture functions as realizations of intricate and nonstationary stochastic processes. When compared to general photographic image analysis, the acquisition devices and protocols in biomedical imaging yield data with well controlled and standardized fundamental properties such as pixel size and intensity, as well as image orientation and sample/patient position. Texture analysis challenges are therefore specific to the domain of biomedical imaging and require adequate methods for obtaining optimal results.

Second, we introduced a general problem formulation for BTA in Section 1.3. It essentially consisted of considering that any biomedical texture analysis approach can be characterized by a series of local texture operators and regional aggregation functions. Operators can be handcrafted to highlight desired properties of the input texture function such as spatial scales in a well-defined frequency band, image gradients along horizontal directions, cooccurrences, local binary patterns, or circular frequencies. They can also be learned from data to yield optimal texture representation for reconstruction or discrimination. Image normalization and operator invariance/equivariance to rigid transformations were found to be fundamental for BTA. Aggregation functions are required to summarize the responses of operators over ROIs. It allows obtaining collections of scalar-valued texture measurements that can be used as quantitative imaging biomarkers. The latter can be further combined with other -omics and patient data to allow precision and personalized medicine by predicting diagnosis, treatment response, as well as to enable biomedical discovery (see Chapter 8).

The challenges of multiscale and multidirectional biomedical texture analysis are further developed in Chapter 2. A qualitative comparison of popular approaches in terms of the proposed general problem formulation introduced in this chapter is discussed in Chapter 3.

ACKNOWLEDGMENTS

This work was supported by the Swiss National Science Foundation (under grant PZ00P2_154891), and the CIBM.

REFERENCES

[1] A.S. Szczesniak, Texture is a sensory property, Food Qual. Prefer. 13 (4) (2002) 215–225.
[2] M. Bourne, Food Texture and Viscosity: Concept and Measurement, Academic Press, 2002.
[3] W. Schiff, E. Foulke, Tactual Perception: A Sourcebook, Paperback Re-issue, Cambridge University Press, 1982.
[4] W. Lin, K. Kuppusamy, E.M. Haacke, H. Burton, Functional MRI in human somatosensory cortex activated by touching textured surfaces, J. Magn. Reson. Imaging 6 (4) (July 1996) 565–572.
[5] T.A. Whitaker, C. Simões-Franklin, F.N. Newell, Vision and touch: independent or integrated systems for the perception of texture?, Brain Res. 1242 (2008) 59–72.
[6] G.D. Tourassi, Journey toward computer-aided diagnosis: role of image texture analysis, Radiology 213 (2) (July 1999) 317–320.
[7] P.J. Robinson, Radiology's Achilles' heel: error and variation in the interpretation of the Rontgen image, Br. J. Radiol. 70 (839) (1997) 1085–1098.
[8] G.D. Tourassi, S. Voisin, V. Paquit, E.A. Krupinski, Investigating the link between radiologists' gaze, diagnostic decision, and image content, J. Am. Med. Inform. Assoc. 20 (6) (November 2013) 1067–1075.
[9] Z.A. Aziz, A.U. Wells, D.M. Hansell, G.A. Bain, S.J. Copley, S.R. Desai, S.M. Ellis, F.V. Gleeson, S. Grubnic, A.G. Nicholson, S.P. Padley, K.S. Pointon, J.H. Reynolds, R.J. Robertson, M.B. Rubens, HRCT diagnosis of diffuse parenchymal lung disease: inter-observer variation, Thorax 59 (6) (June 2004) 506–511.
[10] C.P. Langlotz, RadLex: a new method for indexing online educational materials, Radiographics 26 (6) (2006) 1595–1597.
[11] E. Lazarus, M.B. Mainiero, B. Schepps, S.L. Koelliker, L.S. Livingston, BI-RADS lexicon for US and mammography: interobserver variability and positive predictive value, Radiology 239 (2) (2006) 385–391.
[12] A. Depeursinge, C. Kurtz, C.F. Beaulieu, S. Napel, D.L. Rubin, Predicting visual semantic descriptive terms from radiological image data: preliminary results with liver lesions in CT, IEEE Trans. Med. Imaging 33 (8) (August 2014) 1–8.
[13] D.F. Gleason, G.T. Mellinger, The Veretans Administration Cooperative Urological Research Group, Prediction of prognosis for prostatic adenocarcinoma by combined histological grading and clinical staging, J. Urol. 167 (2) (2002) 953–958.
[14] A. Depeursinge, A. Foncubierta-Rodríguez, D. Van De Ville, H. Müller, Three-dimensional solid texture analysis and retrieval in biomedical imaging: review and opportunities, Med. Image Anal. 18 (1) (2014) 176–196.

[15] K.P. Andriole, J.M. Wolfe, R. Khorasani, Optimizing analysis, visualization and navigation of large image data sets: one 5000-section CT scan can ruin your whole day, Radiology 259 (2) (May 2011) 346–362.

[16] V. Kumar, Y. Gu, S. Basu, A. Berglund, S.A. Eschrich, M.B. Schabath, K. Forster, H.J.W.L. Aerts, A. Dekker, D. Fenstermacher, D.B. Goldgof, L.O. Hall, P. Lambin, Y. Balagurunathan, R.A. Gatenby, R.J. Gillies, Radiomics: the process and the challenges, Magn. Reson. Imaging 30 (9) (2012) 1234–1248.

[17] R.M. Summers, Texture analysis in radiology: does the emperor have no clothes?, Abdom. Radiol. (2016).

[18] F. Orlhac, M. Soussan, J.-A. Maisonobe, C.A. Garcia, B. Vanderlinden, I. Buvat, Tumor texture analysis in 18F-FDG PET: relationships between texture parameters, histogram indices, standardized uptake values, metabolic volumes, and total lesion glycolysis, J. Nucl. Med. 55 (3) (2014) 414–422.

[19] E.A. Kidd, P.W. Grigsby, Intratumoral metabolic heterogeneity of cervical cancer, Clin. Cancer Res. 14 (16) (2008) 5236–5241.

[20] F. Davnall, C.S.P. Yip, G. Ljungqvist, M. Selmi, F. Ng, B. Sanghera, B. Ganeshan, K.A. Miles, G.J. Cook, V. Goh, Assessment of tumor heterogeneity: an emerging imaging tool for clinical practice?, Insights Imaging 3 (6) (2012) 573–589.

[21] F. Orlhac, C. Nioche, M. Soussan, I. Buvat, Understanding changes in tumor textural indices in PET: a comparison between visual assessment and index values in simulated and patient data, J. Nucl. Med. (October 2016), http://dx.doi.org/10.2967/jnumed.116.181859.

[22] M.A. Haidekker, Texture Analysis, John Wiley & Sons, Inc., 2010, pp. 236–275.

[23] M. Petrou, Texture in biomedical images, in: T.M. Deserno (Ed.), Biomedical Image Processing, Springer-Verlag, Berlin, Heidelberg, 2011, pp. 157–176.

[24] B.M. ter Haar Romeny, Multi-scale and multi-orientation medical image analysis, in: T.M. Deserno (Ed.), Biomedical Image Processing, Springer-Verlag, Berlin, Heidelberg, 2011, pp. 177–196.

[25] C. Blakemore, F.W. Campbell, On the existence of neurones in the human visual system selectively sensitive to the orientation and size of retinal images, J. Physiol. 203 (1) (1969) 237–260.

[26] D.P. McGovern, K.S. Walsh, J. Bell, F.N. Newell, Individual differences in context-dependent effects reveal common mechanisms underlying the direction aftereffect and direction repulsion, Vis. Res. (2016), http://dx.doi.org/10.1016/j.visres.2016.08.009.

[27] R.M. Haralick, Statistical and structural approaches to texture, Proc. IEEE 67 (5) (May 1979) 786–804.

[28] M.M. Galloway, Texture analysis using gray level run lengths, Comput. Graph. Image Process. 4 (2) (1975) 172–179.

[29] G. Thibault, B. Fertil, C. Navarro, S. Pereira, P. Cau, N. Levy, J. Sequeira, J.-L. Mari, Texture indexes and gray level size zone matrix application to cell nuclei classification, in: Pattern Recognition and Information Processing, 2009, pp. 140–145.

[30] A.C. Bovik, M. Clark, W.S. Geisler, Multichannel texture analysis using localized spatial filters, IEEE Trans. Pattern Anal. Mach. Intell. 12 (1) (1990) 55–73.

[31] D.G. Lowe, Distinctive image features from scale-invariant keypoints, Int. J. Comput. Vis. 60 (2) (2004) 91–110.

[32] T. Ojala, M. Pietikäinen, T. Mäenpää, Multiresolution gray-scale and rotation invariant texture classification with local binary patterns, IEEE Trans. Pattern Anal. Mach. Intell. 24 (7) (July 2002) 971–987.

[33] J. Bruna, S.G. Mallat, Invariant scattering convolution networks, IEEE Trans. Pattern Anal. Mach. Intell. 35 (8) (2013) 1872–1886.

[34] L. Sifre, S. Mallat, Rigid-motion scattering for texture classification, CoRR, abs/1403.1687, 2014.

[35] A. Krizhevsky, I. Sutskever, G.E. Hinton, ImageNet classification with deep convolutional neural networks, in: F. Pereira, C.J.C. Burges, L. Bottou, K.Q. Weinberger (Eds.), Advances in Neural Information Processing Systems, vol. 25, Curran Associates, Inc., 2012, pp. 1097–1105.

[36] O. Ronneberger, P. Fischer, T. Brox, U-Net: convolutional networks for biomedical image segmentation, in: N. Navab, J. Hornegger, W.M. Wells, A.F. Frangi (Eds.), Medical Image Computing and Computer-Assisted Intervention, MICCAI 2015, in: Lecture Notes in Computer Science, vol. 9351, Springer International Publishing, 2015, pp. 234–241.

[37] M.J. Gangeh, A. Ghodsi, M.S. Kamel, Dictionary learning in texture classification, in: Proceedings of the 8th International Conference on Image Analysis and Recognition, Part I, 2011, pp. 335–343.

[38] J. Mairal, F. Bach, J. Ponce, G. Sapiro, A. Zisserman, Supervised dictionary learning, in: Advances in Neural Information Processing Systems, 2008, pp. 1033–1040.

[39] A. Hyvärinen, J. Hurri, P.O. Hoyer, Energy correlations and topographic organization, in: Natural Image Statistics: A Probabilistic Approach to Early Computational Vision, Springer London, London, 2009, pp. 239–261.

[40] P. Cirujeda, Y. Dicente Cid, H. Müller, D. Rubin, T.A. Aguilera, B.W. Loo Jr., M. Diehn, X. Binefa, A. Depeursinge, A 3-D Riesz-covariance texture model for prediction of nodule recurrence in lung CT, IEEE Trans. Med. Imaging (2016).

[41] J. Scharcanski, Stochastic texture analysis for monitoring stochastic processes in industry, Pattern Recognit. Lett. 26 (11) (2005) 1701–1709.

[42] T. Ojala, T. Mäenpää, M. Pietikäinen, J. Viertola, J. Kyllönen, S. Huovinen, Outex – new framework for empirical evaluation of texture analysis algorithms, in: 16th International Conference on Pattern Recognition, in: ICPR, vol. 1, IEEE Computer Society, August 2002, pp. 701–706.

[43] M. Petrou, P. García Sevilla, Image Processing: Dealing with Texture, Wiley, 2006.

[44] J. Bela, Textons, the elements of texture perception, and their interactions, Nature 290 (5802) (March 1981) 91–97.

[45] J. Bela, Visual pattern discrimination, IRE Trans. Inf. Theory 8 (2) (February 1962) 84–92.

[46] J. Bela, A theory of preattentive texture discrimination based on first-order statistics of textons, Biol. Cybern. 41 (2) (1981) 131–138.

[47] B. Julesz, E.N. Gilbert, L.A. Shepp, H.L. Frisch, Inability of humans to discriminate between visual textures that agree in second-order statistics—revisited, Perception 2 (4) (1973) 391–405.

[48] B. Julesz, R.A. Schumer, Early visual perception, Annu. Rev. Psychol. 32 (1) (1981) 575–627.

[49] S.-C. Zhu, C.-E. Guo, Y. Wang, Z. Xu, What are textons?, in: Special Issue on Texture Analysis and Synthesis, Int. J. Comput. Vis. 62 (1–2) (April 2005) 121–143.

[50] M. Varma, A. Zisserman, A statistical approach to texture classification from single images, Int. J. Comput. Vis. 62 (1–2) (2005) 61–81.

[51] J. Sivic, B.C. Russell, A.A. Efros, A. Zisserman, W.T. Freeman, Discovering objects and their location in images, in: Tenth IEEE International Conference on Computer Vision, 2005, ICCV 2005, vol. 1, October 2005, pp. 370–377.

[52] J.R. Bergen, M.S. Landy, Computational modeling of visual texture segregation, in: Computational Models of Visual Processing, MIT Press, 1991, pp. 253–271.

[53] A. Depeursinge, A. Foncubierta-Rodríguez, D. Van De Ville, H. Müller, Rotation-covariant texture learning using steerable Riesz wavelets, IEEE Trans. Image Process. 23 (2) (February 2014) 898–908.

[54] A. Depeursinge, A. Foncubierta-Rodríguez, D. Van De Ville, H. Müller, Multiscale lung texture signature learning using the Riesz transform, in: Medical Image Computing and Computer-Assisted Intervention, MICCAI 2012, in: Lecture Notes in Computer Science, vol. 7512, Springer, Berlin/Heidelberg, October 2012, pp. 517–524.

[55] A. Dumas, M. Brigitte, M.F. Moreau, F. Chrétien, M.F. Baslé, D. Chappard, Bone mass and microarchitecture of irradiated and bone marrow-transplanted mice: influences of the donor strain, Osteoporos. Int. 20 (3) (2009) 435–443.

[56] T.M. Deserno, Fundamentals of Biomedical Image Processing, Springer, 2011.

[57] Z. Püspöki, M. Storath, D. Sage, M. Unser, Transforms and operators for directional bioimage analysis: a survey, in: W.H. De Vos, S. Munck, J.-P. Timmermans (Eds.), Focus on Bio-Image Informatics, vol. 219, Springer International Publishing, Cham, 2016, pp. 69–93.

[58] A. Ahmadian, A. Mostafa, An efficient texture classification algorithm using Gabor wavelet, in: Proceedings of the 25th Annual International Conference of the IEEE Engineering in Medicine and Biology Society, vol. 1, September 2003, pp. 930–933.

[59] C. Vonesch, F. Stauber, M. Unser, Steerable PCA for rotation-invariant image recognition, SIAM J. Imaging Sci. 8 (3) (2015) 1857–1873.

[60] G. Quellec, M. Lamard, P.M. Josselin, G. Cazuguel, B. Cochener, C. Roux, Optimal wavelet transform for the detection of microaneurysms in retina photographs, IEEE Trans. Med. Imaging 27 (9) (2008) 1230–1241.

[61] S. Mallat, Understanding deep convolutional networks, Philos. Trans. R. Soc. A, Math. Phys. Eng. Sci. 374 (2065) (March 2016).

[62] N. Chenouard, M. Unser, 3D steerable wavelets and monogenic analysis for bioimaging, in: 2011 IEEE International Symposium on Biomedical Imaging: From Nano to Macro, April 2011, pp. 2132–2135.

[63] M. Papadakis, G. Gogoshin, I.A. Kakadiaris, D.J. Kouri, D.K. Hoffman, Nonseparable radial frame multiresolution analysis in multidimensions and isotropic fast wavelet algorithms, in: Wavelets: Applications in Signal and Image Processing X, in: Proc. SPIE, vol. 5207, August 2003, pp. 631–642.

[64] G. Collewet, M. Strzelecki, F. Mariette, Influence of MRI acquisition protocols and image intensity normalization methods on texture classification, Magn. Reson. Imaging 22 (1) (2004) 81–91.

[65] M. Macenko, M. Niethammer, J.S. Marron, D. Borland, J.T. Woosley, X. Guan, C. Schmitt, N.E. Thomas, A method for normalizing histology slides for quantitative analysis, in: Proceedings of the Sixth IEEE International Conference on Symposium on Biomedical Imaging: From Nano to Macro, ISBI'09, IEEE Press, Piscataway, NJ, USA, 2009, pp. 1107–1110.

[66] P. Thévenaz, T. Blu, M. Unser, Image interpolation and resampling, in: I.N. Bankman (Ed.), Handbook of Medical Imaging, Processing and Analysis, Academic Press, San Diego CA, USA, 2000, pp. 393–420, chapter 25.

[67] S. Lazebnik, C. Schmid, J. Ponce, A sparse texture representation using local affine regions, IEEE Trans. Pattern Anal. Mach. Intell. 27 (8) (August 2005) 1265–1278.

[68] M.N. Gurcan, L.E. Boucheron, A. Can, A. Madabhushi, N.M. Rajpoot, B. Yener, Histopathological image analysis: a review, IEEE Rev. Biomed. Eng. 2 (2009) 147–171.

[69] D. Marcos Gonzalez, M. Volpi, N. Komodakis, D. Tuia, Rotation equivariant vector field networks, CoRR, abs/1612.09346, 2016.

[70] A. Depeursinge, Z. Püspöki, J.-P. Ward, M. Unser, Steerable wavelet machines (SWM): learning moving frames for texture classification, IEEE Trans. Image Process. 26 (4) (2017) 1626–1636.

[71] A. Depeursinge, D. Racoceanu, J. Iavindrasana, G. Cohen, A. Platon, P.-A. Poletti, H. Müller, Fusing visual and clinical information for lung tissue classification in high-resolution computed tomography, Artif. Intell. Med. 50 (1) (September 2010) 13–21.

[72] J. Deng, W. Dong, R. Socher, L.-J. Li, K. Li, L. Fei-Fei, ImageNet: a large-scale hierarchical image database, in: IEEE Conference on Computer Vision and Pattern Recognition, CVPR 2009, 2009, pp. 248–255.

[73] A.S. Razavian, H. Azizpour, J. Sullivan, S. Carlsson, CNN features off-the-shelf: an astounding baseline for recognition, in: IEEE Computer Society Conference on Computer Vision and Pattern Recognition Workshops, 2014, pp. 512–519.

[74] J. Yosinski, J. Clune, Y. Bengio, H. Lipson, How transferable are features in deep neural networks?, in: Proceedings of NIPS, in: Advances in Neural Information Processing Systems, vol. 27, 2014, pp. 1–9.

CHAPTER 2

Multiscale and Multidirectional Biomedical Texture Analysis
Finding the Needle in the Haystack

Adrien Depeursinge*,†

*École Polytechnique Fédérale de Lausanne (EPFL), Biomedical Imaging Group, Lausanne, Switzerland
†University of Applied Sciences Western Switzerland (HES-SO), Institute of Information Systems, Sierre, Switzerland

Abstract

This chapter clarifies the important aspects of biomedical texture analysis under the general framework introduced in Chapter 1. It was proposed that any approach can be characterized as the combination of local texture operators and regional aggregation functions. The type of scale and directional information that can or cannot be modeled by categories of texture processing methods is revealed through theoretic analyses and experimental validations. Several key aspects are found to be commonly overlooked in the literature and are highlighted. First, we demonstrate the risk of using regions of interest for aggregation that are regrouping tissue types of different natures. Second, a detailed study of the type of directional information important for biomedical texture characterization suggests that fundamental properties lie in the local organization of image directions. In addition, it was found that most approaches cannot efficiently characterize the latter, and even fewer can do it with invariance to local rotations. We conclude by deriving novel comparison axes to evaluate the relevance of biomedical texture analysis methods in a specific medical or biological applicative context.

Keywords

Texton, Moving frames, Uncertainty principle, Texture analysis

2.1 INTRODUCTION

The diversity of existing Biomedical Texture Analysis (BTA) approaches illustrates the various properties required in different applicative contexts [1,2]. Desired BTA properties are, *e.g.*, ease of use, interpretability, low computational cost, and most importantly high discriminatory performance and specificity. The latter is strongly dependent on the nature of the texture information required for a specific task in hand. The purpose of this work is to dissect the wide range of BTA properties to provide a set of comparison dimensions between approaches. A formal definition of biomedical texture information was proposed in Section 1.2 of Chapter 1. The latter was found to be characterized by the type of spatial transitions and dependencies between pixel values. In particular, it was demonstrated that the scales (*i.e.*, speed of variation or frequency) and directions of the spatial transitions are fundamental properties of biomedical texture functions.

Biomedical Texture Analysis
DOI: 10.1016/B978-0-12-812133-7.00002-8

In Section 1.3.1 of Chapter 1 a general problem formulation for biomedical texture analysis was introduced, considering that any approach can be characterized as a set of texture operators and aggregation functions. The operators allow locally isolating desired texture information in terms of spatial scales and directions of a texture image. The application of the operators over all positions of the image yields translation-equivariant feature maps containing every local response of the latter. Scalar-valued texture measurements are obtained by aggregating feature maps over regions of interest.

In this chapter, clarifications are provided on possible design choices in terms of spatial scales and directions for operators and aggregation functions. An excellent description of the problem can be found in [3], which concerns medical image analysis in general. Our aim is to discuss the particularities of multiscale and multidirectional analysis for biomedical texture characterization. A focus is made on linear operators. However, although the theoretic concepts presented are only valid for the latter, they should also provide intuition for designing nonlinear operators wherever possible. Multiple examples and toy problems will be provided to illustrate the concepts introduced. Among these, the uncertainty principle, a theoretic limitation of the trade-off between operator scale and locality is first recalled to provide guidelines for optimal design of operator scales. The influence of the size and shape of the region of interest used for aggregation on texture classification and segmentation is demonstrated. The latter motivates the creation of digital tissue atlases of organs or tumors, providing powerful models of digital phenotypes. In Section 1.3.3 of Chapter 1 and in Chapter 7 the importance of approaches that are robust to rigid transformations (translation and rotation) was emphasized. In the second part of this chapter (Section 2.4), clarifications are made on directional information types that are important for biomedical texture analysis. In particular, the Local Organization of Image Directions (LOID: how directional structures intersect) are found to be fundamental. Characterizing the latter with invariance to local rotations raises several challenges. In this context, operators that are insensitive to image directions (called *circularly/spherically symmetric*) are compared to their *directional* counterparts. The destructive effect of aggregation on the ability of directional operators to characterize the LOIDs is demonstrated and motivates the use of Moving Frame (MF) texture representations. The latter consist of locally adapting a coordinate frame (*e.g.*, a set of noncollinear operators) based on an alignment criteria that is consistent[1] for all positions in the texture image. We provide evidence that MF representations allow detailed characterizations of the LOIDs with invariance to rigid transformation. A quantitative performance comparison of circularly symmetric, directional, and MF texture representations for 2D texture classification is presented in Section 2.4.4. Finally, most important aspects of operator and aggregation function design are summarized under the form of a checklist matrix in Section 2.5.

---

[1] A simple and reliable alignment criteria is to orient all operators at each image position with the direction that maximizes the local image gradient.

2.2 NOTATION

Additional notation are introduced based on the notations initially defined in Section 1.2.2 of Chapter 1. To further analyze the relationship between scales and directions, let us consider the definition of texture functions in 2D polar coordinates as

$$f(r, \theta) \qquad \overset{\mathcal{F}}{\longleftrightarrow} \qquad \hat{f}(\rho, \vartheta)$$

$$r \in \mathbb{R}^+, \, \theta \in [0, 2\pi) \qquad\qquad \rho \in \mathbb{R}^+, \, \vartheta \in [0, 2\pi)$$

$$\boldsymbol{x} = \begin{pmatrix} x_1 \\ x_2 \end{pmatrix} = r \begin{pmatrix} \cos\theta \\ \sin\theta \end{pmatrix} \qquad\qquad \boldsymbol{\omega} = \begin{pmatrix} \omega_1 \\ \omega_2 \end{pmatrix} = \rho \begin{pmatrix} \cos\vartheta \\ \sin\vartheta \end{pmatrix}$$

and in 3D spherical coordinates as

$$f(r, \theta, \phi) \qquad \overset{\mathcal{F}}{\longleftrightarrow} \qquad \hat{f}(\rho, \vartheta, \varphi)$$

$$r \in \mathbb{R}^+, \, \theta \in [0, \pi], \, \phi \in [0, 2\pi) \qquad\qquad \rho \in \mathbb{R}^+, \, \vartheta \in [0, \pi], \, \varphi \in [0, 2\pi)$$

$$\boldsymbol{x} = \begin{pmatrix} x_1 \\ x_2 \\ x_3 \end{pmatrix} = r \begin{pmatrix} \sin\theta\cos\phi \\ \sin\theta\sin\phi \\ \cos\theta \end{pmatrix} \qquad\qquad \boldsymbol{\omega} = \begin{pmatrix} \omega_1 \\ \omega_2 \\ \omega_3 \end{pmatrix} = \rho \begin{pmatrix} \sin\vartheta\cos\varphi \\ \sin\vartheta\sin\varphi \\ \cos\vartheta \end{pmatrix}$$

Polar and spherical representations allow separating the *angular* part (*i.e.*, image directions) from the *radial* part (*i.e.*, spatial frequencies related to image scales) [4]. For simplifying the notation, all coordinate domains are considered continuous in this chapter. Their discretized versions can be obtained following the notions introduced in Section 1.2.2 of Chapter 1.

2.3 MULTISCALE IMAGE ANALYSIS

The need for multiscale image analysis is motivated throughout various chapters of this book (*e.g.*, Section 1.3 of Chapter 1) as well as Chapters 4, 5, and 7. In this section, we will define more precisely the important aspects of multiscale texture operator and aggregation function design. In particular the discussion on which scales are optimal for biomedical texture measurements hinges on two important facets:

- How to optimally define the spatial support(s) $\boldsymbol{G}_n = G_{1,n} \times \cdots \times G_{D,n}$ and the radial responses of the operator(s) $\mathcal{G}_n$?
- What is the best size and shape of the region of interest $\boldsymbol{M}$ for aggregation?

These two aspects are illustrated in Fig. 2.1 and detailed in the following Subsections 2.3.1 and 2.3.2.

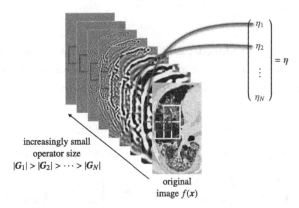

$$\left.\begin{array}{c} \eta_1 \\ \eta_2 \\ \vdots \\ \eta_N \end{array}\right\} = \eta$$

increasingly small
operator size
$|G_1| > |G_2| > \cdots > |G_N|$

original
image $f(x)$

Figure 2.1 Aspects of multiscale texture operator and aggregation function design [5]. How to optimally define the sizes $|G_1| > |G_2| > \cdots > |G_N|$ and the radial responses of a collection of operators $\mathcal{G}_n$? What is the best position, size, and shape of the region of interest M for aggregation?

2.3.1 Spatial versus spectral coverage of linear operators: the uncertainty principle

Linear texture operators are expected to be *band-pass* functions, which means that the covered spatial frequencies are in a range defined by $\rho_{\min} > 0$ and $\rho_{\max} < \pi$ in the Fourier domain. Band-pass operators are commonly used as operators for texture analysis because their geometric behavior, isotropy, or directionality can be well controlled. Since they do not include the zero frequency $\rho = ||\boldsymbol{\omega}|| = 0$, they are only sensitive to transitions between pixel values (*i.e.*, texture) and not to the average regional intensity. In between 0 and π, the *uncertainty principle* allows defining rules to design texture operators with optimal spectrum coverage $[\rho_{\min}, \rho_{\max}]$ (see Eq. (2.1)). Ideal texture operators would be accurately localized both in spatial and Fourier domains. On the one hand, well-localized operators in the spatial domain allow identifying precise local texture properties without including surrounding image structures. On the other hand, optimally localized operators in the Fourier domain can precisely characterize narrow frequency bands without mixing with other neighboring spectral components. Unfortunately, having both properties together is not possible and subject to a theoretic limitation called the *uncertainty principle* [6]. Intuitively, the latter can be understood as follows: it is impossible to measure rich texture information from gray-level transitions between a few pixels only. Likewise, measuring all transitions from a large number of pixels yields detailed texture information, but requires large image neighborhoods. More precisely, the relationship between the spatial support $\boldsymbol{G} = G_1 \times \cdots \times G_D$ and the

$g_n(x)$

$\mathcal{F} \updownarrow$

$\hat{g}_n(\omega)$

Figure 2.2 Uncertainty principle: the function of a linear operator $g_n(x)$ (2D circularly symmetric low-pass Gaussian in this example) cannot be well localized both in spatial and Fourier domains. The theoretic limit observes $G_1^2 \Gamma_1^2 G_2^2 \Gamma_2^2 \geq \frac{1}{16}$.

spectral support $\boldsymbol{\Gamma} = \Gamma_1 \times \cdots \times \Gamma_D$ of a linear texture operator is

$$\prod_{d=1}^{D} G_d^2 \Gamma_d^2 \geq \frac{1}{4^D}. \qquad (2.1)$$

This trade-off is illustrated in Fig. 2.2 for 2D circularly symmetric Gaussian operators $g_n(x)$. Therefore an operator with an accurate spatial localization (*i.e.*, narrow support) yields poor spectrum estimates. This has a direct implication in practice, and can be critical when the texture processes are multispectral and highly nonstationary. The spatial support of the operator needs to be large enough to accurately characterize the frequency components of intricate texture processes, and can potentially be larger than the studied texture region. This is illustrated in Fig. 2.3 for the characterization of local ground glass and reticular regions in lung CT. In addition, controlling the profile of the operators (*i.e.*, the decay slope at their boundaries) is important to avoid extensive *ringing effects* in their dual representation resulting in poorly localized analysis (see Fig. 2.4) [7]. Finding the optimal trade-off between the accurate definitions of operator supports in space and in Fourier requires identifying spatial frequencies that are important for the texture segregation task in hand. The lowest discriminative frequency will determine the smallest operator size needed to differentiate between the various texture classes (see Fig. 2.5). The latter is not straightforward in most cases and machine learning can be used to determine discriminative scales [8].

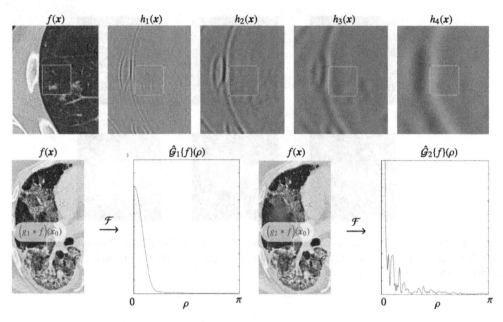

Figure 2.3 Challenges of operator design for complex and nonstationary biomedical texture processes. Top row (peripheral ground glass opacities in chest CT): large influence of proximal objects when the support of operators is larger than the region of interest. The lung boundary has an increasingly important impact in the peripheral region, which can be observed on the response maps $h_n(x)$ of increasingly large LoG linear operators with functions $g_n(x)$ [7]. Bottom row (reticular and normal lung parenchyma in CT): on the left image, a small-sized Gaussian windowed Fourier transform operator g_1 ($\sigma = 3.2$ mm) is precisely located in the reticular pattern but yields a poor characterization of the spectral content inside its support. Conversely, the right image shows a large Gaussian window g_2 ($\sigma = 38.4$ mm) allowing an accurate estimation of spatial frequencies, but its spatial support encroaches upon normal parenchyma and mixes properties from the two distinct texture classes.

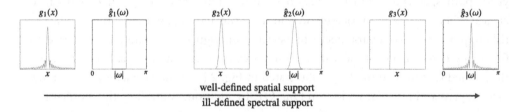

Figure 2.4 Importance of the decay of linear operators on spatial versus spectral support (1D). Smooth operators (center) avoid ringing effects in the dual representation.

2.3.2 Region of interest and response map aggregation

Another critical aspect of scale definition in biomedical texture analysis concerns the design of the ROI for aggregating the operators' response maps (see Fig. 2.1). The fundamental underlying question is: how large must be the ROI M? Addressing this

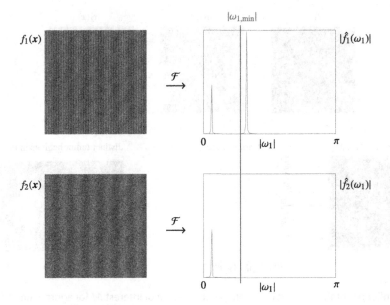

Figure 2.5 A texture operator characterizing spatial frequencies along x_1 in $[\omega_{1,\min}, \pi)$ is enough to discriminate f_1 from f_2. Finding this lower bound in the Fourier domain allow defining texture operators with narrow spatial supports that are optimally localized.

issue requires considering once more the spectral complexity and spatial stationarity of the considered texture processes. On the one hand, M should be large enough to capture the discriminative statistics of the operators' responses. On the other hand, using large M covering several contiguous interleaving nonstationary processes will mix the statistics of the latter and result in meaningless texture measures, even when using appropriate texture operators.

Two examples are developed to illustrate the impact of the size of M on texture classification and segmentation (see Figs. 2.6 and 2.7). For both examples, simple circularly symmetric band-pass and multiscale operators are used. They are based on two consecutive dyadic iterations of Simoncelli wavelet frames [11] $g_n(\boldsymbol{x})$. The 2D version is defined in Fourier in polar coordinates (ρ, ϑ) as

$$
\begin{aligned}
\hat{g}_1(\rho) &= \begin{cases} \cos\left(\frac{\pi}{2}\log_2\left(\frac{2\rho}{\pi}\right)\right) & \text{for } \frac{\pi}{4} < \rho \leq \pi, \\ 0 & \text{otherwise.} \end{cases} \\
\hat{g}_2(\rho) &= \begin{cases} \cos\left(\frac{\pi}{2}\log_2\left(\frac{4\rho}{\pi}\right)\right) & \text{for } \frac{\pi}{8} < \rho \leq \frac{\pi}{2}, \\ 0 & \text{otherwise.} \end{cases}
\end{aligned}
\tag{2.2}
$$

Because it is circularly symmetric, this operator depends on the radial coordinate ρ only and is qualitatively similar to 2D LoGs. It is applied to a texture function

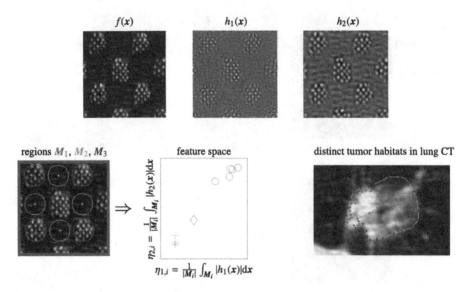

Figure 2.6 Influence of the size and localization of the region of interest M for aggregating the feature maps using the average. Bottom center: each region M_i is represented by a point (*e.g.*, "+") in a feature space spanned by feature averages $(\eta_{1,i}, \eta_{2,i})$. Whereas red and green regions are well separated and regrouped in the feature space, averaging the feature maps over the entire image (blue region) yield texture features that do not correspond to anything visually (see the blue diamond in the feature space). Likewise, averaging texture properties over entire tumor regions including distinct habitats will provide quantitative texture measures that do not correspond to anything biologically [9,10]. (For interpretation of the references to color in this figure legend, the reader is referred to the web version of this chapter.)

f with convolution (*i.e.*, equivalent to a multiplication in the Fourier domain) as $\hat{h}_n(\rho, \vartheta) = \hat{g}_n(\rho) \cdot \hat{f}(\rho, \vartheta)$.

A first example is detailed in Fig. 2.6 and demonstrates the negative impact of averaging feature maps over large ROIs including texture functions from distinct spatial processes. This is a critical issue when texture analysis is used to characterize the structural properties of tumors because it requires defining smaller ROIs based on distinct tumor habitats [9] (*e.g.*, ground glass versus solid tumor components in lung adenocarcinoma [12]). When the component textures are too expensive to delineate (*e.g.*, tumor habitats in 3D imaging) or when they are not known in advance, unsupervised texture segmentation approaches can be used to reveal the diversity of patterns contained in a given ROI. Examples of such methods are superpixels [13], graph cuts [14], or the Pott's model [15]. An advantage of the Pott's model is its ability to handle multiple feature maps for segmenting the subregions and, therefore, can easily run on the registered outputs of several operators. The aggregation function has itself a strong influence on the specificity of the texture measures. In [10], Cirujeda et al. showed that measurements based on the covariances of the operator's responses provided a better characterization

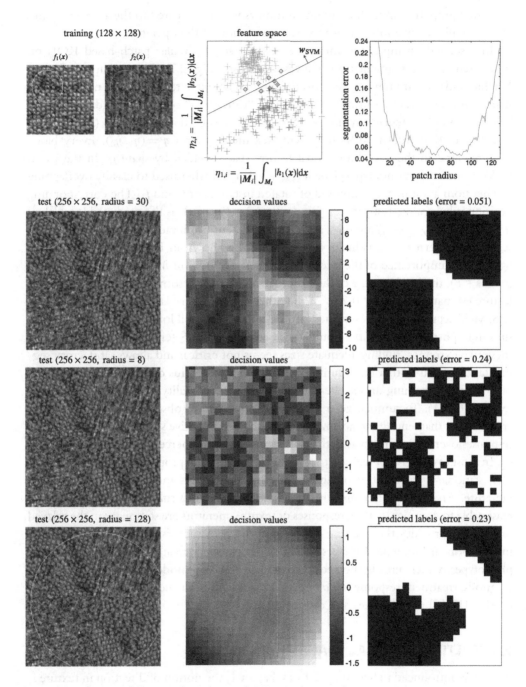

Figure 2.7 Influence of the size of circular patch ROIs on supervised texture segmentation.

of texture properties in multicomponent tumors when compared to the average, thanks to their ability to quantify the pointwise coactivation of the operators.

In a second example, the influence of the size of circular patch-based ROIs on supervised texture segmentation is investigated (see Fig. 2.7). Linear Support Vector Machines (SVM) are trained from overlapping ROIs extracted from two unrotated instances of classes $f_1(\boldsymbol{x})$: "canvas002" and $f_2(\boldsymbol{x})$: "canvas003" of the Outex database [16]. The responses of circularly symmetric Simoncelli wavelet frames (see Eq. (2.2)) are averaged over the circular ROIs to provide texture measures $\boldsymbol{\eta} = (\eta_1, \eta_2)$. Every block is represented in the two-dimensional feature space yielded by span($\boldsymbol{\eta}$). In the latter, SVMs learn a separating hyperplane $\boldsymbol{w}_{\text{SVM}} + b$ that is further used to classify overlapping patches from a test image composed of rotated instances of f_1 and f_2. The corresponding decision values of the test patches (*i.e.*, $\langle \boldsymbol{\eta}, \boldsymbol{w}_{\text{SVM}} \rangle + b$) as well as the predicted local labels (*i.e.*, sgn $(\langle \boldsymbol{\eta}, \boldsymbol{w}_{\text{SVM}} \rangle + b)$) are shown for three different patch radii. The evolution of the segmentation error with radii varying in [0, 128] is shown in Fig. 2.7 top right, highlighting the importance of the size of the aggregation region $\boldsymbol{M}$. When $\boldsymbol{M}$ is too small (radius = 8), the local average of the feature maps is poorly estimated, which yields noisy feature estimates $\boldsymbol{\eta}$ (error = 0.24). At the other extreme, very large regions $\boldsymbol{M}$ (radius = 128) yield accurate estimates of the features, but are not well localized spatially. This results in important errors at the boundaries between f_1 and f_2 (error = 0.23). In between these two extremes, finding adequate sizes seems not critical and allows satisfactory segmentation results with a minimum error of 0.045 for a radius of 90. However, a radius of 30 allows obtaining an excellent trade-off between locality and average estimation (error = 0.051). To summarize, a simple rule of thumb to observe is to use ROIs that are no larger than enough to accurately estimate discriminative statistics of the operators' responses over stationary areas defined in terms of human perception or tissue biology.

In most cases, it is not realistic to assume that the texture properties are homogeneous (*i.e.*, stationary) over entire organs or tumors, which was mostly overlooked in the literature. An interesting approach is to divide organs into subregions for which it is reasonable to consider that the responses of texture operators are stationary in the relaxed sense. This provides the exciting opportunity to construct tissue atlases from texture information in biomedical images. They can be used to create disease-specific digital phenotypes, which already showed to constitute powerful models for predicting disease diagnosis, treatment response, and/or patient prognosis in the context of interstitial lung diseases [17,18] (see Fig. 2.8) and cancer [12,19].

2.4 MULTIDIRECTIONAL IMAGE ANALYSIS

As already introduced in Section 1.2.1 of Chapter 1, the notion of direction in texture is fundamental and complementary to the notion of scale (see Fig. 1.2 of Chapter 1). An important question is then: which image directions are important for deriving texture

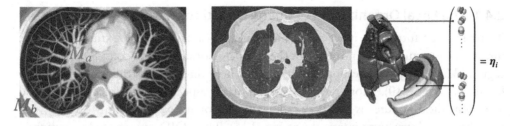

Figure 2.8 Digital phenotypes for interstitial lung diseases. Left: due to the presence of thicker bronchovascular structures in the region M_a that is close to the mediastinium, the texture properties of normal and altered parenchymal tissue cannot be considered similar as the ones in the peripheral region M_b. Right: therefore it is relevant to divide lungs into regions for which it is reasonable to consider that texture properties are homogeneous and create tissue atlases to derive digital phenotypes for interstitial lung diseases [17,18].

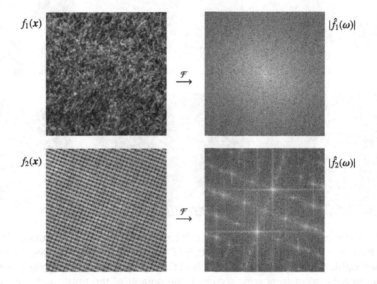

Figure 2.9 Examples of texture functions with weak (*i.e.*, $f_1(x)$) versus strong (*i.e.*, $f_2(x)$) directionality. The modulus of the Fourier representation of $f_2(x)$ shows that there are clear directional spatial frequencies, standing out as bright spots in $|\hat{f}_2(\omega)|$. This also demonstrates that image directionality is defined for a particular scale (*i.e.*, spatial frequency).

measurements and allowing adequate texture segregation? Fig. 2.9 shows textures and their Fourier modulus. For one of them, $f_1(x)$, directionality seems not obvious because the texture contains little structure and is highly stochastic. For the other, $f_2(x)$, dominant directions are clearly visible, creating oriented grid patterns and corresponding peaks in the Fourier domain. Moreover, it appears that texture directionality is defined for a particular position and scale (*i.e.*, spatial frequency). These aspects of texture directionality are developed in the next Subsections 2.4.1, 2.4.2, and 2.4.3.

2.4.1 The Local Organization of Image Directions (LOID)

Thinking even further, it appears that most biomedical and natural textures have clear directional structures or primitives, but the latter are not necessarily consistent over large regions M. Most often the opposite happens where directional structures are defined locally (see Fig. 2.10). More precisely, an important aspect of directionality in natural and biomedical textures is the Local Organization of Image Directions (LOID), *i.e.*, how directional structures intersect (see Fig. 2.11). The LOIDs were already mentioned in the literature as being central in preattentive texture segregation [23] as well as com-

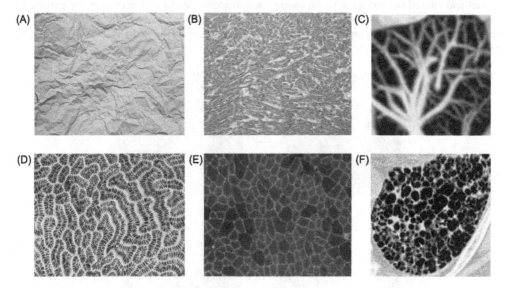

Figure 2.10 Importance of the Local Organization of Image Directions (LOID) in natural and biomedical textures (*i.e.*, how directional structures intersect). (A) Photograph of creased paper. (B) Photomicrograph of hypertrophic cardiomyopathy [20]. (C) Chest CT angiography. (D) Photograph of meandroid coral. (E) Fluorescence microscopy cross-sectional photograph of the tibialis anterior muscle of a mouse [21]. (F) Honeycombing fibrosis in lung CT [22].

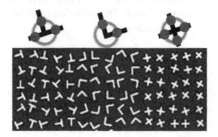

Figure 2.11 Importance of the LOIDs in preattentive texture segregation [23] (see Section 1.2.4 of Chapter 1). The LOIDs can be distinguished by counting the number of endpoints of the primitives (top row).

puterized texture analysis [24,25] (see Section 1.2.4 of Chapter 1). They relate to the primitives or textons of biomedical texture (*i.e.*, the essential "stitches" of the tissue), which often have random local orientations. The various forms of these tissue stitches are even richer in 3D, where the potential complexity of the primitives follows a cubic growth. The number of possible discrete image directions grows as $(2r+1)^3 - 1$ in 3D versus $(2r+1)^2 - 1$ in 2D [1]. Therefore, defining texture operators that are able to characterize the LOIDs in a locally rotation–invariant fashion are required to accurately analyze biomedical texture. Advanced methods to meet these challenging requirements are further developed in Section 2.4.3.

2.4.2 Directional sensitivity of texture operators

It is convenient to consider two distinct categories of operators in terms of directional characterization: directionally sensitive versus insensitive. In the particular case of linear operators, directionally insensitive operators are called *circularly/spherically symmetric* operators and their functions do not depend on the angular coordinate(s):

$$g_n(r) \xleftrightarrow{\mathcal{F}} \hat{g}_n(\rho). \tag{2.3}$$

Examples of such operators in 2D and 3D are Gaussian filters, LoGs, and circularly symmetric wavelets [11,26] (see Fig. 2.12). Examples of nonlinear directionally insensitive operators are max or median filters. Directional operators constitute a vast category where operator functions depend on all polar/spherical coordinates. They include Fourier basis functions, circular and spherical harmonics [27,28], directional filters and wavelets (*e.g.*, Gabor [29], Riesz [30], Simoncelli's steerable pyramid [31], curvelets [32]), Histogram of Oriented Gradients (HOG) [33–35], GLCMs [36], LBPs [37], GLRLMs [38,39], CNNs [40], DL [41,42], and others.

By construction, directionally insensitive operators are locally rotation-invariant, but insensitive to image directions. Texture measures obtained from this category of operators are therefore invariant to local rotations. However, they can hardly differentiate between +-shaped, **L**-shaped or blob-shaped texture primitives (see Fig. 2.13 bottom left). Therefore they cannot characterize the LOIDs and can only be used to distinguish between biomedical tissue types with manifest differences in image scales. Directional counterparts are sensitive to image directions, but may not be locally rotation-invariant even in an approximate sense. They are able to identify the LOIDs only when they have all the same orientation, which is very unlikely in biomedical textures (see Fig. 2.10). Moreover, the characterization of the LOIDs can be challenging even when the latter are all aligned to each other. In fact, the aggregation function plays itself an important role when unidirectional operators[2] are jointly used to characterize the LOIDs [25].

---

[2] Unidirectional operators are "seeing" only one direction.

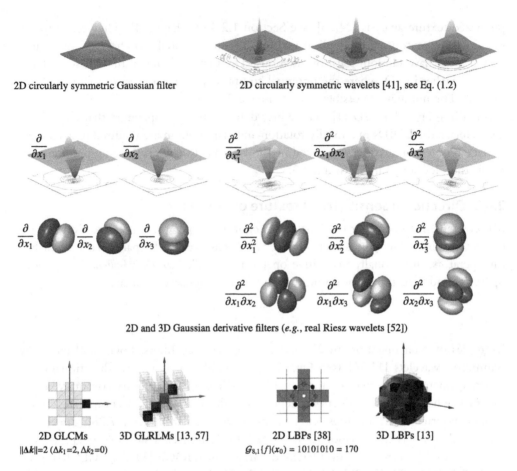

2D circularly symmetric Gaussian filter   2D circularly symmetric wavelets [41], see Eq. (1.2)

2D and 3D Gaussian derivative filters (*e.g.*, real Riesz wavelets [52])

2D GLCMs   3D GLRLMs [13, 57]   2D LBPs [38]   3D LBPs [13]

$\|\Delta \boldsymbol{k}\|=2$ ($\Delta k_1=2, \Delta k_2=0$) $\mathcal{G}_{8,1}\{f\}(\boldsymbol{x}_0) = 10101010 = 170$

Figure 2.12 Directionally sensitive versus insensitive texture operators. Top row: linear circularly symmetric operators that are not sensitive to image directions and, therefore, locally rotation-invariant. Middle rows: linear directionally sensitive operators. Bottom row: nonlinear directionally sensitive operators. Directional operators are sensitive to image directions but not locally rotation-invariant by construction.

When separately integrated, the responses of unidirectional individual operators are not local anymore and their joint responses become only sensitive to the global amount of image directions in the region $\boldsymbol{M}$. For instance, the joint responses of image gradients

$$\mathcal{G}_d\{f\}(\boldsymbol{x}_0) = \frac{\partial f}{\partial x_d}(\boldsymbol{x}_0), \quad d = 1, 2, \tag{2.4}$$

are not able to discriminate between the two textures classes $f_1(\boldsymbol{x})$ and $f_2(\boldsymbol{x})$ shown in Fig. 2.13 when integrated over the full image domain $\boldsymbol{M}$. This loss of information is

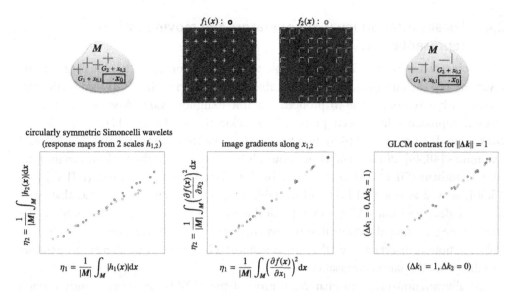

Figure 2.13 $f_1(x)$ and $f_2(x)$ only differ in terms of the LOIDs (*i.e.*, +-shaped versus **L**-shaped). Bottom row: one circle in each feature representation corresponds to one realization (*i.e.*, full image) of f_i, where feature maps are averaged over the entire image. Bottom left: feature vectors η_i obtained from the responses of circularly symmetric operators (two consecutive dyadic scales of Simoncelli wavelets, see Eq. (2.2)) provide poor distinction between the two classes. Bottom center and right: even when the LOIDs are all aligned to each other, the joint responses of directional operators (*e.g.*, image gradients along x_d, cooccurrences along x_d) can hardly discriminate between f_1 and f_2 when integrated over the full image domain M.

detailed by Sifre et al. in terms of separable group invariants [43]. When integrated separately, the responses of unidirectional operators become invariant to a larger family of roto-translations where different orientations are translated by different values. For instance, it can be observed in Fig. 2.13 that f_2 can be obtained from f_1 by vertically translating horizontal bars only and horizontally translating vertical bars only.

Further refinements are required to allow for a true locally rotation-invariant characterization of the LOIDs. Several approaches were proposed to increase the local rotation-invariance of directional operators. GLCMs and GLRLMs are made approximately insensitive to directions either by averaging feature measures or by summing the counts over all directions of the operators [44] (*e.g.*, cooccurrences along x_1 or x_2 are mixed). Likewise, rotation-based data augmentation in CNNs and DL improves invariance to local rotations [45]. Unfortunately, these processes reduce the ability of operators to characterize the LOIDs by making them insensitive to image directions. More advanced approaches were developed to allow enhanced locally rotation-invariant characterization of the LOIDs and are described in Section 2.4.3.

2.4.3 Locally rotation-invariant operators and moving frames representations

It was observed in Section 2.4.2 that simple texture operators are either locally rotation-invariant and directionally insensitive or able to characterize the LOIDs (directionally sensitive), but regrouping both properties is not straightforward. A variety of texture analysis approaches have been proposed to tackle this challenge. Those include the Maximum Response 8 (MR8) filterbank [46], rotation-invariant LBPs [47,37] and extensions [48,49], discrete and continuous HOGs (*e.g.*, used in the Scale-Invariant Feature Transform (SIFT) and in the Rotation-Invariant Feature Transform (RIFT)) [34, 35,50], as well as oriented [51] and steerable [52,25,53] filters and wavelets that were also included in recent CNNs [45,43]. All the aforementioned approaches rely on the same strategy: using directionally-sensitive operators $\mathcal{G}_n$ that are (approximately) locally rotation-equivariant over their own support $\mathbf{G}_n$ and then to align the operators to achieve local rotation-invariance.

Locally rotation-invariant characterization of the LOIDs can be efficiently carried out using Moving Frames (MF) representations [24,25]. The key idea of MFs is to locally adapt a coordinate frame directly to image contours instead of using fixed extrinsic coordinates for all image locations (see Fig. 2.14). The two necessary and sufficient requirements to define MFs are (i) using a set of $N \geq D$ noncollinear texture operators and (ii) having a consistent criteria to define the local orientation of the frame bundle (*e.g.*, using the tangent as the first unit vector of the frame). In 2D, orthonormal MFs are defined as

$$\mathbf{e}_{1,\boldsymbol{x}_0} = \cos\theta_{\boldsymbol{x}_0} \cdot \mathbf{e}_1 + \sin\theta_{\boldsymbol{x}_0} \cdot \mathbf{e}_2, \tag{2.5}$$

$$\mathbf{e}_{2,\boldsymbol{x}_0} = \cos(\theta_{\boldsymbol{x}_0} + \pi/2) \cdot \mathbf{e}_1 + \sin(\theta_{\boldsymbol{x}_0} + \pi/2) \cdot \mathbf{e}_2, \tag{2.6}$$

where $\{\mathbf{e}_1, \mathbf{e}_2\}$ is the canonical basis for $\mathbb{R}^2$, $\{\mathbf{e}_{1,\boldsymbol{x}_0}, \mathbf{e}_{2,\boldsymbol{x}_0}\}$ is a local moving frame bundle for the position $\boldsymbol{x}_0$, and $\theta_{\boldsymbol{x}_0}$ its orientation relatively to $\{\mathbf{e}_1, \mathbf{e}_2\}$. Image representations obtained from MFs are robust to rigid transformations [54] (proof in [25]). They are

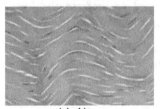

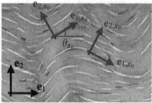

original image   field of dominant directions (gradient)   moving frames (tangent bundles)

Figure 2.14 Construction of MFs in a histopathological image of dense connective tissue (left). Right: MFs are based on local directions $\theta_{\boldsymbol{x}_0}$ maximizing the gradient magnitude (center).

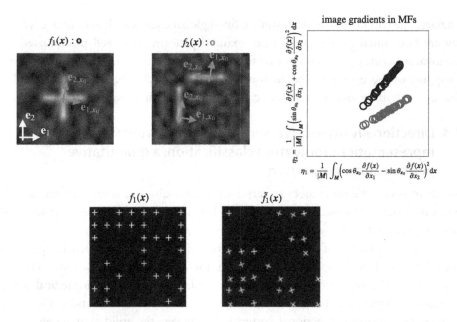

Figure 2.15 Using gradient-based MF representations to discriminate texture classes that only differ in terms of their LOIDs. Because the MF bundle $\{\mathbf{e}_{1,x_0}, \mathbf{e}_{2,x_0}\}$ is locally aligned with the direction θ_{x_0} maximizing the gradient at each position x_0, the energies of gradients along $\mathbf{e}_{2,x_0}$ are null except at the center of +-shaped primitives. The MF representation yields a linearly separable feature representation of f_1 and f_2, while the same unidirectional texture operator pair (*i.e.*, orthogonal image gradients) used in global coordinates could not distinguish between the two (see Fig. 2.13). Bottom row: the flip side of the coin is that MF representation cannot differentiate textures that only differ in terms of the orientations of their primitives (*e.g.*, f_1 versus $\tilde{f}_1$).

invariant to local rotations and equivariant to translations. Moreover, deriving the local orientation of the frame tends to preserve the joint information between positions and orientations even when the operators are integrated (*e.g.*, averaged) over an image domain $\mathbf{M}$. Suitable local orientation measures for defining a consistent MF alignment criterion θ_{x_0} are, *e.g.*, simple pixel differences of a Gaussian-smoothed response map [34] (see Eq. (3.7) of Chapter 3), or the Gaussian-smoothed structure tensor, which can be interpreted as a localized covariance matrix of the gradient [55] (2D [56], 3D [57]). The construction of 2D MFs based on directions maximizing the gradient is illustrated in Fig. 2.14. It can be observed in Fig. 2.15 that when compared to using global coordinates (see Fig. 2.13), texture measures obtained from identical operators (*e.g.*, image gradients) but expressed in MFs can provide very detailed characterizations of the LOIDs.

It is important to note that image representations based on locally rotation-invariant operators $\mathcal{G}_n$ are not preserving image layouts (*i.e.*, large-scale organization of image structure) larger than their spatial supports $G_{1,n} \times \cdots \times G_{D,n}$. They cannot be used alone to characterize natural images with well-defined global layouts such as in ImageNet [58]. They are best suited for discriminating textured patterns with well-pronounced local directional structures, which is the case for most biomedical tissue architectures. Tuning the support of the operators and/or using hybrid approaches based on both local and global image transforms can be required to achieve optimal texture discrimination.

2.4.4 Directionally insensitive, sensitive, and moving frames representations for texture classification: a quantitative performance comparison

In order to reveal the differences in terms of texture discrimination performance between directionally insensitive, sensitive, and MF representations, a quantitative comparison is proposed in this section.

The Outex database of 2D natural textures with highly controlled imaging conditions [16] is chosen to compare between the various representations. The Outex_TC_10 test suite is used, because it has similar properties to biomedical images: it contains no significant changes in terms of image scale and illumination. The texture classes contain strongly directional patterns. Moreover, the validation scheme allows training on unrotated images only, but the testing set contains rotated instances only. This allows evaluating two important properties of the texture representations: rotation-invariance and their ability to characterize directional patterns. The classes are very pure though, where little intraclass variations are present, which differs from most biomedical texture analysis problems. Outex_TC_10 contains 24 classes, which are depicted in Fig. 2.16. It has a total of 4320 ($24 \cdot 20 \cdot 9$) nonoverlapping 128×128 image instances. The training set consists of the 480 ($24 \cdot 20$) unrotated images and the remaining 3840 ($24 \cdot 20 \cdot 8$) images from 8 different orientations are constituting the test set.

Riesz wavelet frames are used as texture operators, because variations in their design allow implementing all three representation types that we want to compare [52]. The latter are detailed in Section 3.2.2 of Chapter 3. Qualitatively, Riesz wavelet frames correspond to multiscale directional image derivatives and evaluate not only the magnitude, but also the type of transitions between image pixels (i.e., derivative order such as the gradient, Hessian). Moreover, Riesz wavelets are steerable, which means that it is relatively inexpensive to locally align every texture operator in order to obtain rich MF representations. Four iterations of Simoncelli's circularly symmetric wavelets (see Eq. (2.2)) are used to define the spatial supports of Riesz-based image derivatives. In order to obtain MF representations, the angle $\theta_{\boldsymbol{x}_0}$ maximizing the response of the first element of the filterbank $\mathcal{G}_{\sigma,L,0}\{f\}(\boldsymbol{x}_0)$ is used to define the MF alignment criteria at each position $\boldsymbol{x}_0$ (see, e.g., Eq. (3.14) of Chapter 3). A collection of texture measurements $\boldsymbol{\eta}$ is obtained by averaging the energies of the responses of each operator over the full 128×128 support $\boldsymbol{M}$ of the texture instances. Simple one-versus-all SVM models

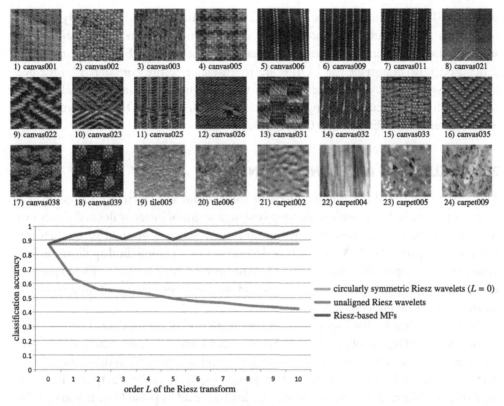

Figure 2.16 Quantitative comparison of directionally insensitive, sensitive, and MF representations for texture classification. Whereas MF representations achieve best performance, directionally insensitive texture operators allow close classification accuracies with a much lower computational complexity. Finally, unaligned directionally sensitive Riesz filters perform poorly because they lack rotation-invariance.

using Gaussian kernels are used to learn decision boundaries in the space spanned by η for texture classification.

The performance comparison for various orders L of the Riesz transform is shown in Fig. 2.16. The performance of directionally sensitive unaligned Riesz wavelets appears to be clearly inferior to both directionally insensitive and MF representations. This can be explained by their lack of rotation invariance, where rotations of the texture instances swaps the responses between filters and results in noisy and diffuse class representations in the feature space. Directionally insensitive representations (*i.e.*, using circularly symmetric Simoncelli wavelets only with Riesz order $L = 0$) benefit from their invariance to local rotations and achieve an honorable classification accuracy of 87.5% with a very simple approach and low computational complexity. Best results are obtained by MF representations with an accuracy of 97.42% for $L = 4$, which highlights the importance of the LOIDs in texture analysis. However, when compared to directionally insensitive operators, MFs involve a much higher computational cost to locally estimate θ_{x_0} and align the Riesz frame accordingly. Therefore the success of MF representations will depend on how important the LOIDs are to characterize the considered biomedical texture classes, which will set the threshold to invest the extra computational cost required. Other techniques allowing the characterization of the LOIDs with invariance to local rotations are discussed in Chapters 3, 5, and 7.

2.5 DISCUSSIONS AND CONCLUSIONS

This chapter explained the essential theoretic foundations and practical consequences of texture operator and aggregation function design in terms of image scales and directions. A set of comparison dimensions was introduced, which can be used to evaluate the relevance of a particular BTA approach or design for a given biomedical application. The most important aspects are recalled in a checklist matrix (see Fig. 2.17). The expression of operators and texture functions in polar and spherical coordinates allowed to clearly separate scale from directional considerations.

Section 2.3 detailed the importance and consequences of appropriate choices of scale for operators and region of interest (see Sections 2.3.1 and 2.3.2, respectively). The uncertainty principle, a fundamental theoretic limitation was recalled to make the relation between operator locality in space and frequency explicit (see Eq. (2.1) and Fig. 2.2). As a rule of thumb, the operators should be kept as small as possible in the spatial do-

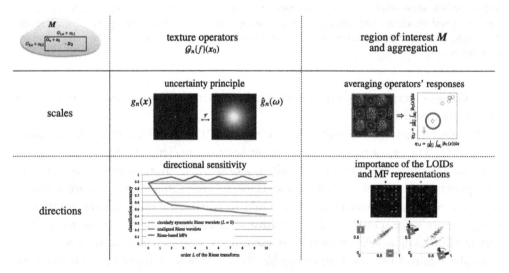

Figure 2.17 Checklist matrix of essential theoretic foundations and practical consequences in terms of choices of scales and directions for texture operator and aggregation function design.

main to allow local texture analysis. However, they should be sufficiently large to allow for an adequate and accurate characterization of texture frequency components (see Figs. 2.3 bottom row and 2.5). The importance of using operators with smooth profiles was highlighted to obtain an optimal trade-off between localization in the spatial and Fourier domains [3] (see Fig. 2.4). In addition, operator smoothness allows limiting the effect of proximal objects surrounding the region of interest for texture analysis, which was illustrated in Fig. 2.3 (top row). The influence of the design and shape of the aggregation region was detailed in Section 2.3.2. In particular, the hazard of using large ROIs encompassing multiple nonstationary texture processes for aggregation with integrative aggregation functions was highlighted. The latter will mix texture properties of distinct tissues, yielding texture measurement that are not corresponding to anything visually or biologically (see Fig. 2.6). This was found to be widely overlooked in the literature and motivated the use of alternate aggregation functions (*e.g.*, covariances [10]) as well as defining digital tissue atlases of tumors or organs containing collections of regions for which it is reasonable to consider that texture properties are homogeneous. The latter provides the exciting opportunity to construct disease-specific digital phenotypes, which already showed to constitute powerful models for predicting disease diagnosis, treatment response and/or patient prognosis [17,18,12,19] (see Fig. 2.8).

The second part of the chapter, Section 2.4, studied the type of directional information that is relevant for BTA. It was found that directional structures are defined locally but are not necessarily consistent over large regions. More precisely, a fundamental aspect of texture directionality is how directional structure intersect, which we called the Local Organization of Image Directions (LOID, see Section 2.4.1). The latter relate to the essential stitches of biomedical tissue, as well as to the texture primitives and texton theory discussed in Sections 1.2.4 of Chapter 1 and 3.2.3 of Chapter 3. Two adversarial categories of operators were analyzed in Section 2.4.2: directionally insensitive versus sensitive (see Fig. 2.12). Designing texture operators that are able to accurately characterize the LOIDs with robustness to rigid transformation was found to be challenging. One the one hand, directionally insensitive operators are invariant to local rotations but are insensitive to image directions. On the other hand, directionally sensitive operators can sense directions but are not invariant to local rotations. In addition, the effect of aggregation using integrative functions kills the ability of unidirectional operators to characterize the LOIDs even when the latter are all aligned to each other. In Section 2.4.3, we provided evidence that using Moving Frame (MF) texture representations consisting of locally aligning sets of noncollinear operators (see Eq. (2.5) and Fig. 2.14) allowed robust recognition of the LOIDs with invariance to local rigid transformations. A quantitative comparison of the classification performance of texture classes with pronounced directional patterns and nonrigid transformations confirmed the superiority of MF representations, but at the expense of a high computational complexity when compared to much simpler directionally insensitive representations (see Section 2.4.4).

Overall, this chapter introduced a new set of comparison dimensions between BTA methods that is specific to biomedical imaging. The latter is further used in Chapter 3 to perform a systematic qualitative comparison of most popular BTA methods, which constitutes a user guide to assess the relevance of each approach for a particular medical or biological task in hand.

ACKNOWLEDGMENTS

This work was supported by the Swiss National Science Foundation (under grant PZ00P2_154891) and the CIBM.

REFERENCES

[1] A. Depeursinge, A. Foncubierta-Rodríguez, D. Van De Ville, H. Müller, Three-dimensional solid texture analysis and retrieval in biomedical imaging: review and opportunities, Med. Image Anal. 18 (1) (2014) 176–196.

[2] M. Petrou, Texture in biomedical images, in: T.M. Deserno (Ed.), Biomedical Image Processing, Springer-Verlag, Berlin, Heidelberg, 2011, pp. 157–176.

[3] B.M. ter Haar Romeny, Multi-scale and multi-orientation medical image analysis, in: T.M. Deserno (Ed.), Biomedical Image Processing, Springer-Verlag, Berlin, Heidelberg, 2011, pp. 177–196.

[4] Q. Wang, O. Ronneberger, H. Burkhardt, Fourier Analysis in Polar and Spherical Coordinates, Technical Report, Albert-Ludwigs-Universität Freiburg, Institut für Informatik, 2008.

[5] A. Depeursinge, D. Van De Ville, A. Platon, A. Geissbuhler, P.-A. Poletti, H. Müller, Near-affine-invariant texture learning for lung tissue analysis using isotropic wavelet frames, IEEE Trans. Inf. Technol. Biomed. 16 (4) (July 2012) 665–675.

[6] M. Petrou, P. García Sevilla, Non-Stationary Grey Texture Images, John Wiley & Sons, Ltd, 2006, pp. 297–606.

[7] A. Depeursinge, P. Pad, A.C. Chin, A.N. Leung, D.L. Rubin, H. Müller, M. Unser, Optimized steerable wavelets for texture analysis of lung tissue in 3-D CT: classification of usual interstitial pneumonia, in: IEEE 12th International Symposium on Biomedical Imaging, ISBI 2015, IEEE, April 2015, pp. 403–406.

[8] A.K. Jain, K. Karu, Learning texture discrimination masks, IEEE Trans. Pattern Anal. Mach. Intell. 18 (2) (February 1996) 195–205.

[9] R.A. Gatenby, O. Grove, R.J. Gillies, Quantitative imaging in cancer evolution and ecology, Radiology 269 (1) (2013) 8–14.

[10] P. Cirujeda, Y. Dicente Cid, H. Müller, D. Rubin, T.A. Aguilera, B.W. Loo Jr., M. Diehn, X. Binefa, A. Depeursinge, A 3-D Riesz-covariance texture model for prediction of nodule recurrence in lung CT, IEEE Trans. Med. Imaging (2016), http://dx.doi.org/10.1109/TMI.2016.2591921.

[11] J. Portilla, E.P. Simoncelli, A parametric texture model based on joint statistics of complex wavelet coefficients, Int. J. Comput. Vis. 40 (1) (2000) 49–70.

[12] A. Depeursinge, M. Yanagawa, A.N. Leung, D.L. Rubin, Predicting adenocarcinoma recurrence using computational texture models of nodule components in lung CT, Med. Phys. 42 (2015) 2054–2063.

[13] R. Achanta, A. Shaji, K. Smith, A. Lucchi, P. Fua, S. Süsstrunk, SLIC superpixels compared to state-of-the-art superpixel methods, IEEE Trans. Pattern Anal. Mach. Intell. 34 (11) (2012) 2274–2281.

[14] Y. Boykov, O. Veksler, R. Zabih, Fast approximate energy minimization via graph cuts, IEEE Trans. Pattern Anal. Mach. Intell. 23 (11) (November 2001) 1222–1239.

[15] M. Storath, A. Weinmann, M. Unser, Unsupervised texture segmentation using monogenic curvelets and the Potts model, in: IEEE International Conference on Image Processing, ICIP, October 2014, pp. 4348–4352.

[16] T. Ojala, T. Mäenpää, M. Pietikäinen, J. Viertola, J. Kyllönen, S. Huovinen, Outex – new framework for empirical evaluation of texture analysis algorithms, in: 16th International Conference on Pattern Recognition, in: ICPR, vol. 1, IEEE Computer Society, August 2002, pp. 701–706.

[17] A. Depeursinge, A.C. Chin, A.N. Leung, D. Terrone, M. Bristow, G. Rosen, D.L. Rubin, Automated classification of usual interstitial pneumonia using regional volumetric texture analysis in high-resolution CT, Invest. Radiol. 50 (4) (April 2015) 261–267.

[18] Y. Dicente Cid, H. Müller, A. Platon, J.-P. Janssens, L. Frédéric, P.-A. Poletti, A. Depeursinge, A lung graph-model for pulmonary hypertension and pulmonary embolism detection on DECT images, in: MICCAI Workshop on Medical Computer Vision: Algorithms for Big Data, MICCAI-MCV, 2016.

[19] S. Torquato, Toward an Ising model of cancer and beyond, Phys. Biol. 8 (1) (2011) 15017.

[20] P. Gattuso, V.B. Reddy, O. David, D.J. Spitz, M.H. Haber, Differential Diagnosis in Surgical Pathology, Elsevier Health Sciences, 2009.

[21] A. Briguet, I. Courdier-Fruh, M. Foster, T. Meier, J.P. Magyar, Histological parameters for the quantitative assessment of muscular dystrophy in the *mdx*-mouse, Neuromuscul. Disord. 14 (10) (2004) 675–682.

[22] A. Depeursinge, A. Vargas, A. Platon, A. Geissbuhler, P.-A. Poletti, H. Müller, Building a reference multimedia database for interstitial lung diseases, Comput. Med. Imaging Graph. 36 (3) (April 2012) 227–238.

[23] J.R. Bergen, M.S. Landy, Computational modeling of visual texture segregation, in: Computational Models of Visual Processing, MIT Press, 1991, pp. 253–271.

[24] O. Ben-Shahar, S.W. Zucker, The perceptual organization of texture flow: a contextual inference approach, IEEE Trans. Pattern Anal. Mach. Intell. 25 (4) (April 2003) 401–417.

[25] A. Depeursinge, Z. Püspöki, J.-P. Ward, M. Unser, Steerable wavelet machines (SWM): learning moving frames for texture classification, IEEE Trans. Image Process. 26 (4) (2017) 1626–1636.

[26] D. Van De Ville, T. Blu, M. Unser, Isotropic polyharmonic B-splines: scaling functions and wavelets, IEEE Trans. Image Process. 14 (11) (November 2005) 1798–1813.

[27] M. Unser, N. Chenouard, A unifying parametric framework for 2D steerable wavelet transforms, SIAM J. Imaging Sci. 6 (1) (2013) 102–135.

[28] J.P. Ward, M. Unser, Harmonic singular integrals and steerable wavelets in $L_2(\mathbb{R}^d)$, Appl. Comput. Harmon. Anal. 36 (2) (March 2014) 183–197.

[29] A.C. Bovik, M. Clark, W.S. Geisler, Multichannel texture analysis using localized spatial filters, IEEE Trans. Pattern Anal. Mach. Intell. 12 (1) (1990) 55–73.

[30] M. Unser, N. Chenouard, D. Van De Ville, Steerable pyramids and tight wavelet frames in $L_2(\mathbb{R}^d)$, IEEE Trans. Image Process. 20 (10) (October 2011) 2705–2721.

[31] E.P. Simoncelli, W.T. Freeman, The steerable pyramid: a flexible architecture for multi-scale derivative computation, in: Proceedings of International Conference on Image Processing, 1995, vol. 3, October 1995, pp. 444–447.

[32] E.J. Candès, D.L. Donoho, Curvelets – a surprisingly effective nonadaptive representation for objects with edges, in: Curves and Surface Fitting, Vanderbilt University Press, Nashville, 2000, pp. 105–120.

[33] D.G. Lowe, Object recognition from local scale invariant features, in: Proceedings of the International Conference of Computer Vision, Corfu, Greece, September 1999.

[34] D.G. Lowe, Distinctive image features from scale-invariant keypoints, Int. J. Comput. Vis. 60 (2) (2004) 91–110.

[35] K. Liu, H. Skibbe, T. Schmidt, T. Blein, K. Palme, T. Brox, O. Ronneberger, Rotation-invariant HOG descriptors using Fourier analysis in polar and spherical coordinates, Int. J. Comput. Vis. 106 (3) (2014) 342–364.

[36] R.M. Haralick, Statistical and structural approaches to texture, Proc. IEEE 67 (5) (May 1979) 786–804.

[37] T. Ojala, M. Pietikäinen, T. Mäenpää, Multiresolution gray-scale and rotation invariant texture classification with local binary patterns, IEEE Trans. Pattern Anal. Mach. Intell. 24 (7) (July 2002) 971–987.

[38] M.M. Galloway, Texture analysis using gray level run lengths, Comput. Graph. Image Process. 4 (2) (1975) 172–179.

[39] D.-H. Xu, A.S. Kurani, J. Furst, D.S. Raicu, Run-length encoding for volumetric texture, in: The 4th IASTED International Conference on Visualization, Imaging, and Image Processing, VIIP 2004, Marbella, Spain, September 2004.

[40] A.S. Razavian, H. Azizpour, J. Sullivan, S. Carlsson, CNN features off-the-shelf: an astounding baseline for recognition, in: IEEE Computer Society Conference on Computer Vision and Pattern Recognition Workshops, 2014, pp. 512–519.

[41] M.J. Gangeh, A. Ghodsi, M.S. Kamel, Dictionary learning in texture classification, in: Proceedings of the 8th International Conference on Image Analysis and Recognition, Part I, 2011, pp. 335–343.

[42] J. Mairal, F. Bach, J. Ponce, G. Sapiro, A. Zisserman, Supervised dictionary learning, in: Advances in Neural Information Processing Systems, 2008, pp. 1033–1040.

[43] L. Sifre, S. Mallat, Rigid-motion scattering for texture classification, CoRR, abs/1403.1687, 2014.

[44] G. Thibault, B. Fertil, C. Navarro, S. Pereira, P. Cau, N. Levy, J. Sequeira, J.-L. Mari, Texture indexes and gray level size zone matrix application to cell nuclei classification, in: Pattern Recognition and Information Processing, 2009, pp. 140–145.

[45] D. Marcos Gonzalez, M. Volpi, D. Tuia, Learning rotation invariant convolutional filters for texture classification, CoRR, abs/1604.06720, 2016.

[46] M. Varma, A. Zisserman, A statistical approach to texture classification from single images, Int. J. Comput. Vis. 62 (1–2) (2005) 61–81.

[47] T. Ahonen, J. Matas, C. He, M. Pietikäinen, Rotation invariant image description with local binary pattern histogram Fourier features, in: A.-B. Salberg, J. Hardeberg, R. Jenssen (Eds.), Image Analysis, in: Lecture Notes in Computer Science, vol. 5575, Springer, Berlin, Heidelberg, 2009, pp. 61–70.

[48] Z. Guo, L. Zhang, D. Zhang, A completed modeling of local binary pattern operator for texture classification, IEEE Trans. Image Process. 19 (6) (June 2010) 1657–1663.

[49] L. Liu, S. Lao, P.W. Fieguth, Y. Guo, X. Wang, M. Pietikäinen, Median robust extended local binary pattern for texture classification, IEEE Trans. Image Process. 25 (3) (March 2016) 1368–1381.

[50] S. Lazebnik, C. Schmid, J. Ponce, A sparse texture representation using local affine regions, IEEE Trans. Pattern Anal. Mach. Intell. 27 (8) (August 2005) 1265–1278.

[51] Y. Xu, X. Yang, H. Ling, H. Ji, A new texture descriptor using multifractal analysis in multi-orientation wavelet pyramid, in: IEEE Conference on Computer Vision and Pattern Recognition, CVPR, June 2010, pp. 161–168.

[52] A. Depeursinge, A. Foncubierta-Rodríguez, D. Van De Ville, H. Müller, Rotation-covariant texture learning using steerable Riesz wavelets, IEEE Trans. Image Process. 23 (2) (February 2014) 898–908.

[53] Y. Dicente Cid, H. Müller, A. Platon, P.-A. Poletti, A. Depeursinge, 3D solid texture classification using locally-oriented wavelet transforms, IEEE Trans. Image Process. 26 (4) (2017) 1899–1910, http://dx.doi.org/10.1109/TIP.2017.2665041.

[54] O.D. Faugeras, Cartan's Moving Frame Method and Its Application to the Geometry and Evolution of Curves in the Euclidean, Affine and Projective Planes, Technical Report, Institut National de Recherche en Informatique et en Automatique (INRIA), 1993.

[55] Z. Püspöki, V. Uhlmann, C. Vonesch, M. Unser, Design of steerable wavelets to detect multifold junctions, IEEE Trans. Image Process. 25 (2) (February 2016) 643–657.

[56] L. Sorensen, M. Nielsen, P. Lo, H. Ashraf, J.H. Pedersen, M. de Bruijne, Texture-based analysis of COPD: a data-driven approach, IEEE Trans. Med. Imaging 31 (1) (January 2012) 70–78.

[57] N. Chenouard, M. Unser, 3D steerable wavelets and monogenic analysis for bioimaging, in: 2011 IEEE International Symposium on Biomedical Imaging: from Nano to Macro, April 2011, pp. 2132–2135.

[58] J. Deng, W. Dong, R. Socher, L.-J. Li, K. Li, L. Fei-Fei, ImageNet: a large-scale hierarchical image database, in: IEEE Conference on Computer Vision and Pattern Recognition, CVPR 2009, 2009, pp. 248–255.

CHAPTER 3

Biomedical Texture Operators and Aggregation Functions
A Methodological Review and User's Guide

Adrien Depeursinge*,†, Julien Fageot*

*École Polytechnique Fédérale de Lausanne (EPFL), Biomedical Imaging Group, Lausanne, Switzerland
†University of Applied Sciences Western Switzerland (HES-SO), Institute of Information Systems, Sierre, Switzerland

Abstract

This chapter reviews most popular texture analysis approaches under novel comparison axes that are specific to biomedical imaging. A concise checklist is proposed as a user guide to assess the relevance of each approach for a particular medical or biological task in hand. We revealed that few approaches are regrouping most of the desirable properties for achieving optimal performance. In particular, moving frames texture representations based on learned steerable operators showed to enable data-specific and rigid-transformation-invariant characterization of local directional patterns, the latter being a fundamental property of biomedical textures. Potential limitations of having recourse to data augmentation and transfer learning for deep convolutional neural networks and dictionary learning approaches to palliate the lack of large annotated training collections in biomedical imaging are mentioned. We conclude by summarizing the strengths and limitations of current approaches, providing insights on key aspects required to build the next generation of biomedical texture analysis approaches.

Keywords

Biomedical texture analysis, Convolutional neural networks, Deep learning, Dictionary learning, Gabor filters, Gray-level cooccurrence matrices, Gray-level run-length matrices, Gray-level size-zone matrices, Local binary patterns, Steerable wavelets

3.1 INTRODUCTION

When starting from scratch to enable quantitative image analysis in a biomedical research project or a clinical setting, the number of available approaches can be disconcerting. Whereas implementing intensity-based features already raises several challenges in terms of reproducibility and relevance to a particular medical or biological applicative context, finding adequate texture measures requires extensive expertise as well as time-consuming iterative validation processes. Knowing precisely the type of textural information sought and further opting for the appropriate analysis technique is as much challenging as crucial for success. To that end, Section 1.2 of Chapter 1 proposed a formal definition of Biomedical Texture (BT) from both perceptual and mathematical

Biomedical Texture Analysis
DOI: 10.1016/B978-0-12-812133-7.00003-X

perspectives. It was suggested that BTs are realizations of intricate and nonstationary spatial stochastic processes, and that spatial scales and directions are fundamental properties of BTs. The relation between scale and directions was further developed in Section 2.4.1 of Chapter 2, suggesting that the Local Organization of Image Directions (LOID) is a fundamental property of BT. The LOIDs relate to the texton theory (see Section 1.2.4 of Chapter 1), where textons are crucial elements of preattentive vision [1]. Section 1.3.1 of Chapter 1 introduced a general framework of texture analysis methods where any BTA approach can be decomposed into a succession of local texture operators and regional aggregation functions. In addition, Section 1.3.3 of Chapter 1 and Chapter 7 provided evidence that the type of geometric invariances required for BTA are invariances to nonrigid transformations,[1] which considerably differs from general purpose texture analysis traditionally used in computer vision for photographic image analysis. These aspects were further clarified and exemplified in Chapter 2, where a set of comparison dimensions between BTA methods was introduced. In particular, nonexclusive categories of texture operators were presented, including directionally insensitive/sensitive, aligned, Moving Frames (MF), and learned. The discrimination abilities of each category was evaluated, where the destructive effects of integrative aggregation were demonstrated. In particular, it was found that some groups of operators lose their ability to characterize the LOIDs when integrated over a Region Of Interest (ROI), the latter being required to obtain scalar-valued texture measurements. The ultimate challenging requirement of combining the ability to describe the LOIDs with robustness to rigid transformations was only fulfilled by the groups of MF operators.

In this work, we compare most popular BTA approaches under the light of the novel comparison dimensions introduced in Chapter 2. A concise checklist is proposed to assess the relevance of each BTA approach for a particular medical or biological task in hand. The main groups of methods considered are (i) convolutional (Section 3.2, further information on CNNs can be found in Chapters 4 and 9), (ii) Gray-Level Matrices (GLM, Section 3.3), (iii) Local Binary Patterns (LBP, Section 3.4), and (iv) fractals (Section 3.5, further information can be found in Chapter 5). The main categories of BTA methods are summarized in Fig. 3.1. Our choice to review these categories is based on their popularity and diversity. The strengths and weaknesses of the methods are summarized and discussed in Section 3.6. The notation used in this chapter are based on Section 1.2.2 of Chapter 1.

---

[1] Optimal operators should be equivariant to translations and locally equivariant to rotations.

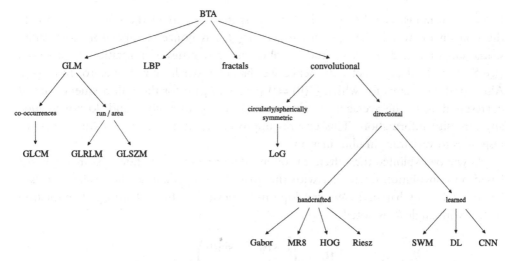

Figure 3.1 Overview and dependence of the BTA approaches reviewed in this chapter. These categories were chosen based on their popularity and diversity.

3.2 CONVOLUTIONAL APPROACHES

A large group of approaches called *convolutional* are based on linear texture operators. In spatial domain, the application of a linear operator $\mathcal{G}_n$ to $f(\boldsymbol{x})^2$ at the position $\boldsymbol{x}_0$ is characterized by the *operator function* $g_n(\boldsymbol{x})$ as

$$\mathcal{G}\{f\}(\boldsymbol{x}_0) = (g_n * f)(\boldsymbol{x}) = \int_{\boldsymbol{x} \in \mathbb{R}^D} f(\boldsymbol{x}) g_n(\boldsymbol{x}_0 - \boldsymbol{x}) \, \mathrm{d}\boldsymbol{x}. \tag{3.1}$$

The operation $(g_n * f)(\boldsymbol{x}_0)$ is called a *convolution*: we slide the operator function $g_n(\boldsymbol{x})$ to the input texture function $f(\boldsymbol{x})$ over all positions $\boldsymbol{x}_0$. When the position $\boldsymbol{x}_0$ is fixed, we remark that the value of $\mathcal{G}\{f\}(\boldsymbol{x}_0)$ is simply the scalar product

$$\mathcal{G}_n\{f\}(\boldsymbol{x}_0) = \langle f(\cdot), g_n(\boldsymbol{x}_0 - \cdot)\rangle. \tag{3.2}$$

A convolution in the spatial domain corresponds to a multiplication in the Fourier domain as

$$(g_n * f)(\boldsymbol{x}) \overset{\mathcal{F}}{\longleftrightarrow} \hat{g}_n(\boldsymbol{\omega}) \cdot \hat{f}(\boldsymbol{\omega}).$$

Convolution operators are the ones for which the response map $h_n(\boldsymbol{x})$ depends linearly on the input texture function $f(\boldsymbol{x})$. The definition of the function $g_n(\boldsymbol{x})$ is a priori

---

[2] In this section, except as otherwise stipulated, we will consider continuous operators and functions indexed by the coordinate vector $\boldsymbol{x}$. Discretized versions can be obtained following the notions introduced in Section 1.2.2 of Chapter 1.

free of constraints. For the convolution, the spatial support of the operator is exactly the domain on which the operator function $g_n(\boldsymbol{x})$ is nonzero. Operators with finite spatial supports are desirable to study local texture properties of nonstationary processes (see Section 2.3.1 of Chapter 2), hence we shall consider localized operator functions. Also, band-pass filters for which $\hat{g}_n(\boldsymbol{0}) = 0$ (*i.e.*, zero gain for the null frequency $\boldsymbol{\omega} = \boldsymbol{0}$ corresponding to the average) are able to focus on texture alone and do not include any intensity information. This ensures improved robustness of the texture operator responses to variations in illumination.

As will be exploited thereafter, it is often interesting to define new texture operators based on convolution operators, with the goal of locally aligning the orientation (see Section 2.4.3 of Chapter 2). We develop this framework in the 2D setting. The rotation matrix with angle θ_0 is noted

$$\mathbf{R}_{\theta_0} = \begin{pmatrix} \cos\theta_0 & -\sin\theta_0 \\ \sin\theta_0 & \cos\theta_0 \end{pmatrix}.$$

Consider a convolution operator $\mathcal{G}_n$ with operator function $g_n(\boldsymbol{x})$. For each location $\boldsymbol{x}_0$, we set

$$\theta_{\boldsymbol{x}_0} = \arg\max_{\theta_0 \in [0, 2\pi)} \langle f(\cdot), g_n(\boldsymbol{x}_0 - \mathbf{R}_{\theta_0}\cdot) \rangle \qquad (3.3)$$

Here, $\theta_{\boldsymbol{x}_0}$ is the angle that maximizes the scalar product between the texture image $f(\boldsymbol{x})$ and the rotated version of $g_n(\boldsymbol{x})$ around the location $\boldsymbol{x}_0$. It is therefore the angle for which the operator function is the most aligned with $f(\boldsymbol{x})$ at position $\boldsymbol{x}_0$. We then define

$$\mathcal{H}_n\{f\}(\boldsymbol{x}_0) = \langle f(\cdot), g_n(\boldsymbol{x}_0 - \mathbf{R}_{\theta_{\boldsymbol{x}_0}}\cdot) \rangle \qquad (3.4)$$
$$= \int_{\boldsymbol{x}\in\mathbb{R}^D} f(\boldsymbol{x}) g_n(\boldsymbol{x}_0 - \mathbf{R}_{\theta_{\boldsymbol{x}_0}}\boldsymbol{x}) \mathrm{d}\boldsymbol{x}.$$

The response map $h_n(\boldsymbol{x}) = \mathcal{H}_n\{f\}(\boldsymbol{x})$ is no longer linear in $f(\boldsymbol{x})$, therefore the texture operator $\mathcal{H}_n$ is not a convolution operator in the sense of Eq. (3.1). We include the angle alignment in this section since it is based on the convolutional framework and characterized by the operator function $g_n(\boldsymbol{x})$ according to Eq. (3.4). Convolution operators and local angle alignment together allow to design texture operators equivariant to local rotations that can be used to study the local orientation in the texture image (see Section 1.3.3 of Chapter 1).

The most popular convolutional approaches in texture analysis are detailed in the following subsections while distinguishing three main categories of convolutional operator functions (see Fig. 3.1): circularly/spherically symmetric filters (Section 3.2.1), directional filters (Section 3.2.2), and learned filters (Section 3.2.3). It is worth noting that directional and learned filters are nonexclusive categories.

3.2.1 Circularly/spherically symmetric filters

Circularly/spherically symmetric filters are convolutional texture operators with functions that only depend on the radial polar coordinate r (see Section 2.2 of Chapter 2): $g_n(\boldsymbol{x}) = g_n(||\boldsymbol{x}||) = g_n(r)$. A direct consequence of this is their complete lack of directional sensitivity and their invariance to local rotations.

A simple example of such convolutional texture processing is based on Laplacian of Gaussian (LoG) filters.[3] Their handcrafted operator function $g_\sigma(\boldsymbol{x})$ is a radial second-order derivative of a D–dimensional Gaussian filter as

$$g_\sigma(\boldsymbol{x}) = -\frac{1}{\pi \sigma^2}\left(1 - \frac{||\boldsymbol{x}||^2}{2\sigma^2}\right)e^{-\frac{||\boldsymbol{x}||^2}{2\sigma^2}}, \tag{3.5}$$

where the standard deviation of the Gaussian σ controls the scale of the operator. LoGs are band-pass and circularly/spherically symmetric. It is straightforward to notice in Eq. (3.5) that $g_\sigma(\boldsymbol{x})$ only depends on the norm of $\boldsymbol{x}$ and is therefore circularly symmetric: we have $g_\sigma(\boldsymbol{x}) = g_\sigma(r)$. Their 2D profiles and response maps obtained with synthetic tumors are depicted in Fig. 1.12 of Chapter 1. LoGs can also be approximated by a difference of two Gaussians.[4] LoGs were mentioned to be important in biological visual processing by Marr in [2].

Multiscale texture measurements η can be obtained by averaging the absolute values or the energies of the response maps (i.e., $|h_{\sigma_n}(\boldsymbol{x})|$ or $h_{\sigma_n}^2(\boldsymbol{x})$, respectively) of a series of operators with increasing values of $\sigma_1 < \sigma_2 < \cdots < \sigma_N$ (see Fig. 1.12 of Chapter 1). The average is computed over a ROI $\boldsymbol{M}$. They are implemented in 2D in the TexRAD[5] commercial medical research software [4]. The properties of LoG filters are summarized in Table 3.1.

3.2.2 Directional filters

Several handcrafted approaches were proposed for texture analysis using directional filters. The latter are sensitive to image directions (see Section 2.4.2 of Chapter 3). The most important and popular methods are discussed in this section, including Gabor wavelets (Section 3.2.2.1), Maximum Response 8 (MR8, Section 3.2.2.2), Histogram of Oriented Gradients (HOG, Section 3.2.2.3), and the Riesz transform (Section 3.2.2.4).

---

[3] They are also called *Mexican hat* filters due their 2D shape.

[4] The Difference of Gaussians (DoG) best approximates LoG when the ratio of the two variances of the Gaussians are $\sigma_1 = \frac{\sigma_2}{\sqrt{2}}$.

[5] http://texrad.com, University of Sussex, Brighton, UK, as of November 21, 2016.

Table 3.1 Properties LoG filters

Operator linearity	Linear.
Handcrafted	Yes.
3D extension	Trivial as the second-order derivative of a 3D Gaussian function. The LoGs are circularly/spherically symmetric functions which only depend on the radial coordinate r.
Coverage of image directions	Complete: the angular part is constant for a fixed radius r.
Directionality and local rotation-invariance	Directionally insensitive and locally rotation-invariant.
Characterization of the LOIDs	No.
Coverage of image scales	Incomplete for a single value of σ. However, their wavelet extension allows a full coverage of the spatial spectrum [3].
Band-pass	Yes.
Gray-level reduction	Not required.
Illumination-invariance	No. However, they are robust to changes in illumination because LoGs are band-pass filters.
Aggregation function	Integrative (typically): computes the average of the absolute values or the energies of the response maps over $\boldsymbol{M}$.

3.2.2.1 Gabor wavelets

A popular example of directional texture operators is the family of Gabor wavelets [5–7], which consists in a systematic parcellation of the Fourier domain with elliptic Gaussian windows (see Fig. 3.2). The D-dimensional filter function of the Gabor operator $g_{\rho,\sigma,\theta}(x)$ is the most general function that minimizes the uncertainty principle [8] (see Section 2.3.1 of Chapter 2). Moreover, it was suggested in [9] that 2D Gabor filters are appropriate models of the transfer function of simple cells in the visual cortex of mammalian brains and thus mimicking the early layers of human visual perception. In the spatial domain, they correspond to Gaussian-windowed oscillatory functions as

$$g_{\rho,\sigma,\theta}(\boldsymbol{x}) = \frac{\sigma_1\sigma_2}{2\pi} \, e^{-\sigma_1^2\tilde{x}_1^2+\sigma_2^2\tilde{x}_2^2} \, e^{j2\pi\rho\tilde{x}_1}, \qquad (3.6)$$

where ρ is the center of the Gaussian window in Fourier, $\boldsymbol{\sigma} = (\sigma_1, \sigma_2)$ contains the radial and orthoradial standard deviations of the Gaussian window in Fourier, $\tilde{\boldsymbol{x}} = (\tilde{x}_1, \tilde{x}_2) = \mathbf{R}_\theta \boldsymbol{x}$, and $\tilde{\boldsymbol{\omega}} = (\tilde{\omega}_1, \tilde{\omega}_2) = \mathbf{R}_\theta \boldsymbol{\omega}$ (see Fig. 3.2).

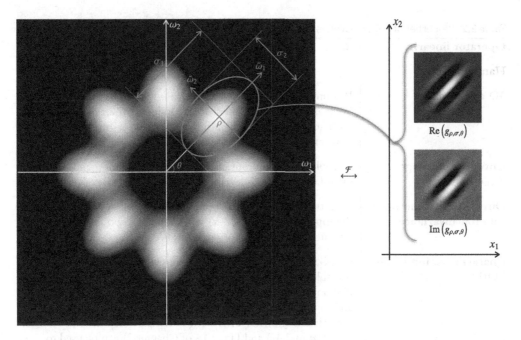

Figure 3.2 One isolated scale ρ of Gabor wavelets with 4 orientations θ and a spectral dispersion of $\sigma = (\sigma_1, \sigma_2)$. Note that the spatial angular polar coordinate θ is equivalent to the Fourier angular coordinate ϑ.

Multiscale and multiorientation texture measurements $\boldsymbol{\eta}$ are obtained by averaging the absolute values or the energies of the response maps (*i.e.*, $|h_{\rho,\sigma,\theta}(\boldsymbol{x})|$ or $h_{\rho,\sigma,\theta}^2(\boldsymbol{x})$, respectively) for various values of ρ, $\boldsymbol{\sigma}$, and θ. The properties of Gabor wavelets are summarized in Table 3.2.

3.2.2.2 Maximum Response 8 (MR8)

Another popular example of handcrafted directional filterbank is the MR8 approach [11]. The latter includes a filterbank with a collection of 38 operator functions $g_n(\boldsymbol{x})$ (see Fig. 3.3). Two of them are circularly symmetric (one Gaussian and one LoG). Eighteen are directional multiscale edge detectors based on oriented first-order Gaussian derivatives. Another 18 are directional multiscale ridge detectors based on oriented second-order Gaussian derivatives. However, the approach yields a total of eight response maps. The first two come from the convolution of the image with the circularly symmetric Gaussian and LoG filters. The remaining six are computed as follows. For each directional detector types (*i.e.*, first- and second-order Gaussian derivatives), only one detector per scale is kept. Among the six different filter orientations per scale, the detector kept at the position $\boldsymbol{x}_0$ is the one that maximizes the detection in the sense of Eq. (3.3), where θ_0 is coarsely discretized. This results in six non-linear texture operators

Table 3.2 Properties of Gabor wavelets

Operator linearity	Yes.
Handcrafted	Yes.
3D extension	Requires using 3D elliptic Gaussian windows in the volumetric Fourier domain and systematically indexing their orientations with angles θ, ϕ (corresponding to angles ϑ, φ in spherical Fourier, see Section 2.2 of Chapter 2).
Coverage of image directions	Complete with appropriate choices of directions θ and orthoradial standard deviations σ_2 respecting Parseval's identity.
Directionality and local rotation-invariance	Directional and not locally rotation-invariant. A local rotation of the input image $f(\mathbf{R}_{\theta_0, \boldsymbol{x}_0} \cdot \boldsymbol{x})$ will swap the responses of the operators with various orientations θ.
Characterization of the LOIDs	No. Gabor filters/wavelets are unidirectional operators that are not able to characterize the LOIDs when used with an integrative aggregation function (*e.g.*, average, see Section 2.4.2 and Fig. 2.13 of Chapter 2). MF representations based on a consistent alignment criteria (*e.g.*, Hessian-based structure tensor, see Section 2.4.3 and Fig. 2.14 of Chapter 2) can be used to locally align Gabor operators (using, *e.g.*, steerability [10]) and allow characterizations of the LOIDs.
Coverage of image scales	Complete for filters with appropriate choices of radial frequencies ρ and standard deviation σ_1 respecting Parseval's identity. This is the case for Gabor wavelets.
Band-pass	Yes.
Gray-level reduction	Not required.
Illumination-invariance	No. However, they are robust to changes in illumination because Gabor filters are band-pass.
Aggregation function	Integrative (typically): computes the average of the absolute values or the energies of the response maps over $\boldsymbol{M}$.

$\mathcal{G}_{n=1,\dots,6}$ that are "aligned" at every position and achieve approximated local rotation-invariance. This process is illustrated in Fig. 3.3.

The aggregation function usually consists in pixel-wise clustering of the maximum filter responses to create a texton dictionary (see Section 1.2.4 of Chapter 1). Class–wise models can then be created as texton occurrence histograms, which can be used to compare texture instances by measuring distances between them (*e.g.*, Euclidean, χ^2). The vector of texture measurements $\boldsymbol{\eta}$ contains the bin values of the texton occurrences. The properties of the MR8 approach are summarized in Table 3.3.

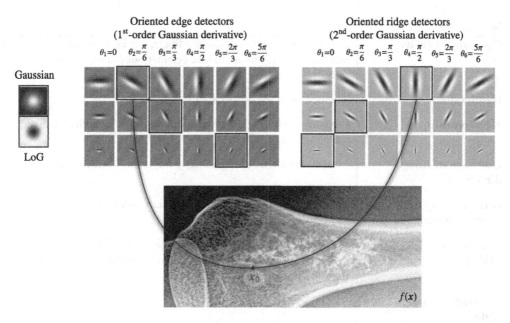

Figure 3.3 Top row: the 38 operators of the MR8 filterbank. At a given position x_0, the collection of operators $\mathcal{G}_{n=1,\dots,8}\{f\}(x_0)$ consists of a subset of the six directional operators with maximum response over all orientations $\theta_{1,\dots,6}$ plus the two circularly symmetric operators. This results in eight responses for each position (hence the name of the approach), which are marked in red in the example above.

3.2.2.3 Histogram of Oriented Gradients (HOG)

An efficient gradient-based MF representation is the local image descriptor used in the Scale Invariant Feature Transform (SIFT) approach [12] called the Histogram of Oriented Gradients (HOG). The texture function $f(x)$ is first filtered with circularly symmetric multiscale DoG filters $g_{\sigma_i}(x)$ (see Section 3.2.1) using a dyadic scale progression (i.e., $\sigma_{i+1} = 2\sigma_i$), which yields a collection of response maps $h_{\sigma_i}(x)$. At a fixed scale i, corresponding to the standard deviation $\sigma_i = \sigma$, the gradient orientation map $h_\sigma^\theta(x)$ is computed from the response map $h_\sigma(x)$ as

$$h_\sigma^\theta(x) = \arctan\left(\frac{\nabla_{x_2}\{h_\sigma\}(x)}{\nabla_{x_1}\{h_\sigma\}(x)}\right), \tag{3.7}$$

where $\nabla_{x_1}\{h_\sigma\}(x)$ and $\nabla_{x_2}\{h_\sigma\}(x)$ yield gradients maps. Equation (3.7) provides local angle values maximizing the gradient magnitude in the spirit of Eq. (3.3). For a fixed orientation θ, we set $u(\theta) = R_\theta(1,0) = (\cos\theta, \sin\theta)$ and $\nabla_{u(\theta)} = \cos\theta \cdot \nabla_{x_1} + \sin\theta \cdot \nabla_{x_2}$ the oriented derivative with direction $u(\theta)$. We define the oriented response map as $h_{\sigma,\theta}\{x_0\} = \nabla_{u(\theta)}\{h_\sigma\}(x_0)$. When using discretized image functions, operators and response maps $h_\sigma(\xi)$ (see Section 1.2.2 of Chapter 1), the latter can be estimated using

Table 3.3 MR8 properties

Operator linearity	No. The filtering operations are linear, but keeping the maximum value among the six orientations is a nonlinear operation.
Handcrafted	The operators are handcrafted.[a]
3D extension	Requires extending the filter orientations to 3D with systematic sampling of angles θ, ϕ.
Coverage of image directions	Near to complete, but not strictly respecting Parseval's identity.
Directionality and local rotation-invariance	Directional and approximate local rotation-invariance.
Characterization of the LOIDs	No, all MR8 operators are unidirectional. In addition, they are "aligned" independently at a position x_0 and do not yield MF representations.
Coverage of image scales	Incomplete.
Band-pass	Yes, with the exception of the circularly symmetric Gaussian filter (low-pass).
Gray-level reduction	Not required.
Illumination-invariance	No, mostly because of the circularly symmetric Gaussian filter.
Aggregation function	Consists of two consecutive aggregation functions: a first piece-wise integrative function[b] is used to construct the texton dictionary and second integrative function counts the texton occurrences and organizes them in a histogram.

[a] The texton dictionary is derived from the data, but the operators are handcrafted.
[b] Clustering results in averaging within local homogeneous regions in the feature space.

pixel differences as

$$\nabla_{x_1}\{h_\sigma\}(\boldsymbol{\xi}) = h_\sigma(\xi_1 + \Delta\xi_1, \xi_2) - h_\sigma(\xi_1 - \Delta\xi_1, \xi_2),$$
$$\nabla_{x_2}\{h_\sigma\}(\boldsymbol{\xi}) = h_\sigma(\xi_1, \xi_2 + \Delta\xi_2) - h_\sigma(\xi_1, \xi_2 - \Delta\xi_2). \tag{3.8}$$

For each position x_0, the dominant gradient direction θ_{x_0} is obtained by maximizing Eq. (3.7). To define the HOG operators, we align and sample the orientation $\theta = \theta_q - \theta_{x_0} = \frac{2\pi}{q} - \theta_{x_0}$ with $q = 1, \dots, 8$. The final collection of HOG operators $\mathcal{G}_{\sigma,q}$ is given at location x_0 by

$$\mathcal{G}_{\sigma,q}\{f\}(x_0) = h_{\sigma,q}(x_0).$$

The value of $h_{\sigma,q}(\boldsymbol{x}_0)$ can be efficiently evaluated using the steerability of the gradient operator [13] as

$$h_{\sigma,q}(\boldsymbol{x}_0) = \cos(\theta_q - \theta_{\boldsymbol{x}_0}) \cdot \nabla_{x_1}\{h_\sigma\}(\boldsymbol{x}_0) + \sin(\theta_q - \theta_{\boldsymbol{x}_0}) \cdot \nabla_{x_2}\{h_\sigma\}(\boldsymbol{x}_0). \tag{3.9}$$

The collection of HOG operators $\mathcal{G}_{\sigma,q}$ provides a MF representation oriented with $\theta_{\boldsymbol{x}_0}$ and containing eight redundant frame components (see Section 2.4.3 of Chapter 2). Whereas the response maps $h_\sigma(\boldsymbol{x})$ depend linearly on the texture image $f(\boldsymbol{x})$, it is not anymore the case for the HOG operators due to the local alignment of the angle: the HOG operators are not convolution operators themselves, but take advantage of the convolution framework since they are based on the DoG filters.

For obtaining texture measurements $\boldsymbol{\eta}$, the responses of HOG operators can be aggregated over regions $\boldsymbol{M}$ using component-wise averages (see Fig. 2.15 of Chapter 2). This allows building scale-wise histograms of oriented gradients, where each bin q corresponds to the average response of the response map $h_{\sigma,q}(\boldsymbol{x})$ of its corresponding operator $\mathcal{G}_{\sigma,q}$ over $\boldsymbol{M}$. The properties of HOGs are summarized in Table 3.4.

3.2.2.4 Riesz transform

A more elegant approach to compute directional transitions between pixel values is to compute them in the Fourier domain instead of using pixel differences in Gaussian-smoothed images as presented in Eq. (3.8). This also provides the opportunity to easily compute higher-order image derivatives of order l as

$$\frac{\partial^l}{\partial x_d^l} f(\boldsymbol{x}) \xleftrightarrow{\mathcal{F}} (\mathrm{j}\omega_d)^l \hat{f}(\boldsymbol{\omega}), \tag{3.10}$$

where $1 \le d \le D$. It can be noticed that differentiating an image along the direction x_d only requires multiplying its Fourier transform by $\mathrm{j}\omega_d$. Computing lth-order derivatives has an intuitive interpretation (*e.g.*, texture gradient for $l = 1$, curvature for $l = 2$), which makes them attractive for understanding the meaning of the texture measures in a particular medical or biological applicative context. However, a pure image derivative filter as computed in Eq. (3.10) is high-pass (because multiplied by ω_d) and accentuates high frequencies along x_d. Therefore it is desirable to implement image derivatives as all-pass filters, which is provided with the real Riesz transform[6] $\mathcal{R}\{f\}(\boldsymbol{x})$ as [15]

$$\mathcal{R}\{f\}(\boldsymbol{x}) = \begin{pmatrix} \mathcal{R}_1\{f\}(\boldsymbol{x}) \\ \vdots \\ \mathcal{R}_D\{f\}(\boldsymbol{x}) \end{pmatrix} \xleftrightarrow{\mathcal{F}} -\mathrm{j}\frac{\boldsymbol{\omega}}{||\boldsymbol{\omega}||}\hat{f}(\boldsymbol{\omega}). \tag{3.11}$$

---

[6] The Riesz transform is the multidimensional extension of the Hilbert transform.

Table 3.4 HOG properties

Operator linearity	No. The initial filtering operations are linear, but "aligning" the HOGs with the dominant directions θ_{x_0} obtained from Eq. (3.9) is a nonlinear operation.
Handcrafted	Yes.
3D extension	Requires extending the gradient orientations to 3D with systematic sampling of angles θ, ϕ, which is proposed in [14].
Coverage of image directions	Complete.
Directionality and local rotation-invariance	Directional and locally rotation-invariant.
Characterization of the LOIDs	Yes. The HOGs provide MF representations oriented with θ_{x_0} (see Eq. (3.7)).
Coverage of image scales	Incomplete and depends on the choices of DoG scales σ.
Band-pass	Yes.
Gray-level reduction	Not required.
Illumination-invariance	No. However, they are robust to changes in illumination because gradient operators computed on top of DoGs coefficients are band-pass.
Aggregation function	The average of the response maps of each HOG operator's responses can be used to build the scale-wise histogram of oriented gradients. This yields gradient-based MF representations (see Section 2.4.3 of Chapter 2).

It can be noticed by dividing the Fourier representation with the norm of $\boldsymbol{\omega}$ transforms Eq. (3.10) in D all-pass operators $\mathcal{R}_d$. For a fixed order L, the collection of higher-order all-pass image derivatives are defined in Fourier as

$$\widehat{\mathcal{R}^l}\{f\}(\boldsymbol{\omega}) = (-\mathrm{j})^L \sqrt{\frac{L!}{l_1! \cdots l_D!}} \frac{\omega_1^{l_1} \cdots \omega_D^{l_D}}{\left(\omega_1^2 + \cdots + \omega_D^2\right)^{L/2}} \hat{f}(\boldsymbol{\omega}), \qquad (3.12)$$

which yields a total of $\binom{L+D-1}{D-1} = \frac{(L+D-1)!}{L!(D-1)!}$ all-pass filters for all combinations of the elements l_d of the vector $\boldsymbol{l}$ as $|\boldsymbol{l}| = l_1 + \cdots + l_D = L$. The collection of Riesz operators of order L is denoted by $\mathcal{R}^L$. A set of band-pass, multiscale, and multiorientation operator functions $g_{\sigma,l}(\boldsymbol{x})$ can be obtained by simply applying the Riesz transform to circularly

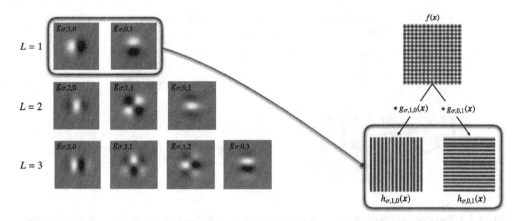

Figure 3.4 Two-dimensional Lth-order Riesz texture operators with functions $g_{\sigma,l}(x)$ providing Lth-order image derivatives at a fixed scale σ. Applying $L = 1$ texture operators with functions $g_{\sigma,1,0}$ and $g_{\sigma,0,1}$ to the input texture f with convolution yields response maps $h_{\sigma,1,0}(x)$ and $h_{\sigma,0,1}(x)$ decomposing f into its vertical and horizontal directions, respectively. Qualitatively, $L = 1$ corresponds to gradient estimation whereas $L = 2$ estimates the Hessian. They both have intuitive interpretations (*e.g.*, texture slope for $L = 1$, curvature for $L = 2$), which makes them attractive for understanding the meaning of the texture measures in a particular medical or biological applicative context.

symmetric wavelets or multiscale filters, *e.g.*, the LoG filter g_σ (see Eq. (3.5)) as

$$g_{\sigma,l}(x) = \mathcal{R}^l\{g_\sigma\}(x).$$

Examples of 2D real Riesz operators and their application to a directional texture are shown in Fig. 3.4. They are implemented as a plugin for quantitative image analysis on the ePAD radiology platform (see Chapter 13). 3D real Riesz filters are depicted in Fig. 2.12 of Chapter 2 and are implemented in the QuantImage radiomics web platform[7] (see Chapter 12).

Riesz operators as defined in Eq. (3.12) are not locally rotation–invariant/equivariant. However, local rotation–invariance/equivariance and rich MF representations can be achieved in a convenient fashion through the most interesting property of Riesz texture operators, which is steerability. The Riesz operator functions $g_{\sigma,l}(x)$ are steerable, which will be detailed in 2D in the following text. 2D steerability means that the responses of $g_{\sigma,l}$ rotated by an angle θ_0 can be very efficiently computed with a linear combination of a finite number of basis elements, which is parameterized by a steering matrix $\mathbf{A}_{\theta_0}^L$ as

$$\mathcal{R}^L\{g_\sigma\}(\mathbf{R}_{\theta_0}x) = \mathbf{A}_{\theta_0}^L\mathcal{R}^L\{g_\sigma\}(x), \tag{3.13}$$

[7] https://radiomics.hevs.ch, as of March 1, 2017.

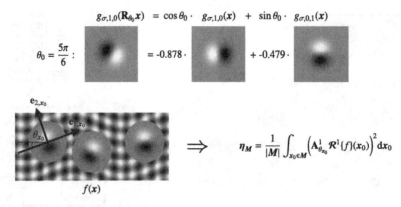

Figure 3.5 Top row: example of the steering of the first-order Riesz operator $g_{\sigma,1,0}$ at the position $x_0 = 0$ with an angle $\theta_0 = \frac{5\pi}{6}$. Bottom row: rich MF representations of $f(x)$ can be obtained from locally steered Riesz filterbanks (exemplified for $L = 1$).

where g_σ is the circularly symmetric function used to control the spatial support of the operators. For order 1 (*i.e.*, $L = 1$), $\mathbf{A}_{\theta_0}^1$ is equal to the 2D rotation matrix $\mathbf{R}_{\theta_0}$. For each scale σ, rich D-dimensional MF texture representations can be obtained by finding the local angle θ_{x_0} maximizing the response of one chosen Riesz operator function $g_{\sigma,l_1,\ldots,l_D}$ for each position x_0 using steerability [16–20]. For instance, 2D angle maps corresponding to a texture function f can be obtained for $L = 1$ as

$$\theta_{x_0} := \arg\max_{\theta_0 \in [0, 2\pi)} \Big(\cos\theta_0 \cdot (g_{\sigma,1,0} * f)(x_0) + \sin\theta_0 \cdot (g_{\sigma,0,1} * f)(x_0) \Big). \qquad (3.14)$$

This is a particular case of the angle alignment framework introduced in Eqs. (3.3) and (3.4).

A collection of MF texture measurements η can be obtained from component-wise averages of the absolute values (or energies) of the Riesz wavelet coefficients steered with angle maps obtained from, *e.g.*, Eq. (3.14). Steerability of the real Riesz transform and its use for the construction of MF representations is illustrated in Fig. 3.5. The properties of real Riesz–based texture analysis are summarized in Table 3.5.

By slightly modifying the definition of the Riesz transform (see Eq. (3.12)) into its complex form, it is also possible to obtain two-dimensional texture descriptors that can linearly quantify the amount of local circular frequencies. The latter are implicitly modeled by LBP operators, which unfortunately require a binarization and discretization of the circular neighborhoods resulting in a potentially large loss of information (see Section 3.4). This is not the case when using the complex Riesz transform. The 2D *l*th order complex Riesz transform $\mathcal{R}_{\mathbb{C}}^l$ is defined in polar coordinates in Fourier as [27]

$$\widehat{\mathcal{R}_{\mathbb{C}}^l\{f\}}(\rho, \vartheta) = e^{jl\vartheta}\hat{f}(\rho, \vartheta). \qquad (3.15)$$

Table 3.5 Properties of real and complex Riesz texture representations, as well as Steerable Wavelet Machines (SWM)

Operator linearity	No. The filtering operations are linear, but "aligning" the filters either using operator steering or the complex magnitudes of CHs are nonlinear operations.
Handcrafted	Yes for real and complex Riesz. No for learned SWMs operators (see Section 3.2.3.1).
3D extension	Real Riesz representations are extended to 3D by considering the subspace of filters spanned by all combinations of partial derivatives relatively to $\{x_1, x_2, x_3\}$ [16,17] (see Eq. (3.12)). The extension of complex Riesz transforms (*i.e.*, CH) to three dimensions is not straightforward. Spherical harmonics-based representations can be considered [21,22].
Coverage of image directions	Complete.
Directionality and local rotation-invariance	Directional and not locally rotation-invariant in their initial form. However, rich directional and locally rotation-invariant representations can be obtained at a low computational cost by either using the complex magnitudes (complex Riesz) or steerability (real and complex Riesz as well as SWMs).
Characterization of the LOIDs	Yes. Moreover, rich and learned MF representations can be obtained with steerability and SWMs.
Coverage of image scales	Complete when used with circularly symmetric wavelet representations [18] (*e.g.*, Simoncelli [23], Meyer [24], Shannon [25]).
Band-pass	Yes.
Gray-level reduction	Not required.
Illumination-invariance	No. However, they are robust to changes in illumination when Riesz operators are based on band-pass circularly symmetric primal functions.
Aggregation function	The average or covariances of the absolute values (or energies) of each Riesz operator response can be used [19,26].

Using a process similar to construct real Riesz filters, applying the complex Riesz transform to circularly symmetric filters or wavelets (*e.g.*, LoG filters, Eq. (3.5), or circularly symmetric Simoncelli wavelets, Eq. (2.2) of Chapter 2) yields steerable, complex, and band-pass Circular Harmonic (CH) filters or wavelets $g_{\sigma,l}(\boldsymbol{x}) = \mathcal{R}_{\mathbb{C}}^l\{g_\sigma\}(\boldsymbol{x}) \in \mathbb{C}$ with $l = 0, \ldots, L$ [28,29]. CHs stand out as the canonical representation of steerability [27],

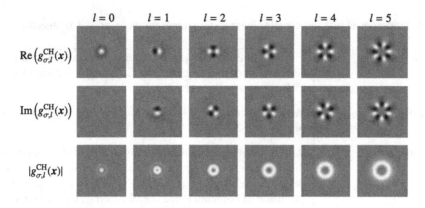

Figure 3.6 Impulse responses of Circular Harmonics (CH) of order $l = 0, \ldots, 5$.

which implies that every steerable representation $g_{\sigma,L,c}$ with $c = 1, \ldots, C$ components per scale can be obtained with a $C \times (L+1)$ complex shaping matrix $\mathbf{U}$ as

$$\begin{pmatrix} g_{\sigma,L,1} \\ \vdots \\ g_{\sigma,L,C} \end{pmatrix} = \mathbf{U} \begin{pmatrix} g_{\sigma,0} \\ \vdots \\ g_{\sigma,L} \end{pmatrix}. \tag{3.16}$$

As a consequence, any steerable representation can be obtained with a specific shaping matrix $\mathbf{U}$ (*e.g.*, gradient and Hessian real Riesz, Simoncelli's steerable pyramid). Examples of CH filters are shown in Fig. 3.6. Rich and locally rotation-invariant descriptions of the LOIDs can be obtained with a very cheap computational cost by computing the complex magnitudes of the response maps $h_{\sigma,l=0,\ldots,L}(\mathbf{x})$. The collection of CH with various orders $l = 0, 1, \ldots, L$ defines an orthonormal system corresponding to a Fourier basis for circular frequencies up to a maximum order L. The process is similar to LBP as presented in [30] but with the desirable properties of a fully linear approach (*e.g.*, no gray-level transformation required) besides the final step computing the complex magnitudes. However, absolute values of CH filters do not define MF representations since the interharmonic phase is lost by taking the magnitude of the responses of the operators. Rich MF representations can be obtained by steering all harmonics with a unique local orientation $\theta_{\mathbf{x}_0}$, as proposed in [26]. Although more computationally expensive than using the magnitudes of the operators, steering them is relatively cheap because CH are self-steerable, resulting in block-diagonal steering matrices $\mathbf{A}_{\theta_0}^L$ for multiorder representations.

Texture measures $\boldsymbol{\eta}$ can be obtained from the averages or covariances of absolute values (or energies) of steered or nonsteered Riesz response maps over a ROI $\boldsymbol{M}$ [19]. The properties of complex Riesz-based texture analysis are summarized in Table 3.5.

3.2.3 Learned filters

All texture operators described and discussed in the previous sections are handcrafted. This means that the type of texture information extracted by theses operators is assumed to be relevant for the texture analysis task in hand. Therefore the design of these operators in terms of the coverage of image scales and directions was based on prior assumptions (*e.g.*, ad hoc or based on theoretic guidelines, see Section 1.3.1 of Chapter 1). Whereas classical approaches use machine learning on top of handcrafted representations, more recent approaches proposed to derive the design of the operators from data to identify the combinations of scales and directions that are optimal for the texture analysis task in hand. When compared to handcrafted operators, learned ones reduce the risk of unnecessary modeling of texture properties that are not related to the targeted application (see Fig. 2.5 of Chapter 2). They eliminate human bias to include or exclude arbitrary operator scales and directions. It is worth noting that learned approaches still predominantly use handcrafted components of the aggregation function (*e.g.*, ReLU, sigmoid, pooling, see Section 3.2.3.3). Three important approaches for learning convolutional texture filters are discussed in this section, including Steerable Wavelet Machines (SWM, Section 3.2.3.1), Dictionary Learning (DL, Section 3.2.3.2), and deep Convolutional Neural Networks (CNNs, Section 3.2.3.3).

3.2.3.1 Steerable Wavelet Machines (SWM)

In [31–33] and [26], we proposed to use SVMs to learn optimally discriminant linear combinations of real or complex Riesz operators. The most interesting property of this approach called Steerable Wavelet Machines (SWM) is to combine the flexibility of learned representations with steerability,[8] *i.e.*, we learned the lines u_c of the shaping matrix $\mathbf{U}$ (see Eq. (3.16)) using one-versus-all classification configurations. This yields class-specific steerable signatures of the essential stitches of biomedical tissue (allowing locally rotation-invariant/equivariant descriptions of the LOIDs) that can be used for building data-specific MF representations. This approach is based on handcrafted representations of image scales based on dyadic circularly symmetric wavelet functions and we are currently extending the learning to image scales as well. The learned operators are limited to the span of Riesz representations. However, it was observed that a limited number of circular harmonics (*e.g.*, $L = 3, \dots, 10$) yields optimal texture representations for classification. In addition, the span of CH becomes less and less restricted with large values of L and can represent any function for $L \rightarrow \infty$ since it can be interpreted as a Fourier transform for circular frequencies. The properties of SWMs are summarized in Table 3.5.

---

[8] Linear combinations of steerable subspaces are themselves steerable.

3.2.3.2 *Dictionary Learning (DL)*

Recent popular approaches propose to fully learn operators from data (*e.g.*, a collection of bounded discretized texture functions $\mathcal{I} = \{f_1(\boldsymbol{\xi}), f_2(\boldsymbol{\xi}), \ldots, f_I(\boldsymbol{\xi})\}$) with very little constraints besides their maximal spatial support $G_1 \times \cdots \times G_D$. A first notable example is unsupervised Dictionary Learning (DL). The learned set of operators will depend on the learning criteria used [34]. Basic dimensionality reduction methods such as Principal Component Analysis (PCA), Independent Component Analysis (ICA), or K-means clustering based on the $|\boldsymbol{G}|$ pixel values of $G_1 \times \cdots \times G_D$ patches $\gamma_p(\boldsymbol{\xi})$ can be used to derive N essential *atoms* $g_{n,\mathcal{I}}(\boldsymbol{\xi})$, which can be further used as operator functions [35]. PCA requires that the atoms are orthogonal to each other (*i.e.*, it removes correlation between them). ICA minimizes the correlation as well as higher-order dependence between atoms, which are not necessarily orthogonal. K-means finds N prototype atoms $g_{n,\mathcal{I}}(\boldsymbol{\xi})$ from clustering of the $|\boldsymbol{G}|$-dimensional space spanned by the pixel values of the patches $\gamma_p(\boldsymbol{\xi})$. The atoms are also called *textons* [36,37], which were identified as the elementary units of preattentive human texture perception [1]. Textons relate to texture primitives (see Section 1.2.4 of Chapter 1) and to the LOIDs (see Section 2.4.1 of Chapter 2). The process for extracting 3×3 textons from a texture function $f(\boldsymbol{\xi})$ is illustrated in Fig. 3.7.

Another successful unsupervised approach was based on creating a collection of atoms from which we can reconstruct the patches $\gamma_p(\boldsymbol{\xi})$ with a vector of coefficients

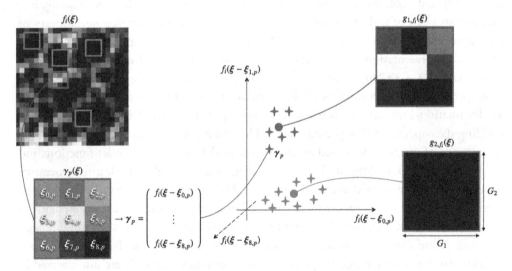

Figure 3.7 Dictionary learning (DL) process from one single image $f_i(\boldsymbol{\xi})$. A collection of 3×3 patches $\gamma_p(\boldsymbol{\xi})$ are extracted from f_i. The patch matrices are vectorized to create vectors $\boldsymbol{\gamma}_p$. K-means clustering is used in the space spanned by instance vectors $\boldsymbol{\gamma}_p$ to find N atoms $g_{n,f_i}(\boldsymbol{\xi})$ that are representative of f_i. These atoms are called *textons* and can be further used as texture operator functions.

$\boldsymbol{\alpha}_p$ containing a minimum number of nonzero elements [38,39]. Every atoms $g_{n,\mathcal{I},\lambda}$ are vectorized, transposed and piled up to create the $|\mathbf{G}| \times N$ dictionary matrix $\mathbf{D}$ with N atoms. Solving the following optimization problem for a collection of P patches can be used to compute both $\mathbf{D}$ and α as

$$\mathbf{D} := \underset{\mathbf{D},\alpha}{\arg\min} \sum_{p=1}^{P} ||\boldsymbol{\gamma}_p - \mathbf{D}\boldsymbol{\alpha}_p||_2^2 + \lambda ||\boldsymbol{\alpha}_p||_1, \tag{3.17}$$

where $|| \cdot ||_1$ is the ℓ_1-norm[9] and λ is controlling the sparsity of the atom coefficients $\boldsymbol{\alpha}$. Additional topological constraints were added in Topographic Independant Component Analysis (TICA) to take into account the important knowledge that patches that are spatially close to each other have similar statistical properties [40]. All texture operators dictionaries $\mathbf{D}$ learned with unsupervised approaches (e.g., K-means, PCA, ICA, TICA, reconstruction, see Eq. (3.17)) are not directly optimizing a discriminative criteria, which means that they are not necessarily optimal for biomedical texture classification of a set of considered tissue types. To tackle this issue, supervised dictionary learning was proposed [38], where $\mathbf{D}$ is obtained as

$$\mathbf{D} := \underset{\mathbf{D},b,\alpha}{\arg\min} \sum_{p=1}^{P} \mathcal{C}\left(\gamma_p, \boldsymbol{\gamma}_p^T \mathbf{D}\boldsymbol{\alpha}_p + b\right) + \lambda ||\mathbf{D}||_F^2, \tag{3.18}$$

where $\boldsymbol{\gamma}_p$ are the vectorized training patches and their labels $\left(y_p \in \{-1,+1\}\right)_{p=1,\dots,P}$, $\mathcal{C}(\gamma_p, \check{y}_p)$ is the cost function,[10] b the bias, and $|| \cdot ||_F$ the Frobenius matrix norm.[11] An inherent challenge of DL methods is the large number of free parameters (i.e., features or variables) to learn which is equal the size (i.g., measure) of the dictionary $|\mathbf{G}| \times N$, resulting in a very high-dimensional feature space. This typically requires a very large collection of training patches to respect the recommended ratio between feature dimensionality and number of training instances equal to ten [41]. As an order of magnitude, 1,210,000 training patches are required to learn 1000 11×11 atoms while respecting this ratio. A collection of N texture measurements $\boldsymbol{\eta}$ can be obtained by computing the averages of the absolute values or energies of the response maps $h_n(\boldsymbol{\xi}) = (g_n * f)(\boldsymbol{\xi})$ within a ROI $\boldsymbol{M}$. The properties of DL-based texture analysis are summarized in Table 3.6.

---

[9] The ℓ_1-norm of a vector $\boldsymbol{x}$ is $||\boldsymbol{x}||_1 = \sum_{d=1}^{D} |x_d|$, whereas the ℓ_2-norm is $||\boldsymbol{x}||_2 = \sqrt{\sum_{d=1}^{D} x_d^2}$.

[10] The cost function defines how label prediction errors of the estimations $\check{y}_p$ are penalized. The logistic loss function is used in [38] and is defined as $\mathcal{C}(\gamma_p, \check{y}_p) = \log(1 + e^{-\gamma_p \check{y}_p})$.

[11] The Frobenius matrix norm is defined as $||\mathbf{A}||_F = \sqrt{\sum_{i=1}^{I} \sum_{j=1}^{J} a_{i,j}^2}$, where $a_{i,j}$ are the elements of $\mathbf{A}$.

Table 3.6 Properties DL texture operators

Operator linearity	Yes.
Handcrafted	No. The texture operators functions $g_n(\xi)$ (*i.e.*, the dictionary atoms) are fully derived from a collection of training texture functions $\mathcal{I}$.
3D extension	Straightforward. It requires vectorizing 3D image patches $\gamma_p(\xi)$. However, the number of free parameters will grow cubically.
Coverage of image directions	Complete, but only over the restricted spatial support $\mathbf{G}$ of the atoms.
Directionality and local rotation-invariance	Directional and not locally rotation-invariant. Data augmentation can be used to improve robustness to input rotation, but it has undesirable effects (see Section 3.2.3.4).
Characterization of the LOIDs	Yes, but not with local rotation-invariance.
Coverage of image scales	Incomplete. Typical spatial supports $\mathbf{G}$ of the atoms are 3×3, 5×5, 11×11, which is much smaller than the spatial supports $\mathbf{F}$ of biomedical texture functions.
Band-pass	No.
Gray-level reduction	Not required.
Illumination-invariance	No. The texture operators functions are not band-pass filters.
Aggregation function	The average of the absolute values (or energies) of each atom response can be used.

3.2.3.3 Deep Convolutional Neural Networks (CNN)

One major and very successful work on filter learning are CNNs and their deep architectures[12] [43–47]. Review and applications of deep learning in texture analysis for tissue image classification are further detailed in Chapters 4, 9, and 10. In a nutshell, deep CNNs consist of a cascade of Q convolutional layers, where each of the latter typically contains:

(i) a multichannel convolution of the input $f_{n',q}(\xi)$[13] with a set of N $G_1 \times \cdots \times G_D$ multichannel operator functions $g_{n,n',q}(\xi)$, called *receptive fields*,

(ii) a simple pointwise nonlinear gating $\mathcal{V}(x)$ of the response map $h_{n,q}(\xi)$,

(iii) an optional cross-channel normalization [45] and,

---

[12] A notable exception is the Scattering Transform (ST) [42], which is a handcrafted deep CNN.

[13] The first layer $q = 1$ will use the original image $f_{n',1}(\xi)$ as input, whereas the next layers $q = 2, \ldots, Q$ will use the $n' = 0, \ldots, N'$ outputs (*i.e.*, *channels*) $f_{n',q-1}(\xi)$ of the previous layer as input for the convolution.

(iv)   a pooling operation resulting in a down- or up-sampling of $h_{n,q}(\xi)$.

In 2D, the multichannel convolution in a layer q between the input $f_{n',q-1}(\xi)$ and the set of N multichannel filters $g_{n,n',q}(\xi)$ is

$$h_{n,q}(\xi) = \sum_{n'=1}^{N'} (g_{n,n',q} * f_{n',q-1})(\xi). \tag{3.19}$$

(3.19) yields a multichannel response map $h_{n,q}(\xi)$ with a dimensionality of $F_1 \times F_2 \times N'$ for a 2D input function of domain $\boldsymbol{F}$. The convolution is two-dimensional, where no cross-channel convolution is carried out. Popular examples of nonlinear gating functions $\mathcal{V}(x)$ are the Rectified Linear Unit (ReLU)

$$\mathcal{V}_{\text{ReLU}}(x) = \max(0, x),$$

the sigmoid function

$$\mathcal{V}_{\text{sig}}(x) = \frac{1}{1 + e^{-x}},$$

the absolute value and the energy. The aim of the pooling operation is to down- or up-sample feature maps to achieve a multiscale analysis throughout the cascade of convolutional layers. Whereas the upsampling operation is often straightforward (no specific superresolution approach [46]), the downsampling requires using a criteria on which value to keep over local patches (*e.g.*, maximum, minimum, median, sum, average). The structure of one convolutional layer is depicted in Fig. 3.8.

Deep CNN architectures consist of a cascade of convolutional layers containing a large collection of operator functions $g_{n,n',q}(\xi)$ learned to minimize a cost function $\mathcal{C}(y_p, \check{y}_p)$. The total number of free parameters is equal to the product of the number of channels per operator N', the number of operator per layer N, the size of the spatial support $|\boldsymbol{G}|$ of the operators, and the number of layers Q, if we assume that N, N', and $|\boldsymbol{G}|$ are identical for all layers. It is very common for CNN architectures designed for classification to have a final *Fully Connected* (FC) layer that computes the decision value based on a linear combination of all elements of the final feature maps $h_{n,Q}(\xi)$. As an example, the complete forward function $u_{\text{forward}}(f(\xi))$ of a classification CNN transforms the input texture $f_p(\xi)$ into an estimated class label $\check{y}_p$[14] is a composition of all functions u_q from all layers as

$$u_{\text{forward}}(f(\xi); g) : \quad \boldsymbol{F} \to \mathbb{R}, \tag{3.20}$$

$$\check{y}_p = u_{\text{forward}}(f_p(\xi); g) = u_{\text{FC}}(\cdots u_2(u_1(f_p(\xi); g_1); g_2) \cdots); g_{\text{FC}}), \tag{3.21}$$

---

[14] $\check{y}_p$ can be binary (*e.g.*, $\check{y}_p \in \{-1, +1\}$) or continuous (*e.g.*, probability $\check{y}_p \in [0, 1]$, or $\check{y}_p \in \mathbb{R}$). We will assume that $\check{y}_p \in \mathbb{R}$ in our example.

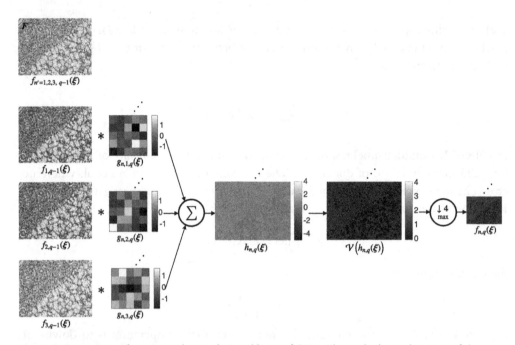

$f_{n'=1,2,3,\,q-1}(\boldsymbol{\xi})$

$f_{1,q-1}(\boldsymbol{\xi})$ $g_{n,1,q}(\boldsymbol{\xi})$

$f_{2,q-1}(\boldsymbol{\xi})$ $g_{n,2,q}(\boldsymbol{\xi})$ $h_{n,q}(\boldsymbol{\xi})$ $\mathcal{V}\big(h_{n,q}(\boldsymbol{\xi})\big)$ $f_{n,q}(\boldsymbol{\xi})$

$f_{3,q-1}(\boldsymbol{\xi})$ $g_{n,3,q}(\boldsymbol{\xi})$

Figure 3.8 Structure of one typical convolutional layer of CNNs. The multichannel output of the previous layer $q-1$ constitutes the input $f_{n',q-1}(\boldsymbol{\xi})$ of the layer q. A multichannel convolution is carried out between the channels of $f_{n',q-1}$ and the collection of $n=1,\dots,N$ operator functions $g_{n,n',q}$ and yields a collection of response maps $h_{n,q}(\boldsymbol{\xi})$ as detailed in Eq. (3.19). The latter undergo a simple pointwise nonlinear gating $\mathcal{V}(x)$, followed by a pooling operation to output the final feature maps $f_{n,q}(\boldsymbol{\xi})$. In the example above the 5×5 operator functions $g_{n,n',q}$ are initialized with random values following a normal distribution with zero mean. The nonlinear gating function $\mathcal{V}$ is a ReLU, and the pooling operation is a $4\times$ downsampling process (noted as $\downarrow 4$) where the maximum value is kept over 4×4 patches and a distance (or *stride*) of 4 between their centers.

where $\boldsymbol{g}_q$ is the collection of free parameters of the operator functions $g_{n,n',q}(\boldsymbol{\xi})$ in the layer q, and $\boldsymbol{g}$ is the total collection of free parameters of the model. When CNN architectures are designed for segmentation (*e.g.*, U–Net [46]), the output of the forward function is an estimated segmentation map $\breve{y}_p(\boldsymbol{\xi})$. Based on a set of training textures from which the true labels y_p are known, it is possible to compute the prediction errors between $\breve{y}_p$ and y_p. The cost of this error is determined by the cost function $\mathcal{C}(y_p,\breve{y}_p)$. Typical $\mathcal{C}(y_p,\breve{y}_p)$ (also called *loss* function) are the logistic loss function for classification, or Euclidean loss for regression. The total cost function $\mathcal{C}_{\text{tot}}$ over the entire training set (called *epoch*) is accumulating errors over instances p as

$$\mathcal{C}_{\text{tot}}(\boldsymbol{g}) = \frac{1}{P}\sum_{p=1}^{P}\mathcal{C}\left(y_p, u_{\text{forward}}(f_p(\boldsymbol{\xi}); \boldsymbol{g})\right).$$

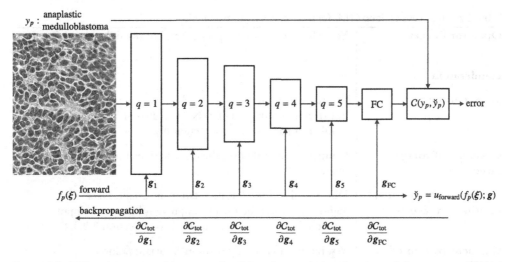

Figure 3.9 CNN architecture with a cascade of $Q = 5$ convolutional and one fully connected layer. The forward function $u_{\text{forward}}(f(\xi); g)$ is a composition of layer-wise functions u_g and transforms an input image $f_p(\xi)$ into a predicted label $\check{y}_p$. The cost function $\mathcal{C}_{\text{tot}}$ of accumulated errors over the entire training set is minimized through several epochs using backpropagation to update the free parameters g. The latter allows learning the optimal profile of the convolutional operator functions $g_{n,n',q}$.

After computing the total cost for one epoch t, the parameters of the model g_t are modified to minimize the total loss as

$$g_{t+1} = g_t - \beta_t \frac{\partial \mathcal{C}_{\text{tot}}}{\partial g}(g_t), \tag{3.22}$$

where β_t is the learning rate at the epoch t. Solving Eq. (3.22) learns the operator functions $g_{n,n',q}(\xi)$ and is called *backpropagation*. $\frac{\partial \mathcal{C}_{\text{tot}}}{\partial g}$ is computed using the chain rule, which allows combining the derivatives of the loss with respect to the parameters g_q of each consecutive layer [43]. Solving Eq. (3.22) through a total number T of epoch can be efficiently carried out with gradient descent optimization techniques. An example of a CNN architecture, its forward function and backpropagation is illustrated in Fig. 3.9. CNN-based texture analysis do not yield texture measurements *per se* as it directly outputs a tissue class probability or segmentation map. The aggregation function results from a cascade of pooling operations and, if applicable, the FC layer. The number of free parameters can be extremely large. For instance, a network with ten layers, ten 11×11 operators per layer with ten channels and with a 50×50 FC layer leads to more than 10^8 free parameters. It therefore requires very large training sets to correctly learn the model, which is most often difficult to get in biomedical imaging. Approaches called *transfer learning* or *fine tuning* are often

Table 3.7 Properties of deep CNNs for biomedical texture analysis

Operator linearity	Yes. The operators $\mathcal{G}_{n,n',q}$ are linear, but the full forward function is not.
Handcrafted	No.
3D extension	Requires using 3D operators, which was recently proposed in [51–53]. Training 3D CNNs is challenging because the number of parameters grows cubically.
Coverage of image directions	Complete, but only over the discrete spatial support $\mathbf{G}$ of the operators.
Directionality and local rotation-invariance	Directional and not locally rotation-invariant. Data augmentation can be used to improve robustness to input rotation, but it has undesired effects (see Section 3.2.3.4).
Characterization of the LOIDs	Yes, but not in a locally rotation-invariant fashion.
Coverage of image scales	Near-to-complete with the spatial support of the operators and the cascade of down- or up-sampling pooling operations. The coverage is not systematic though and do not respect Parseval's identity. A truly multiscale CNNs was proposed in [54].
Band-pass	No.
Gray-level reduction	Not required.
Illumination-invariance	No. The texture operator functions are not band-pass filters.
Aggregation function	The aggregation function results from a cascade of pooling operations and, if applicable, the FC layer.

used is to tackle this problem [48,49] (see Section 4.4.3.3 of Chapter 4). It consists of reusing and slightly adapting models trained with millions of images from other domains (*e.g.*, photographs in ImageNet [50]). In the particular case of biomedical texture analysis, reusing these models is risky because the types of invariances learned by networks based on photographic imagery resulting from scene captures obtained with varying viewpoints is very different from the ones desirable in BTA (see Section 1.3.3 of Chapter 1). An interesting observation is that when large datasets are available, the type of operators learned in the first layers of deep CNNs share very similar properties with handcrafted convolutional filters such as LoG, Gabor, MR8, and Riesz. The properties of CNNs for biomedical texture analysis are summarized in Table 3.7.

3.2.3.4 Data augmentation

Data augmentation aims to tackle two important limitations of filter learning approaches relying on a large number of free parameters: DL and CNNs [45,48,49,46]. First, both approaches require very large training sets to respect an acceptable ratio between the number of free parameters and training instances and allow suitable generalization performance (*i.e.*, limiting the risk of overfitting the training data). Second, neither of the methods are invariant to local scalings or local rotations. Whereas invariance to image scale is most often not desirable for biomedical texture analysis, invariance to local rotations is fundamental (see Section 1.3.3 of Chapter 1). Data augmentation consists of generating additional training instances from geometric transformations (including nonrigid) of the available ones, and to further train the models with these new instances. It provided very important performance gain in various applications [45,48], including biomedical [46]. However, it is obvious that forcing convolutional operators to recognize sheared, scaled, and rotated versions of the texture classes without augmenting the degrees of freedom will result in a strong decrease of the specificity of the model. In particular, forcing invariance to local rotations will make operators insensitive to directions, which was illustrated in [55]. A quantitative comparison between circularly symmetric and locally aligned operators (*e.g.*, MF representations) is detailed in Section 2.4.4 of Chapter 2. In this context, learning steerable filters allows adapting the texture representation while keeping local rotation-invariance and high specificity of the operators [31,26].

3.3 GRAY-LEVEL MATRICES

Gray-Level Matrices (GLM) constitute a large and popular group of non-linear texture operators. There are mainly three approaches based on GLMs: Gray-Level Cooccurrence Matrices (GLCM), Gray-Level Run-Length Matrices (GLRLM), and Gray-Level Size Zone Matrices (GLSZM). Their simplicity is at the origin of their popularity and many implementations can be found (*e.g.*, MaZda,[15] see Chapter 11, LIFEx,[16] Matlab,[17] Scikit-image,[18] the QuantImage web platform,[19] see Chapter 12, ePAD's Quantitative Feature Explore (QFExplore) texture plugins,[20] see Chapter 13). They are often used together. They are based on discrete texture functions $f(\xi)$ and non-linear operators $\mathcal{G}_n$. Their extensions to 3D are straightforward. However, they suffer from several imperfections. A major one is the nonsystematic coverage and poor preservation of image scales

---

[15] http://www.eletel.p.lodz.pl/programy/mazda/, as of March 4, 2017.

[16] http://www.lifexsoft.org, as of November 20, 2016.

[17] http://www.mathworks.com/help/images/ref/graycomatrix.html, as of November 20, 2016.

[18] http://scikit-image.org/docs/stable/api/skimage.feature.html, as of February 26, 2017.

[19] https://radiomics.hevs.ch, as of March 4, 2017.

[20] https://epad.stanford.edu/plugins, as of June 20, 2017.

and directions, especially when the spatial supports $G_{1,n} \times \cdots \times G_{D,n}$ of their operators are large. Therefore, they are best suited for applications where the size of the ROIs are small (*e.g.*, small lesions in low-resolution medical images). They also require drastic reductions of gray levels, where the values of $f(\boldsymbol{\xi}) \in \mathbb{R}$ are quantized with Λ values as $f_\Lambda(\boldsymbol{\xi}) \in \{1, \ldots, \Lambda\}$. The quantization is typically based on $\Lambda = 8, 16, 32$, which results in an important potential loss of information when analyzing rich image contents coded with 12 to 16 bits.[21] Their properties are reviewed and detailed in the three following Subsections 3.3.1, 3.3.2, and 3.3.3.

3.3.1 Gray-Level Cooccurrence Matrices (GLCM)

GLCMs [56] can be seen as a collection of operators mapping the discretized and quantized image function $f_\Lambda(\boldsymbol{\xi})$ to binary output values at a position $\boldsymbol{\xi}_0$ as

$$\mathcal{G}_{\Delta \boldsymbol{k}}^{\lambda_i, \lambda_j}\{f_\Lambda\}(\boldsymbol{\xi}_0) = \begin{cases} 1 & \text{if } f_\Lambda(\boldsymbol{\xi}_0) = \lambda_i \text{ and } f_\Lambda(\boldsymbol{\xi}_0 + \Delta \boldsymbol{k} \circ \Delta \boldsymbol{\xi}) = \lambda_j, \\ 0 & \text{otherwise,} \end{cases} \quad (3.23)$$

where λ_i and λ_j are the pixel values at positions $\boldsymbol{\xi}_0$ and $(\boldsymbol{\xi}_0 + \Delta \boldsymbol{k} \circ \Delta \boldsymbol{\xi})$, respectively. $\Delta \boldsymbol{k}$ and $\Delta \boldsymbol{\xi}$ contains the dimension-wise shifts and sampling steps, respectively (see Section 1.2.2 of Chapter 1). $\Delta \boldsymbol{k} \circ \Delta \boldsymbol{\xi}$ denotes the element-wise product[22] between the two vectors as

$$\Delta \boldsymbol{k} \circ \Delta \boldsymbol{\xi} = \begin{pmatrix} \Delta k_1 \cdot \Delta \xi_1 \\ \vdots \\ \Delta k_D \cdot \Delta \xi_D \end{pmatrix}.$$

The aggregation function is integrative. It counts the responses of the operators and organizes them in square *cooccurrence* matrices of dimensions Λ^2 indexed by (λ_i, λ_j), where Λ is the number of gray-levels. A series of scalar texture measurements[23] η is obtained by computing statistics (*e.g.*, contrast, correlation, entropy) from the cooccurrence matrices. Their properties are summarized in Table 3.8. GLCMs are not invariant to local rotations, but the later is often approximated by either regrouping the counts of operators over all directions in a shared matrix, or by averaging scalar texture measurements from cooccurrences matrices obtained with different directions (see Fig. 3.10).

---

[21] Image pixels encoded with 16 bits can take more than 65,000 possible values.
[22] It is also called the *Hadamard* product.
[23] They are often called *Haralick features* [56].

Table 3.8 GLCM properties

Operator linearity	Nonlinear.
Handcrafted	Yes.
3D extension	Straightforward: displacements $\Delta \boldsymbol{k} \circ \Delta \boldsymbol{\xi}$ between pixels can live in subsets of either $\mathbb{R}^2$ or $\mathbb{R}^3$.
Coverage of image directions	Incomplete: typical directions used are $\theta_{1,...,4} = 0, \frac{\pi}{4}, \frac{\pi}{2}, \frac{3\pi}{4}$ in 2D.
Directionality and local rotation-invariance	Unidirectional and not rotation-invariant. However, local rotation-invariance is often approximated by either regrouping the counts of operators over all directions in a shared matrix, or by averaging scalar texture measurements from cooccurrences matrices obtained with different directions (see Fig. 3.10).
Characterization of the LOIDs	No.
Coverage of image scales	Incomplete: typical displacements values are $\|\Delta \boldsymbol{k}\| \approx 1, 2, 3$. Moreover, displacements along image diagonals (*e.g.*, $\theta_2 = \frac{\pi}{4}$ and $\theta_4 = \frac{3\pi}{4}$) are often considered to have integer norms (*e.g.*, $1, 2, 3$) instead of their actual values (*e.g.*, $\sqrt{2}, 2\sqrt{2}, 3\sqrt{2}$), resulting in anisotropic descriptions of image scales.
Band-pass	Qualitatively similar in the sense that the mean value of the image is not influencing the output value of the operator. It is worth noting that the transfer function is not defined in the Fourier domain because the operator is nonlinear.
Gray-level reduction	A reduction is required to avoid obtaining very large and sparse cooccurrence matrices. Typical gray-level reductions are $\Lambda = 8, 16, 32$.
Illumination-invariance	No, although it is approximated by reducing the number of gray-levels Λ.
Aggregation function	Integrative: counts the binary responses of each GLCM operator $\mathcal{G}_{\Delta \boldsymbol{k}}^{\lambda_i, \lambda_j}$ over $\boldsymbol{M}$ and organizes them in a corresponding cooccurrence matrix indexed by λ_i and λ_j. Scalar texture measurements $\boldsymbol{\eta}$ called *Haralick features* are obtained from statistics of the cooccurrence matrices.

3.3.2 Gray-Level Run-Length Matrices (GLRLM)

GLRLMs [57] count the number of aligned pixels (called stride) with equal gray-level value λ, length $\gamma \in \mathbb{N}^*$ and direction θ. Their operators $\mathcal{G}_{\lambda, \gamma, \theta}\{f_\Lambda\}(\boldsymbol{\xi}_0)$ yield a value of 1

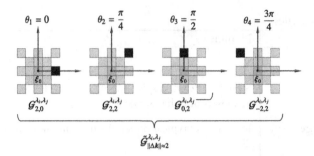

Figure 3.10 Approximate local rotation-invariance with GLCMs: the responses of GLCM operators $\mathcal{G}^{\lambda_i,\lambda_j}_{\Delta k_1,\Delta k_2}$ are combined over four directions $\theta_{1,\dots,4}$ to reduce the directional sensitivity of the operators and approximate locally rotation-invariant texture analysis. In 2D, the angle between the vectors Δk and $e_1 = (1, 0)$ is noted θ (see Section 2.2 of Chapter 2).

$$\mathcal{G}_{\lambda=7,\gamma=4,\theta=0}\{f_\Lambda\}(\xi_0)$$

5	1	1	4	1	0	7	6
2	3	7	7	7	7	2	4
6	7	0	2	6	3	7	1
3	5	1	0	7	0	1	4

$$f_\Lambda(\xi)$$

$\longrightarrow$

0	0	0	0	0	0	0	0
0	0	1	0	0	0	0	0
0	0	0	0	0	0	0	0
0	0	0	0	0	0	0	0

$$h_{\lambda=7,\gamma=4,\theta=0}(\xi)$$

Figure 3.11 Application of a GLRLM operator $\mathcal{G}_{\lambda=7,\gamma=4,\theta=0}$ to an input image f_Λ. Its response map $h_{\lambda=7,\gamma=4,\theta=0}(\xi)$ highlights the presence of the sought stride with gray-level 7, length 4 and orientation 0 at the position ξ_0.

if a stride starting at ξ_0, of length γ, and direction θ is detected, and 0 otherwise. An example of a GLRLM operator and its response map is depicted in Fig. 3.11.

The aggregation function counts the number of strides detected with the corresponding operator and organizes them in a *run-length* matrix indexed by λ and γ. The size of this matrix is the number Λ of gray-levels considered, times the number of stride lengths tested. The texture measures η are obtained from statistics of the run–length matrices (*e.g.*, short run emphasis, run–length nonuniformity, see [58]). Their properties are summarized in Table 3.9.

3.3.3 Gray-Level Size Zone Matrices (GLSZM)

GLSZMs [59] are extending the concept of GLRLM to zone areas or volumes. Their operators $\mathcal{G}_{\lambda,\zeta}\{f_\Lambda\}(\xi_0)$ yield a binary value of 1 if ξ_0 belongs to a uniform zone with gray-level λ, *i.e.*, any zone with area equal to ζ containing ξ_0, and area $\zeta \in \mathbb{N}^*$, and ξ_0 does not belong to a uniform zone with larger area.

The aggregation function counts the number of zones detected with the corresponding operator and organizes them in a *size-zone* matrix indexed by λ and ζ. The size of

Table 3.9 GLRLM properties

Operator linearity	Nonlinear.
Handcrafted	Yes.
3D extension	Straightforward: run directions can be extended to 3D and indexed with angles θ, ϕ.
Coverage of image directions	Incomplete: typical directions used are $\theta_{1,\dots,4} = 0, \frac{\pi}{4}, \frac{\pi}{2}, \frac{3\pi}{4}$ in 2D.
Directionality and local rotation-invariance	Unidirectional and not rotation-invariant. However, local rotation-invariance is often approximated by either regrouping the counts of operators over all directions, or by averaging scalar texture measurements from run-length matrices obtained with different directions.
Characterization of the LOIDs	No.
Coverage of image scales	Complete if the maximum run-length γ is equal to the size of the image. However, run length along image diagonals (*e.g.*, $\theta_2 = \frac{\pi}{4}$ and $\theta_4 = \frac{3\pi}{4}$) are often considered to have integer γ values (*e.g.*, 1, 2, 3) instead of their actual values (*e.g.*, $\sqrt{2}, 2\sqrt{2}, 3\sqrt{2}$), resulting in anisotropic descriptions of image scales.
Band-pass	Qualitatively similar in the sense that the mean value of the image is not influencing the output value of the operator. It is worth noting that the transfer function is not defined in the Fourier domain because the operator is nonlinear.
Gray-level reduction	A reduction is required to avoid obtaining very large and sparse run-length matrices. Typical gray-level reductions are $\Lambda = 8, 16, 32$.
Illumination-invariance	No, although it is approximated by reducing the number of gray-levels Λ.
Aggregation function	Integrative: counts the binary responses of each GLRLM operator $\mathcal{G}_{\lambda,\gamma,\theta}$ over M and organizes them in a corresponding run-length matrix indexed by integer values of λ and γ. Scalar texture measurements η are obtained from statistics of the run-length matrices.

this matrix is the number Λ of gray-levels considered, times the number of zone areas tested. The texture measures η are obtained from statistics of the matrices, which are the same as for the GLRLM (see [58]) plus two additional measures introduced in [59]. Their properties are summarized in Table 3.10.

Table 3.10 GLSZM properties

Operator linearity	Nonlinear.
Handcrafted	Yes.
3D extension	Straightforward: the search for contiguous pixels with identical gray-level values can be extended from 8-connected 2D neighborhoods to their 26-connected 3D counterparts.
Coverage of image directions	Complete.
Directionality and local rotation-invariance	Insensitive to image directions. Elongated and circular zones are mixed.
Characterization of the LOIDs	No.
Coverage of image scales	Complete if the maximum zone surface ζ is equal to the area of the image. However, the notion of scale is ill-defined because a fixed zone area ζ can correspond to both elongated or circular regions.
Band-pass	Qualitatively equivalent in the sense that the mean value of the image is not influencing the output value of the operator. It is worth noting that the transfer function is not defined in the Fourier domain because the operator is nonlinear.
Gray-level reduction	A reduction is required to avoid obtaining very large and sparse size-zone matrices. Typical gray-level reductions are $\Lambda = 8, 16, 32$.
Illumination-invariance	No, although it is approximated by reducing the number of gray-levels Λ.
Aggregation function	Integrative: counts the binary responses of each GLSZM operator $\mathcal{G}_{\lambda,\zeta}$ over M and organizes them in a corresponding size-zone matrix indexed by integer values of λ and ζ. Scalar texture measurements η are obtained from statistics of the size-zone matrices.

3.4 LOCAL BINARY PATTERNS (LBP)

Rotation-invariant Local Binary Patterns (LBP) were first introduced by Ojala et al. in 2002 [60] and many extensions were proposed later on (*e.g.*, [30,61,62]). At a position $\boldsymbol{\xi}_0$ of the image f, the LBP operator $\mathcal{G}_{\gamma,r}\{f\}(\boldsymbol{\xi}_0)$ describes the organization of binarized pixels over circular neighborhoods $\Upsilon(\gamma, r, \boldsymbol{\xi}_0)$ of radius r containing γ equally spaced points. The gray value of points that do not fall exactly in the center of pixels are estimated by interpolation (see Fig. 3.12, top). The decimal value of the operator is given by the binary sequence of the circular neighborhood (*e.g.*, $10101010 = 170$, see Fig. 3.12, middle). The binary value of a point $\boldsymbol{p} \in \Upsilon(\gamma, r, \boldsymbol{\xi}_0)$ is 1 if $f(\boldsymbol{\xi}_1) > f(\boldsymbol{\xi}_0)$, and 0 otherwise, with $\boldsymbol{\xi}_1$ the pixel location where $\boldsymbol{p}$ is located. An example of approximately locally rotation–invariant LBP texture analysis is depicted in Fig. 3.12 middle and bottom.

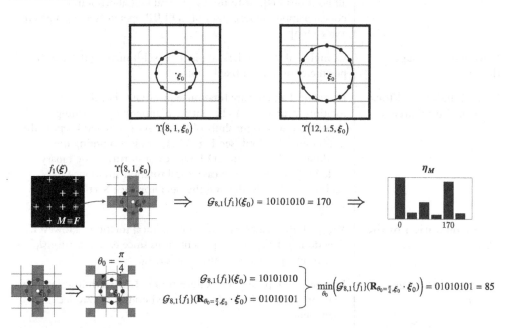

Figure 3.12 Extraction of LBP texture measures on $f_1(\boldsymbol{\xi})$. Top row: example of two different circular neighborhoods $\Upsilon(\gamma, r, \boldsymbol{\xi}_0)$. Middle row: a LBP operator $\mathcal{G}_{8,1}\{f_1\}(\boldsymbol{\xi}_0)$ encodes the LOIDs at the position $\boldsymbol{\xi}_0$ over a circular neighborhood $\Upsilon(8, 1, \boldsymbol{\xi}_0)$ with $\gamma = 8$ equally-spaced points and a radius $r = 1$ pixel. Its binary response for a +-shaped primitive is 10101010, which correspond to a decimal value of 170. The response maps of the operators are aggregated over the region M by counting the binary codes and organizing them into a histogram of texture measurements η_M. Bottom row: local rotation-invariance with LBPs. Local image rotations $\mathbf{R}_{\theta_0,\boldsymbol{\xi}_0} \cdot \boldsymbol{\xi}$ correspond to bit-wise circular shifts of the binary codes. Local rotation-invariance can be achieved by minimizing the decimal values of the binary codes over all possible discrete circular shifts $\theta_0 = \frac{q\pi}{4}$, $\forall q \in \{0, 1, \ldots, 7\}$.

The responses of the operators are aggregated over a ROI M by counting the binary sequences (or decimal values) and organizing them in a histogram. The latter can be used for extracting texture measures η. The properties of LBPs are summarized in Table 3.11.

Table 3.11 LBP properties

Operator linearity	Nonlinear.
Handcrafted	Yes.
3D extension	Not trivial: The ordering of points is straightforward in 2D on circular neighborhoods, but is undefined in 3D for spherical neighborhoods [63]. Approaches were proposed to either define an arbitrary ordering for each (γ, r) over (θ, ϕ) and to use it for all positions [64], or to use cylindrical neighborhoods by concatenating the responses of 2D LBP operators along a given axis ξ_d [65].
Coverage of image directions	Complete if γ considers all pixels/voxels touching the 2D/3D perimeter of Υ for a fixed radius r.
Directionality and local rotation-invariance	Directional and locally rotation-equivariant. Local rotation-invariance can be obtained either by performing circular bit-wise right shifts of the binary codes and keeping the minimum value [60] (see Fig. 3.12), or by computing the modulus of the discrete 1D Fourier transforms of the binary code [30]. In the former case, local rotation-invariance is achieved by locally aligning the operators (see Section 2.4.3 of Chapter 2).
Characterization of the LOIDs	Yes, with invariance/equivariance to local rotations. However, they do not define MF representations since each operator $\mathcal{G}_{\gamma,r}$ is aligned independently at the position ξ_0.
Coverage of image scales	Incomplete: typical radius values are $r = 1, 2, 3$. Multiscale LBPs were proposed in [66] when extracted on top of wavelet coefficients.
Band-pass	Qualitatively similar in the sense that the mean value of the image is not influencing the output value of the operator. It is worth noting that the transfer function is not defined in the Fourier domain because the operator is nonlinear.
Gray-level reduction	Not required. However, the local binarization operation results in an important reduction of the values analyzed.
Illumination-invariance	Yes.
Aggregation function	Integrative: counts the responses of each LBP operator $\mathcal{G}_{\gamma,r}$ over M and organizes them in a histogram. The collection of scalar texture measures η contains the bins of the histogram.

3.5 FRACTALS

Another popular method in BTA is to estimate the fractal properties of biomedical tissue, *e.g.*, how structures are similar across a series of monotonously increasing scales [67,68]. The latter is measured with the *Fractal Dimension* (FD) η_{FD}. The larger the FD, the most regular structures are through multiple scales, which corresponds closely to our intuitive notion of roughness [69]. The maximum FD is equal to the dimensionality D of the texture function. For 2D texture functions indexed by the spatial coordinates $(x_1, x_2) \in \mathbb{R}^2$, we typically consider the set $S = \{(x_1, x_2) \in \mathbb{R}^2 : f(x_1, x_2) = \gamma\} \subset \mathbb{R}^2$ for a fixed value $\gamma \in \mathbb{R}$. Note that S is a level set of f. Then the fractal dimension can take values $0 \leq \eta_{FD} \leq 2$. A comprehensive description and review of fractal analysis for BTA is presented in Chapter 5.

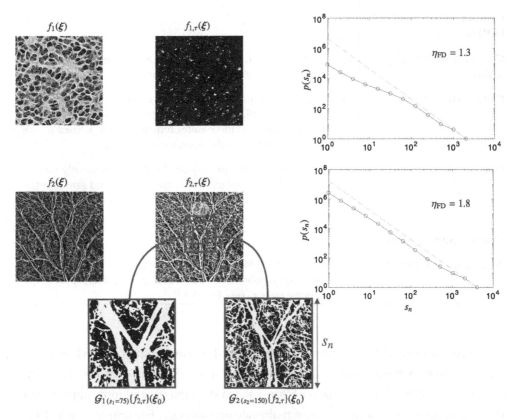

Figure 3.13 2D box-counting method used to estimate the fractal dimension η_{FD} of a texture. Although not recommended in general, a binarization of the texture images is used in this example. The latter is carried out with a thresholding operation, where the threshold τ was chosen to highlight cell nucleus in f_1 and the vascular structure in f_2. The visualization of the profiles $p(s_n)$ reveals the fractal nature of the vascular structure with a FD of 1.8. The maximum FD bound of 2 corresponding to a fully white image corresponds to the dashed red line in the profiles $p(s_n)$.

Table 3.12 Properties of fractal-based texture analysis using the box-counting method

Operator linearity	In principle yes, but it depends on the type of transformation of pixel values. In the example, the binarization of f is a nonlinear operation.
Handcrafted	Yes.
3D extension	Straightforward by using multiscale 3D boxes as operators' supports.
Coverage of image directions	Complete.
Directionality and local rotation-invariance	Insensitive to image directions.
Characterization of the LOIDs	No.
Coverage of image scales	Complete. However, the FD is invariant to image scale, which entails the risk of regrouping tissue structures of different natures.
Band-pass	Not qualitatively equivalent in the sense that the mean value of the image has an influence on the output value of the operator. It is worth noting that the transfer function is not defined in the Fourier domain because the operator is nonlinear.
Gray-level reduction	No. A binarization step or gray-level reduction is usually not recommended as this would both arbitrarily discard important texture information and degrade the stability of the calculation of η_{FD} (see [68,70]).
Illumination-invariance	No.
Aggregation function	Constructs the log-log profile of the averages of operator response maps h_n obtained with operators of spatial supports with varying sizes s_n (see Fig. 3.13).

A popular method for computing the FD is the *box-counting* approach, while several other methods exist (*e.g.*, box counting and probabilities [71], based on Fourier analysis [69] or wavelets [72], see Chapter 5 for detailed descriptions and more approaches). Box-counting relies on a collection of multiscale operators $\mathcal{G}_n$. For simplification, let us assume that the latter have square or cubic spatial support $G_{1,n} = G_{2,n} = \cdots = G_{D,n} = s_n$. The collection of operators have monotonously increasing spatial supports $\boldsymbol{G}_n \subset \boldsymbol{G}_{n+1}$. In order to simply exemplify the box-counting algorithm, let us consider binary images $f_\tau(\boldsymbol{\xi}) \in \{0, 1\}$ obtained from the binarization of an input image $f(\boldsymbol{\xi})$ with a threshold τ^{24}. At a position $\boldsymbol{\xi}_0$, each fractal operator $\mathcal{G}_n$ counts the number p of pixels with value equals to 1 (*e.g.*, $f_\tau(\boldsymbol{\xi}_0) = 1$) over its support $\mathbf{G}_n$. This yields N response maps $h_n(\boldsymbol{\xi}_0) = \mathcal{G}_n\{f_\tau\}(\boldsymbol{\xi}_0)$.

The aggregation function consists of fitting a log-log profile of the averages of h_n through the responses of the collection of operators with spatial supports of varying sizes s_n, where $|\mathbf{G}_n| = s_n^D$. The average number p of pixels with value equals to 1 over a ROI $\boldsymbol{M}$ is computed as

$$p(s_n) = \frac{1}{|\boldsymbol{M}|} \sum_{\boldsymbol{\xi} \in M} h_n(\boldsymbol{\xi}).$$

The FD is obtained with the value of η_{FD} that best fits the following relation:

$$\log\left(p(s_n)\right) \simeq \eta_{\text{FD}} \log\left(s_n\right). \tag{3.24}$$

In this particular case, $\eta_{\text{FD}} = 2$ corresponds to a perfectly regular image filled with ones. The box counting method is illustrated in Fig. 3.13, where two biomedical textures with different FD are compared. The properties of fractal-based texture analysis are summarized in Table 3.12 and are further developed in Chapter 5. The FD is invariant to image scale, which entails the risk of regrouping tissue structures of different natures (see Section 1.3.3 of Chapter 1).

3.6 DISCUSSIONS AND CONCLUSIONS

A qualitative comparison of most popular BTA approaches was proposed under the light of the general framework introduced in Section 1.3.1 of Chapter 1, and based on the comparison dimensions presented in Chapter 2. Our aim is to both provide a user guide for choosing a BTA method that is relevant to the problem in hand, as well as to provide insights on key aspects required to build the next generation of BTA approaches. The review focused on most popular group of methods and was not exhaustive. It included (i) convolutional methods and their operator subtypes: circularly/spherically symmetric, directional and learned, (ii) gray-level matrices and their subtypes: GLCMs, GLRLMs, GLSZMs, (iii) LBPs, and (iv) fractals based on the box-counting method. All approaches' operators were found to be equivariant to translations, which is an inherited property from the general framework introduced in Section 1.3.1 of Chapter 1. However, very few methods were combining both the ability to characterize the LOIDs and with invariance to local rotations. In particular, most approaches are either invariant to local rotations because they are insensitive to directions (*e.g.*, combined GLCM or

---

[24] The binarization step or gray-level reduction is usually not recommended for box-counting as this would both arbitrarily discard important texture information and degrade the stability of the calculation of η_{FD} (see [68,70] and Chapter 5).

GLRLM operators, LoGs, fractals) and they cannot characterize the LOIDs, or they are directional but not locally rotation-invariant (*e.g.*, unidirectional GLCM or GLRLM operators, Gabor wavelets, unaligned real Riesz, DL, CNNs). Within this dilemma, one may favor directional versus directionally insensitive approaches depending on the expected importance of the LOIDs versus invariance to local rotations, respectively. The only approaches that are able to characterize the LOIDs with invariance to (local) rigid transformations are LBPs as well as CH, steered real Riesz wavelets and SWMs. Other factors must be taken into account to evaluate the relevance of the method. An important property that is responsible for the success of approaches such as DL and CNNs is the ability to derive the texture operators[25] from the data in a supervised or unsupervised fashion. A challenge for the success of the learning-based methods is to have sufficient representation of each intra-class variants, which are most often not available in focused and innovative biomedical applications (see Chapter 6). Quick fixes have been extensively used to tackle this challenge. A first one is to introduce implicit handcrafted invariances with data augmentation, which has the undesirable effects of both increasing the training computational load (minor issue), and more importantly to decrease the specificity of the model [55] (see Section 3.2.3.4). A second quick fix is to use transfer learning and fine tuning in order to recycle models trained on other image types. This is not exempt of risks as borrowing models from general computer vision with strong invariance to scale (*e.g.*, models trained on ImageNet) are not fulfilling the requirement of BTA (see Section 3.2.3). An inherent risk of methods requiring important gray-level reductions or binarization (*e.g.*, GLM, LBP, fractals based on box-counting) is to miss or mix important texture properties. It was found that methods based on learned steerable wavelets (*i.e.*, SWMs [26,31]) were regrouping several desirable properties such as the ability to learn optimally discriminant sets of LOIDs with invariance to rigid transformation, with a small amount of training data, and without requiring neither data augmentation nor gray-level reductions. Few methods were found to provide easily interpretable texture measurements. Notable examples among them are the contrast and energy measures of GLCMs, the fractal dimension and roughness [69] (see Chapter 5), as well as 2D and 3D steered real Riesz wavelets that are measuring image derivatives of various orders in a locally rotation–invariant fashion [18,20,17,16] (see Section 12.2.4.4 of Chapter 12).

We recognize several limitations of the current review, including the absence of quantitative comparison of performance, computational complexity and time of the approaches. Detailed reviews of specific BTA properties, approaches or applications are presented in Chapter 7 (invariance to rigid transformations), Chapters 4 and 9 (deep learning), Chapter 6 (machine learning), and Chapter 10 (digital histopathology).

---

[25] CNNs with FC layers are also learning the aggregation function.

ACKNOWLEDGMENTS

This work was supported by the Swiss National Science Foundation (under grant PZ00P2_154891) and the CIBM.

REFERENCES

[1] B. Julesz, Textons, the elements of texture perception, and their interactions, Nature 290 (5802) (mar 1981) 91–97.

[2] D. Marr, E. Hildreth, Theory of edge detection, Proc. R. Soc. Lond. B, Biol. Sci. 207 (1167) (feb 1980) 187–217.

[3] D. Van De Ville, M. Unser, Complex wavelet bases, steerability, and the Marr-like pyramid, IEEE Trans. Image Process. 17 (11) (November 2008) 2063–2080.

[4] B. Ganeshan, E. Panayiotou, K. Burnand, S. Dizdarevic, K.A. Miles, Tumour heterogeneity in non-small cell lung carcinoma assessed by CT texture analysis: a potential marker of survival, Eur. Radiol. 22 (4) (2012) 796–802.

[5] T. Sing Lee, Image representation using 2D Gabor wavelets, IEEE Trans. Pattern Anal. Mach. Intell. 18 (10) (oct 1996) 959–971.

[6] A.G. Ramakrishnan, S.K. Raja, H.V.R. Ram, Neural network-based segmentation of textures using Gabor features, in: Proceedings of the 12th IEEE Workshop on Neural Networks for Signal Processing, 2002, pp. 365–374.

[7] M. Haghighat, S. Zonouz, M. Abdel-Mottaleb, CloudID: trustworthy cloud-based and cross-enterprise biometric identification, Expert Syst. Appl. 42 (21) (2015) 7905–7916.

[8] M. Petrou, P. García Sevilla, Non-Stationary Grey Texture Images, John Wiley & Sons, Ltd, 2006, pp. 297–606.

[9] J.G. Daugman, Uncertainty relation for resolution in space, spatial frequency, and orientation optimized by two-dimensional visual cortical filters, J. Opt. Soc. Am. A, Opt. Image Sci. Vis. 2 (7) (1985) 1160–1169.

[10] W. Pan, T.D. Bui, C.Y. Suen, Rotation-invariant texture classification using steerable Gabor filter bank, in: M. Kamel, A. Campilho (Eds.), Image Analysis and Recognition: Second International Conference, Proceedings, ICIAR 2005, Toronto, Canada, September 28–30, 2005, Springer, Berlin, Heidelberg, ISBN 978-3-540-31938-2, 2005, pp. 746–753.

[11] M. Varma, A. Zisserman, A statistical approach to texture classification from single images, Int. J. Comput. Vis. 62 (1–2) (2005) 61–81.

[12] D.G. Lowe, Distinctive image features from scale-invariant keypoints, Int. J. Comput. Vis. 60 (2) (2004) 91–110.

[13] W.T. Freeman, E.H. Adelson, The design and use of steerable filters, IEEE Trans. Pattern Anal. Mach. Intell. 13 (9) (sep 1991) 891–906.

[14] P. Scovanner, S. Ali, M. Shah, A 3-dimensional SIFT descriptor and its application to action recognition, in: Proceedings of the 15th ACM International Conference on Multimedia, MM'07, ACM, New York, NY, USA, 2007, pp. 357–360.

[15] M. Unser, D. Van De Ville, Wavelet steerability and the higher-order Riesz transform, IEEE Trans. Image Process. 19 (3) (March 2010) 636–652.

[16] N. Chenouard, M. Unser, 3D steerable wavelets in practice, IEEE Trans. Image Process. 21 (11) (2012) 4522–4533.

[17] Y. Dicente Cid, H. Müller, A. Platon, P.-A. Poletti, A. Depeursinge, 3-D solid texture classification using locally-oriented wavelet transforms, IEEE Trans. Image Process. 26 (4) (2017) 1899–1910, http://dx.doi.org/10.1109/TIP.2017.2665041.

[18] A. Depeursinge, P. Pad, A.C. Chin, A.N. Leung, D.L. Rubin, H. Müller, M. Unser, Optimized steerable wavelets for texture analysis of lung tissue in 3-D CT: classification of usual interstitial pneumonia, in: IEEE 12th International Symposium on Biomedical Imaging, ISBI 2015, IEEE, apr 2015, pp. 403–406.

[19] P. Cirujeda, Y. Dicente Cid, H. Müller, D.L. Rubin, T.A. Aguilera, B.W. Loo Jr., M. Diehn, X. Binefa, A. Depeursinge, A 3-D Riesz-covariance texture model for prediction of nodule recurrence in lung CT, IEEE Trans. Med. Imaging 35 (12) (2016) 2620–2630.

[20] A. Depeursinge, A. Foncubierta-Rodríguez, D. Van De Ville, H. Müller, Lung texture classification using locally-oriented Riesz components, in: G. Fichtinger, A. Martel, T. Peters (Eds.), Medical Image Computing and Computer Assisted Intervention, MICCAI 2011, in: Lecture Notes in Computer Science, vol. 6893, Springer, Berlin/Heidelberg, sep 2011, pp. 231–238.

[21] J.P. Ward, M. Unser, Harmonic singular integrals and steerable wavelets in $L_2(\mathbb{R}^d)$, Appl. Comput. Harmon. Anal. 36 (2) (mar 2014) 183–197.

[22] H. Skibbe, M. Reisert, T. Schmidt, T. Brox, O. Ronneberger, H. Burkhardt, Fast rotation invariant 3D feature computation utilizing efficient local neighborhood operators, IEEE Trans. Pattern Anal. Mach. Intell. 34 (8) (aug 2012) 1563–1575.

[23] J. Portilla, E.P. Simoncelli, A parametric texture model based on joint statistics of complex wavelet coefficients, Int. J. Comput. Vis. 40 (1) (2000) 49–70.

[24] I. Daubechies, Ten Lectures on Wavelets, vol. 61, SIAM, 1992.

[25] M. Unser, N. Chenouard, D. Van De Ville, Steerable pyramids and tight wavelet frames in $L_2(\mathbb{R}^d)$, IEEE Trans. Image Process. 20 (10) (oct 2011) 2705–2721.

[26] A. Depeursinge, Z. Püspöki, J.-P. Ward, M. Unser, Steerable wavelet machines (SWM): learning moving frames for texture classification, IEEE Trans. Image Process. 26 (4) (2017) 1626–1636.

[27] M. Unser, N. Chenouard, A unifying parametric framework for 2D steerable wavelet transforms, SIAM J. Imaging Sci. 6 (1) (2013) 102–135.

[28] G. Jacovitti, A. Neri, Multiresolution circular harmonic decomposition, IEEE Trans. Signal Process. 48 (11) (2000) 3242–3247.

[29] K. Liu, H. Skibbe, T. Schmidt, T. Blein, K. Palme, T. Brox, O. Ronneberger, Rotation-invariant HOG descriptors using Fourier analysis in polar and spherical coordinates, Int. J. Comput. Vis. 106 (3) (2014) 342–364.

[30] T. Ahonen, J. Matas, C. He, M. Pietikäinen, Rotation invariant image description with local binary pattern histogram Fourier features, in: A.-B. Salberg, J. Hardeberg, R. Jenssen (Eds.), Image Analysis, in: Lecture Notes in Computer Science, vol. 5575, Springer, Berlin, Heidelberg, 2009, pp. 61–70.

[31] A. Depeursinge, A. Foncubierta-Rodríguez, D. Van De Ville, H. Müller, Rotation-covariant texture learning using steerable Riesz wavelets, IEEE Trans. Image Process. 23 (2) (February 2014) 898–908.

[32] A. Depeursinge, M. Yanagawa, A.N. Leung, D.L. Rubin, Predicting adenocarcinoma recurrence using computational texture models of nodule components in lung CT, Med. Phys. 42 (2015) 2054–2063.

[33] A. Depeursinge, C. Kurtz, C.F. Beaulieu, S. Napel, D.L. Rubin, Predicting visual semantic descriptive terms from radiological image data: preliminary results with liver lesions in CT, IEEE Trans. Med. Imaging 33 (8) (August 2014) 1–8.

[34] I. Tosic, P. Frossard, Dictionary learning, IEEE Signal Process. Mag. 28 (2) (mar 2011) 27–38.

[35] R. Rigamonti, V. Lepetit, Accurate and efficient linear structure segmentation by leveraging ad hoc features with learned filters, in: N. Ayache, H. Delingette, P. Golland, K. Mori (Eds.), Medical Image Computing and Computer-Assisted Intervention, MICCAI 2012, in: Lecture Notes in Computer Science, vol. 7510, Springer, Berlin, Heidelberg, 2012, pp. 189–197.

[36] Y.-J. Lee, O.L. Mangasarian, SSVM: a smooth support vector machine for classification, Comput. Optim. Appl. 20 (1) (2001) 5–22.

[37] S.-C. Zhu, C.-E. Guo, Y. Wang, Z. Xu, What are textons?, in: Special Issue on Texture Analysis and Synthesis, Int. J. Comput. Vis. 62 (1–2) (apr 2005) 121–143.

[38] J. Mairal, F. Bach, J. Ponce, G. Sapiro, A. Zisserman, Supervised dictionary learning, Adv. Neural Inf. Process. Syst. (2008) 1033–1040.

[39] M.J. Gangeh, A. Ghodsi, M.S. Kamel, Dictionary learning in texture classification, in: Proceedings of the 8th International Conference on Image Analysis and Recognition, Part I, 2011, pp. 335–343.

[40] A. Hyvärinen, J. Hurri, P.O. Hoyer, Energy correlations and topographic organization, in: Natural Image Statistics: A Probabilistic Approach to Early Computational Vision, Springer London, London, 2009, pp. 239–261.

[41] I. Guyon, A. Elisseeff, An introduction to variable and feature selection, J. Mach. Learn. Res. 3 (mar 2003) 1157–1182.

[42] J. Bruna, S.G. Mallat, Invariant scattering convolution networks, IEEE Trans. Pattern Anal. Mach. Intell. 35 (8) (2013) 1872–1886.

[43] I. Goodfellow, Y. Bengio, A. Courville, Deep Learning, MIT Press, 2016, http://www.deeplearningbook.org.

[44] A. Vedaldi, K. Lenc, MatConvNet: convolutional neural networks for MATLAB, in: Proceedings of the 23rd ACM International Conference on Multimedia, MM '15, ACM, New York, NY, USA, 2015, pp. 689–692.

[45] A. Krizhevsky, I. Sutskever, G.E. Hinton, ImageNet classification with deep convolutional neural networks, in: F. Pereira, C.J.C. Burges, L. Bottou, K.Q. Weinberger (Eds.), Advances in Neural Information Processing Systems, vol. 25, Curran Associates, Inc., 2012, pp. 1097–1105.

[46] O. Ronneberger, P. Fischer, T. Brox, U-Net: convolutional networks for biomedical image segmentation, in: N. Navab, J. Hornegger, W.M. Wells, A.F. Frangi (Eds.), Medical Image Computing and Computer-Assisted Intervention, MICCAI 2015, in: Lecture Notes in Computer Science, vol. 9351, Springer International Publishing, 2015, pp. 234–241.

[47] B. Sahiner, H.-P. Chan, N. Petrick, D. Wei, M.A. Helvie, D.D. Adler, M.M. Goodsitt, Classification of mass and normal breast tissue: a convolution neural network classifier with spatial domain and texture images, IEEE Trans. Med. Imaging 15 (5) (oct 1996) 598–610.

[48] A.S. Razavian, H. Azizpour, J. Sullivan, S. Carlsson, CNN features off-the-shelf: an astounding baseline for recognition, in: IEEE Computer Society Conference on Computer Vision and Pattern Recognition Workshops, 2014, pp. 512–519.

[49] M. Oquab, L. Bottou, I. Laptev, J. Sivic, Learning and transferring mid-level image representations using convolutional neural networks, in: 2014 IEEE Conference on Computer Vision and Pattern Recognition, jun 2014, pp. 1717–1724.

[50] J. Deng, W. Dong, R. Socher, L.-J. Li, K. Li, L. Fei-Fei, ImageNet: a large-scale hierarchical image database, in: IEEE Conference on Computer Vision and Pattern Recognition, CVPR 2009, 2009, pp. 248–255.

[51] Q. Dou, H. Chen, Y. Jin, L. Yu, J. Qin, P.-A. Heng, 3D Deeply Supervised Network for Automatic Liver Segmentation from CT Volumes, Springer International Publishing, Cham, 2016, pp. 149–157.

[52] F. Milletari, N. Navab, S.-A. Ahmadi, V-Net: fully convolutional neural networks for volumetric medical image segmentation, CoRR, abs/1606.04797, 2016.

[53] Ö. Çiçek, A. Abdulkadir, S.S. Lienkamp, T. Brox, O. Ronneberger, 3D U-Net: learning dense volumetric segmentation from sparse annotation, in: Medical Image Computing and Computer Assisted Intervention, MICCAI 2016, in: Lecture Notes in Computer Science, vol. 9901, Springer International Publishing, 2016, pp. 424–432.

[54] J. Wang, J.D. MacKenzie, R. Ramachandran, D.Z. Chen, A deep learning approach for semantic segmentation in histology tissue images, in: S. Ourselin, L. Joskowicz, M.R. Sabuncu, G. Unal, W. Wells (Eds.), Medical Image Computing and Computer-Assisted Intervention – MICCAI 2016: 19th International Conference, Proceedings, Part II, Athens, Greece, October 17–21, 2016, Springer International Publishing, Cham, ISBN 978-3-319-46723-8, 2016, pp. 176–184.

[55] D. Marcos Gonzalez, M. Volpi, D. Tuia, Learning rotation invariant convolutional filters for texture classification, CoRR, abs/1604.06720, 2016.

[56] R.M. Haralick, Statistical and structural approaches to texture, Proc. IEEE 67 (5) (may 1979) 786–804.

[57] M.M. Galloway, Texture analysis using gray level run lengths, Comput. Graph. Image Process. 4 (2) (1975) 172–179.

[58] D.-H. Xu, A.S. Kurani, J. Furst, D.S. Raicu, Run-length encoding for volumetric texture, in: The 4th IASTED International Conference on Visualization, Imaging, and Image Processing, VIIP 2004, Marbella, Spain, sep 2004.

[59] G. Thibault, B. Fertil, C. Navarro, S. Pereira, P. Cau, N. Levy, J. Sequeira, J.-L. Mari, Texture indexes and gray level size zone matrix application to cell nuclei classification, in: Pattern Recognition and Information Processing, 2009, pp. 140–145.

[60] T. Ojala, M. Pietikäinen, T. Mäenpää, Multiresolution gray-scale and rotation invariant texture classification with local binary patterns, IEEE Trans. Pattern Anal. Mach. Intell. 24 (7) (jul 2002) 971–987.

[61] Z. Guo, L. Zhang, D. Zhang, A completed modeling of local binary pattern operator for texture classification, IEEE Trans. Image Process. 19 (6) (jun 2010) 1657–1663.

[62] L. Liu, S. Lao, P.W. Fieguth, Y. Guo, X. Wang, M. Pietikäinen, Median robust extended local binary pattern for texture classification, IEEE Trans. Image Process. 25 (3) (mar 2016) 1368–1381.

[63] A. Depeursinge, A. Foncubierta-Rodríguez, D. Van De Ville, H. Müller, Three-dimensional solid texture analysis and retrieval in biomedical imaging: review and opportunities, Med. Image Anal. 18 (1) (2014) 176–196.

[64] J. Fehr, H. Burkhardt, 3D rotation invariant local binary patterns, in: 19th International Conference on Pattern Recognition, 2008, ICPR 2008, December 2008, pp. 1–4.

[65] A. Burner, R. Donner, M. Mayerhoefer, M. Holzer, F. Kainberger, G. Langs, Texture bags: anomaly retrieval in medical images based on local 3D-texture similarity, in: H. Greenspan, H. Müller, T. Syeda-Mahmood (Eds.), Medical Content-Based Retrieval for Clinical Decision Support, MCBR-CDS 2011, in: Lecture Notes in Computer Sciences (LNCS), vol. 7075, September 2012, pp. 116–127.

[66] Y.D. Wang, Q.Y. Yan, K.F. Li, Hand vein recognition based on multi-scale LBP and wavelet, in: 2011 International Conference on Wavelet Analysis and Pattern Recognition, jul 2011, pp. 214–218.

[67] B.B. Mandelbrot, The Fractal Geometry of Nature, W.H. Freeman and Company, 1977.

[68] O.S. Al-Kadi, D. Watson, Texture analysis of aggressive and nonaggressive lung tumor CE CT images, IEEE Trans. Biomed. Eng. 55 (7) (July 2008) 1822–1830.

[69] A.P. Pentland, Fractal-based description of natural scenes, IEEE Trans. Pattern Anal. Mach. Intell., PAMI-6 (6) (nov 1984) 661–674.

[70] N. Sarkar, B.B. Chaudhuri, An efficient differential box-counting approach to compute fractal dimension of image, IEEE Trans. Syst. Man Cybern. 24 (1) (jan 1994) 115–120.

[71] J.M. Keller, S. Chen, R.M. Crownover, Texture description and segmentation through fractal geometry, Comput. Vis. Graph. Image Process. 45 (2) (1989) 150–166.

[72] B. Pesquet-Popescu, J.L. Vehel, Stochastic fractal models for image processing, IEEE Signal Process. Mag. 19 (5) (sep 2002) 48–62.

CHAPTER 4

Deep Learning in Texture Analysis and Its Application to Tissue Image Classification

Vincent Andrearczyk, Paul F. Whelan
Vision Systems Group, Dublin City University, Dublin, Ireland

Abstract

In the last decade, artificial intelligence has been revolutionized by deep learning, outperforming human prediction on a wide range of problems. In particular Convolutional Neural Networks (CNNs) are particularly well suited to texture analysis. The trainable filter banks make an excellent tool in the analysis of repetitive texture patterns. Most applications of CNNs on texture analysis share the idea of using the weight sharing and local connectivity while discarding the global shape analysis. Texture analysis is a common issue in biomedical imaging applications and it is no surprise that CNNs have been applied to this field. This chapter introduces key concepts of deep learning and convolutional neural networks and presents a review of the literature on texture-specific CNNs. A novel CNN architecture (available online) that is specific for texture analysis is introduced and evaluated for general and biomedical texture classification with comparison to other methods.

Keywords

Convolutional neural network, Texture, Filter banks, Dense orderless pooling, Tissue images, Ensemble learning

4.1 INTRODUCTION

Convolutional Neural Networks (CNNs) are excellent at analyzing images by learning abstract representations with high levels of semantics. They are also naturally well designed for texture analysis as they learn filter banks with weight sharing and local connectivity which detect patterns at all locations in the image. The first convolution layer of a CNN learns to detect simple features mainly composed of edges which are very similar to Gabor filters [1] commonly used in texture analysis. The following layers learn to detect larger and more complex features from simple nonlinear combinations of the previous ones. CNNs can therefore learn to recognize texture patterns of various complexity and scales, largely improving traditional filter banks approaches by replacing handcrafted filters by a hierarchical architecture combined with a powerful learning algorithm. Convolutional networks largely exploit the texture content present in images with this filter bank approach. However, in deeper layers, with the increase of the receptive fields of the neurons (*i.e.*, area covered in the input image, see Section 4.2.4) and

Biomedical Texture Analysis
DOI: 10.1016/B978-0-12-812133-7.00004-1

95

of the complexity and level of abstraction of the features, CNNs start to detect global structures and shapes over simple texture patterns. CNNs are introduced in Section 4.2 and the reader should refer to [2] for background details on neural networks and [1] for more details on deep learning and CNNs.

This chapter introduces several methods to adapt a classic CNN architecture to the analysis of texture by exploiting the powerful trainable filter banks and discarding the overall shape analysis. We focus on our Texture CNN (T-CNN) approach that densely pools simple texture descriptors from convolution layers and allows end-to-end training with backpropagation. The complexity of our method is reduced while the accuracy on texture datasets is considerably increased when compared to classic CNN approaches [3,4]. We also introduce a method to combine a texture and shape analyses in a single CNN architecture which is trained end-to-end.

Texture is often predominant in biomedical images and its analysis plays a key role in various computer aided diagnostic systems. In particular microscopic tissue images exhibit high texture contents with large pixel intensity variation and repetitive patterns across the images. Texture oriented deep learning methods are therefore legitimate candidates for the analysis of such textured biomedical images. We present an approach to tissue classification based on our T-CNN with an ensemble model method to combine the classification of patches extracted from large tissue images. The proposed approach largely outperforms classic machine learning approaches.

The rest of the chapter is organized as follows. We introduce the fundamental concepts of CNNs and the methods to train them in Section 4.2. In Section 4.3, we present a literature review of CNNs applied to texture and biomedical texture analysis. In Section 4.4, we introduce our Texture CNN approach and experiment on classic texture datasets. Finally, an application to the classification of microscopic tissue images is presented in Section 4.5.

4.2 INTRODUCTION TO CONVOLUTIONAL NEURAL NETWORKS

This section introduces several key concepts necessary to the understanding of CNNs, including artificial neurons, neural networks, and backpropagation, followed by an introduction to CNNs with several important architectures.

4.2.1 Neurons and nonlinearity

An artificial neuron, which we will refer to as a neuron for simplicity, is a mathematical function partly inspired by biological neurons. Neural networks are designed by connecting multiple neurons together. A neuron computes a nonlinear weighted sum

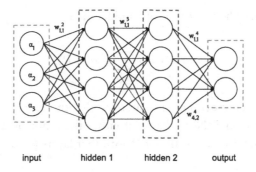

input   hidden 1   hidden 2   output

Figure 4.1 Three-layer Multilayer Perceptron (biases not represented).

as:

$$a = \sigma(h) = \sigma(\sum_{i=1}^{m} w_i \alpha_i + b),\tag{4.1}$$

where $\boldsymbol{w}$ and $\boldsymbol{\alpha}$, respectively the weights and inputs, are both vectors of size m. b is the bias and $\sigma(\cdot)$ is a nonlinear function inspired by the "firing" of biological neurons. The use of nonlinearity is necessary when using multiple layers of neurons as explained in Section 4.2.2. In many representations, including Fig. 4.1, the neuron encapsulates the computation of h and its activation with a nonlinear function $\sigma(h)$ (Eq. (4.1)). Several nonlinearity functions have been used in neural networks including sigmoid, hyperbolic tangent, and more commonly, Rectified Linear Unit (ReLU) (*e.g.*, the ReLU of a real value x is simply obtained as $\max(0, x)$) and its variants.

4.2.2 Neural network

Neurons are organized in an Artificial Neural Network (ANN) into multiple layers. A basic example of such network is the Multilayer Perceptron (MLP). An MLP consists of an input layer (bottom of the network), an output layer (top of the network) and at least one hidden layer. An example of a three layer MLP (+ input) is illustrated in Fig. 4.1. An MLP is fully connected, *i.e.*, each neuron of a given hidden layer is connected to all the neurons of the previous and next layers. If only linear neurons were used, a network of any depth could be reduced to a single layer network which would perform a linear regression [2]. Instead, the output of the neurons are "activated" by nonlinearity to generate more complex functions.

We can generalize to other architectures with L layers including the input layer. In such architectures, h_j^l is the output of neuron j at layer l (with $l = 1$ being the input layer) before activation. The activation of a hidden neuron is computed as

$$a_j^l = \sigma(h_j^l) = \sigma\left(\sum_{i=1}^{m_{l-1}} w_{ij}^l \alpha_i^l + b_j^l\right), \tag{4.2}$$

where w_{ij}^l is the weight connecting the input α_i^l to the jth neuron at layer l and m_l is the number of neurons at layer l. Note that $\boldsymbol{\alpha}^l$ is either the input of the network for $l = 2$ or the output of the previous activation $\boldsymbol{a}^{l-1}$ for other layers $l > 2$.

4.2.3 Training

An ANN can be trained to infer a classification or perform a regression task. We focus in the section on a classification problem in which the network is trained with categorical labels (as opposed to continuous labels for regression), although most concepts introduced here are shared by both approaches. In this context, the output of the last layer is a vector of size N where $N = m_L$ is the number of classes. The value of an output neuron represents a class score or a (nonnormalized) probability of the input data to belong to a certain class. Labeled training data is fed into the network to infer a function whose outputs are close to the given labels, and ideally, generalizes to unknown data. This approach is called *supervised learning*. A key factor of success of neural networks is the automatic learning of the weights. Backpropagation [5] is a training method to iteratively update the weights in order to minimize the error between known training values and the predictions of the network. To this end, the error is propagated backward through every layer of the network by calculating the gradients of a defined loss function with respect to all the weights. These gradients are a measure of how the error will change if we vary each weight individually. An optimization function then uses these gradients to update the weights and reduce the error. The backpropagation algorithm can be summarized as a succession of the following operations: forward pass, error calculation, backpropagation, and optimization of the weights. We discuss these operations in more detail in addition to other key concepts required to understand how to efficiently train neural networks.

4.2.3.1 Forward pass

In the forward pass, the outputs of all the neurons are computed from the input layer to the output layer. The hidden neurons are computed with Eq. (4.2), where $\boldsymbol{\alpha}^l = \boldsymbol{a}^{l-1}$ for $l > 2$ and the output of the network is $y(\boldsymbol{\alpha}) = \boldsymbol{h}^L$.

4.2.3.2 Error

The output of the last layer is a vector of real values that represent a class score. With a training set or subset of input vectors $\boldsymbol{\alpha}^{(k)} \in \boldsymbol{A}$, and associated ground truth vectors

$t^{(k)} = t(\boldsymbol{\alpha}^{(k)}) \in \boldsymbol{T}$, we want to train the network so that it computes a function $y(\boldsymbol{A})$ that maps $\boldsymbol{A}$ to $\boldsymbol{T}$. This function should also generalize well to unknown input data. To know how well the network is performing, the errors between the ground truths $t^{(k)}$ and the predictions of the network $y(\boldsymbol{\alpha}^{(k)})$ are calculated and averaged across the multiple K samples. Common error functions (also known as loss or cost functions) include the *mean squared* and the *cross-entropy* errors.

The (multiclass) cross-entropy is often used for predicting targets interpreted as probabilities by calculating the difference between a predicted probability vector $\tilde{p}(\boldsymbol{\alpha})$ and its ground truth $p(\boldsymbol{\alpha})$. A vector of predictions of the network $y(\boldsymbol{\alpha}) = \boldsymbol{h}^L$ is generally interpreted (or squashed) as a probability vector using the softmax function:

$$\tilde{p}_n^{(k)} = \frac{e^{h_n^{L(k)}}}{\sum_{j=1}^N e^{h_j^{L(k)}}} \text{ for } n = 1, ..., N, \tag{4.3}$$

where $h_j^{L(k)}$ is the jth value of the output vector of the network with input $\boldsymbol{\alpha}^{(k)}$ and N is the dimension of the output vector. The values of the probability prediction vector $\tilde{\boldsymbol{p}}^{(k)}$ are in the range $(0, 1)$ and sum to 1. The cross-entropy loss is then computed as follows:

$$E^{(k)} = -\sum_{n=1}^N p_n^{(k)} log \left(\frac{\tilde{p}_n^{(k)}}{p_n^{(k)}} \right). \tag{4.4}$$

Note that for a regression task with continuous labels t^k, the mean squared error is generally used as loss function and is computed as follows:

$$E^{(k)} = \frac{1}{2} \sum_{n=1}^N (h_n^{L(k)} - t_n^{(k)})^2. \tag{4.5}$$

4.2.3.3 Backpropagation of the error

The gradients of the error with regard to all the weights in the network can now be calculated using the chain rule which enables to compute the derivative of the composition of functions in the network (*e.g.*, weighted sum, activation, etc.).

Using the chain rule and simplifications, the partial derivative of E with respect to the weights of the last layer are expressed as

$$\frac{\partial E^{(k)}}{\partial w_{ij}^L} = \frac{\partial E^{(k)}}{\partial h_j^{L,(k)}} \frac{\partial h_j^{L,(k)}}{\partial w_{ij}^L} = \delta_j^{L(k)} a_i^{L-1(k)} = (\tilde{p}_j^{(k)} - p_j^{(k)}) a_i^{L-1(k)}. \tag{4.6}$$

For all other layers $l < L$, the neurons compute their output with Eq. (4.2) and the gradients are computed as follows:

$$\frac{\partial E^{(k)}}{\partial w_{ij}^{l(k)}} = \delta_j^{l(k)} \alpha_j^{l(k)}, \tag{4.7}$$

where δ^l's is referred to as the error at a certain layer (not to be confused with the Kronecker delta):

$$\delta_j^{l(k)} = \sigma'(h_j^{l(k)}) \sum_{n=1}^{m_{l+1}} w_{jn}^{l+1} \delta_n^{l+1}, \tag{4.8}$$

where σ' is the derivative of the activation function.

4.2.3.4 Stochastic gradient descent

Gradient Descent (GD) is an optimization method to find a local (preferably global) minimum of a function. In backpropagation, it is used to iteratively update the weights in order to minimize the error function. In a GD optimization, all the training samples are used for each update of the weights whereas in Stochastic Gradient Descent (SGD), only one or a small batch of training samples are used for each step. With large training sets (common to CNN implementations), the computation time of the GD optimization can become extremely long as one must compute the outputs, errors, and gradients of all the samples at each iteration. SGD is therefore almost always preferred to GD in neural networks. The weights are updated using

$$w_{ij}^{\tau+1} = w_{ij}^{\tau} + \Delta w_{ij}^{\tau}, \tag{4.9}$$

where τ is the gradient descent iteration and Δw_{ij} is the weight update. Two variants of SGD are generally used, namely *on-line learning* and *batch learning*. In on-line learning, the weights are updated for every input $\boldsymbol{\alpha}^{(k)}$, i.e., in Eq. (4.9) we have

$$\Delta w_{ij} = -\eta \frac{\partial E^{(k)}}{\partial w_{ij}} = -\eta \delta_j^l a_j^{l-1(k)}, \tag{4.10}$$

where the learning rate η is a hyperparameter of the network that determines the speed of change.

In batch learning, the error is calculated for a batch of K inputs $\boldsymbol{A}$ and the weights gradients are computed as

$$\Delta w_{ij} = -\eta \frac{\partial E}{\partial w_{ij}} = -\frac{\eta}{K} \sum_{k=1}^{K} \delta_j^{l(k)} a_j^{l-1(k)}. \tag{4.11}$$

A common issue with gradient based optimization methods is being trapped in a local minimum far from the desired global minimum. With on-line learning the stochastic error surface is noisy which helps escaping such local minima. Batch learning, by averaging the gradients, removes noise and is more prone to being caught in local minima. However, batch training is often preferred as it can be very efficiently implemented with modern computers.

Moreover, training methods have been developed to escape from local minima, making on-line learning obsolete. For example, the momentum method [5] (and variant Nesterov momentum [6]) helps escape from a local minimum by diminishing the fluctuations in weights updates over consecutive iterations. Its effect can be thought of as a ball rolling down a hill in the weights space and picking up pace to roll up the opposite slope and potentially escape a local minimum. Other popular gradient descent methods used in deep learning include the adaptive gradient (Adagrad) [7], adaptive moment estimation (Adam) [8], and adaptive learning rate (Adadelta) [9].

4.2.3.5 Weights initialization

The weights initialization is the starting point of the learning algorithm and requires particular attention. While it might be a reasonable first guess to initialize all the weights to zero, it does not work in practice as they all compute the same gradients and do not enable gradient descent. Indeed, asymmetry is required in the weights initialization. A common method is the Gaussian initialization in which the weights are drawn from a Gaussian distribution of zero mean and small variance. The Xavier initialization [10] is similar to a Gaussian initialization except that the variance is a function of the number of input and output connections, keeping the initial weight values in a desired range. Note that due to the lack of labeled training data and good domain transferability of deep networks, the weights are often pretrained on a large dataset (ImageNet [11]) and fine-tuned on the target task as explained in Section 4.4.3.3.

4.2.3.6 Regularization

Overfitting is a frequent problem in machine learning when training a model. It occurs when an excessively complex model describes the noise in the training data and becomes too specific to the latter. The overfitted model will not generalize well to unknown test data. Several regularization methods are used in neural networks to prevent overfitting by generally penalizing the complexity. Commonly used regularization methods include weight decay, early stopping, dropout [12], and artificial data expansion.

4.2.4 CNN

A CNN is a supervised feedforward ANN (although unsupervised and recurrent variations also exist). The analysis of the visual cortex [13] inspired the arrangement of neurons which respond to overlapped regions tiling the input grid–like data, *e.g.*, images. The neurons behave as a bank of filters with local receptive fields in the input image and enable the exploitation of the local correlations present in natural images (*i.e.*, correlation of pixel values in image neighborhoods) in a deep learning approach. A basic CNN approach can be summarized by a succession of convolution, pooling, nonlinear, and Fully Connected (FC) layers. The convolution filters and fully connected weights are trained by backpropagation and SGD (see Section 4.2.3). Once successfully trained, the filters at multiple layers of a convolutional network respond to features of different complexity. The bottom layer has been shown to detect simple patterns such as edge-like features similar to a Gabor filter set, while deeper layers detect more complex and abstract features with larger receptive fields [1]. Note that in this chapter we refer to a receptive field as the area in the input image to which a neuron is directly or indirectly connected. The receptive fields in the top layers are in that sense larger than in early layers due to the consecutive convolution and pooling layers, although the kernel size is not necessarily larger. This complexity in high layers arises from the combination of simpler features with linear and nonlinear operations.

4.2.4.1 Main building blocks

The *convolution* layer is the most important building block of a CNN. It incorporates several key concepts which are well suited to the analysis of images and, as will be shown in Section 4.3, to the analysis of texture images. In the general case (2D images), the input and output data of a convolution layer is three dimensional with dimensions $C \times H \times W$ for respectively the number of channels, the height, and the width. Note that the number of channels in the input image is generally three for color or one for grayscale data. The inputs of a convolution layer are called input channels while the outputs are referred to as feature maps after activation (nonlinearity). A convolution layer is mainly defined by a set of small filters (or kernels) by which the input data is convolved in the forward pass. The outputs of a convolution layer are obtained as

$$h_j(\boldsymbol{x}) = \sum_{i=1}^{C} (f_i * g_{ij})(\boldsymbol{x}), \tag{4.12}$$

where $h_j(\boldsymbol{x})$ is the jth (nonactivated) output feature map at spatial location $\boldsymbol{x} = (x_1, x_2)$, g_{ij} is the kernel between $\boldsymbol{h}_j$, and the ith input channel f_i. The sum runs over the C input channels. The filters scan the input channels with a certain stride by which the filters slide (a large stride results in smaller output feature maps). The operator $*$ is a 2D convolution defined as follows:

$$(f_i * g_{ij})(x_1, x_2) = \sum_{m} \sum_{n} f_i(m, n) g_{ij}(x_1 - m, x_2 - n). \tag{4.13}$$

The input channels can be zero padded with a desired size to convolve the edges of the input channels. A fully-connected architecture (*e.g.*, MLP) does not consider the local spatial correlation of pixels and the fact that features can be sought across the image. It is also too computationally expensive to connect every pixel when dealing with large input images. The convolution layer deals with these problems specific to grid-like data with the following concepts.

Local connectivity (or sparse connectivity) is a first key concept of the convolution layer. Similar to the local receptive field of the visual cortex discovered in [13], convolution layers use small filters to capture local dependencies in the image. Unlike MLPs, convolution layers ensure that each neuron of the network is connected to a small number of neurons in the input channel equal to the size of the filters. On top of capturing local dependencies, the local connectivity also enables a large reduction of trainable parameters when compared to standard MLPs. The high dimensionality of the input images (number of pixels) makes the fully-connected approach of MLPs very impractical. The second key concept of the convolution layer is the weight sharing. All the hidden neurons within a feature map share the same parameters (*i.e.*, filters) and each hidden neuron within a feature map covers a particular local part of the input channels. This can be interpreted in the sense that we seek the same features (edges, corners, and more complex patterns) at different locations across the image. This is particularly true in the analysis of textures with repetitive patterns. The backpropagation at a convolution layer is similar to the MLP (Section 4.2.3), keeping in mind the local connectivity and weight sharing.

Nonlinearity functions are generally used after convolution and FC layers to activate the neurons. The reader should refer to Section 4.2.1 for an explanation of the nonlinearity in ANNs. The most popular activation functions used in convolutional networks are the ReLU [14] and its variants as they limit the vanishing gradients problem often encountered in deep learning. The latter refers to the exponential decrease of the error gradient with the number of layers due to the successive multiplication of small gradients in each layer, causing a slow training process.

Another important layer in CNNs is the *pooling* layer with a threefold benefit. First, it is used to reduce the size of the data throughout the network. Secondly, it helps the network to learn small transformation invariances (*e.g.*, translation, scale, and rotation) [1]. Thirdly, it increases the receptive field of the neurons. Note that unlike a convolution layer, there is no trainable parameter in a pooling layer. The most popular pooling approach is "max" pooling, which outputs the maximum values of small nonoverlapping patches in the input channels. In this method, the gradients are backpropagated through the neuron with maximum value (maximally activated in the forward pass) as changing nonmaximum neurons do not affect the output. Another approach is "average" pooling in which the average of each nonoverlapping patch is outputted. In an average pooling layer, all the input neurons covered by a pooling kernel are given the same gradient value, *i.e.*, the output gradients divided by the kernel size. Note that several recent approaches discard the pooling operations [15] by using convolution layers with increased stride.

Fully-Connected (FC) layers are often used after a cascade of convolution, nonlinear and pooling layers. A FC layer is similar to an MLP layer, *i.e.*, the neurons are connected to all the input neurons. In a classification scheme the last FC layer contains a number of neurons equal to the number of classes. In a trained classifier network a high activation of the *i*th output value of the last FC layer reflects a high probability of the input image being of class *i*. Note that several recent architectures replace the FC layers by 1×1 convolution layers [16] to obtain a spatial map of output score vectors from input images of varying sizes.

The *loss* layer as explained in Section 4.2.3.2 measures the error between the output of a network and target values. The loss is used during training for calculating the gradients and propagating them backward through the network for SGD optimization. As mentioned previously, a commonly used loss in CNNs is the cross–entropy. The reader should refer to Section 4.2.3 for the forward and backward computation of this loss function.

Several *regularization* methods are used in CNNs to better generalize to unknown data. The most common ones include Dropout, Dropconnect, stochastic pooling, data augmentation, weight decay, early stopping, and ℓ_1 and ℓ_2 regularization [1]. Also batch normalization [17] is used in many recent deep architectures for regularization and avoiding vanishing gradients and internal covariate shift (change in the distribution of the inputs of intermediate layers amplified throughout the network).

4.2.4.2 CNN architectures

A wide range of CNN architectures have been developed in the last decade, some with a continuously growing depth. The most important architectures for this chapter are AlexNet [3] and VGG networks [4,18], primarily used for the ImageNet classification.

AlexNet [3] is a CNN architecture which includes five convolution layers and three FC layers as well as max–pooling, Local Response Normalization (LRN), and dropout. More details on this architecture can be found in the original paper [3].

The VGG architectures [4,18] vary from AlexNet in terms of depth with up to 19 layers [18]. The main idea of VGG networks is to replace convolution layers of large filter sizes by multiple convolution layers of smaller (3×3) kernel size. This approach enables increasing the depth of the network and reducing the number of parameters while keeping the same receptive fields of the neurons. For instance, three convolution layers with 3×3 filters have the same receptive field as a single convolution layer with 7×7 filters and less trainable parameters. However, the large number of channels (width of the convolution layers) in the VGG architectures results in a large number of trainable parameters, making them difficult to train.

Recent architectures were developed with new methods to enable going deeper including network in network [19], GoogleNet [20], and deep residual networks [21].

4.2.4.3 Visualization

Visualizing what CNNs learn is crucial in interpreting the learned features and understanding why the networks perform so well and how they might be improved [22]. Several methods have been developed to gain an insight of what and how CNNs learn. The simplest method is to visualize the feature maps and weights at different layers in the network. However, these are mainly interpretable for the first layer as the feature maps of intermediate layers are a result of a combination of multiple filters in previous layers [22,23]. We will present three more advanced methods that were developed to visualize what the neurons in intermediate layers actually perceive from the input images. The first method simply finds which input image maximally activates a certain neuron [23]. In other words, we separately input all the images from a dataset (generally unknown to the network) to the CNN and find the one for which the activation of the neuron of interest is maximal. This is a very basic method, which is most relevant for neurons in high layers. Another approach, named activation maximization, generates an image (a receptive field) which maximally activates a neuron [23] (see Fig. 4.3). To that end, the input values are updated in a gradient ascent optimization to maximize the activation of a neuron as a function of the inputs of all layers up to the input image. The last visualization method introduced here is the deconvolution approach [22]. The idea is to reconstruct an input image by reversing the operations of the CNN. To visualize what a neuron at a certain layer has learned to detect, we find which activation fires most with unknown input images and reverse the operations to project the activation back into the pixel space. In particular, deconvolution, reversed max-pooling, and reversed ReLU layers are used. Note that the same convolution kernels and pooling "switches" are used for the deconvolution pass as for the forward pass.

4.3 DEEP LEARNING FOR TEXTURE ANALYSIS: LITERATURE REVIEW

4.3.1 Early work

Textures obey statistical properties and exhibit repeated patterns (see Section 1.2 of Chapter 1). Statistics of dense local descriptors have been commonly used in texture analysis in many forms including filter banks [24], Local Binary Patterns (LBP) [25], bag of visual words [26], and feature encoding and orderless pooling methods such as Fisher Vector (FV) [27] and Vector of Locally Aggregated Descriptors (VLAD) [28]. CNNs have therefore naturally attracted attention in texture analysis as they learn to detect features of multiple complexity everywhere in the image which can be densely pooled into powerful descriptors.

4.3.2 Texture specific CNNs

In [29], Basu et al. argue that the dimensionality of texture datasets is too large to classify them with deep neural networks without explicitly extracting handcrafted features beforehand, as opposed to digits or object datasets which lie on a lower dimensional lattice. They affirm that neural networks must therefore be redesigned to learn texture features similar to GLCM and Haralick features. However, recent work on CNNs applied to texture analysis illustrates that they cope very well with texture datasets [30–33]. It was shown that convolutional architectures are, subject to minor modifications, well designed for the analysis of texture and that they largely improve the state-of-the-art in this field.

Basic CNN architectures have been applied to texture recognition such as [34], in which a simple four layers network was used in the early stages of deep learning to classify simple texture images. More recent CNNs were applied to forest species images with high texture content [35]. Transfer learning between texture datasets was studied in [36] with classic CNN architectures and applied to the forest species classification, similar to a texture recognition problem. While more complex and more accurate than [34], these approaches still do not take the characteristics of texture images (*i.e.*, statistical properties, repeated patterns and irrelevance of the overall shape analysis) into consideration as they are simple applications of a standard CNN architecture to a texture dataset. A recurrent neural network approach, *i.e.*, neural network with internal memory developed for sequential data [37], was used for texture classification [38] and texture segmentation [39]. This approach, however, does not benefit from the weight sharing and local connectivity of CNNs to efficiently detect texture patterns and requires data augmentation to learn simple invariances.

Following these early applications of CNN to texture images, several methods have been developed to discard the global shape or structure analysis by an orderless pooling of feature maps (activations of convolution layer) with excellent results.

Cimpoi et al. [31] demonstrated the relevance of densely extracting texture descriptors from a convolutional network with the Fisher Vector CNN (FV-CNN). In this approach, a CNN pretrained on ImageNet [11] is used as a feature extractor. The output of the last convolution layer is used in a FV encoding and SVM classifier. The overall shape (or global spatial) information is discarded in this analysis by replacing the fully-connected layers by this orderless pooling and classification of features. As the network is only being used for feature extraction, the convolutional network is not fine-tuned and does not learn from the texture dataset. However, due to the domain transferability of filters pretrained on ImageNet, they obtain impressive results on both texture recognition and texture recognition in clutter datasets. The FV-CNN is also combined with shape analysis by extracting the outputs of the penultimate fully-connected layer, demonstrating the complementarity of the shape and texture analysis. Finally, the FV-

CNN can efficiently be used in a region based approach [40] as it requires computing the convolution output once and pooling the desired regions with FV encoding.

In [33], the FV-CNN is improved with a dimensionality reduction to reduce the redundancy and increase the discriminative power of the features extracted from the CNN prior to SVM classification. This approach involves extracting the output of a pretrained CNN, FV encoding, training an ensemble of fully connected neural networks for dimensionality reduction and training an SVM for classification.

A bilinear CNN model is developed in [41] for fine-grained recognition (*i.e.*, visually and semantically very similar classes), combining two CNN streams to extract and classify translationally invariant local pairwise features in a neural network framework. One stream (*e.g.*, convolution and pooling layers) works as an object recognition network, *i.e.*, the "what," while the other stream analyses the spatial location of the object in the image, *i.e.*, the "where." Their output feature maps are multiplied using outer product and densely pooled across the image by summing the extracted features. They obtain robust image descriptors which capture translational invariant local feature interactions. The developed architecture can also generalize several classic orderless pooling descriptors including VLAD [28] and FV [27] in a deep learning framework. This method was successfully applied to texture classification in [32] and obtains slightly better results than the FV-CNN while being trained end-to-end. A Fully Convolutional Network [16] is adapted to the segmentation of texture images in [42], combining local shallow features with deeper and more abstract information while discarding the overall shape and layout analysis.

In [43], several deep texture descriptors are evaluated such as FV-CNN, and compared to several variants of LBP descriptors [25]. As expected, the deep convolutional descriptors [31] obtain the best performance, at the cost of a much higher computational complexity as opposed to LBP variants.

Finally, rotation invariance is embedded in a shallow CNN in [44] by tying the weights of multiple rotated versions of filters for texture classification. While rotation invariance of simple texture descriptors can be learned and pooled in a similar manner as [30] (average pooling), the benefit over hand-crafted rotation invariant descriptors is limited by the use of a single layer.

4.3.3 CNNs for biomedical texture classification

CNNs and deep learning have also been applied to biomedical texture images [45–47] with generally shallow networks and an emphasis on combining generative and discriminative learning. In [46], shallow CNNs are used for the classification of lung image patches with interstitial lung disease. The authors argue that the texture patches do not exhibit high level features and, therefore, do not require abstract representations of deep convolutional networks. Therefore they state that a single layer CNN performs as well as deeper ones, taking into account the low number of training samples and the small size of the patches (32×32). Recent research on CNNs applied to texture analysis, however, has shown that using multiple layers in an appropriate architecture achieves better results on larger texture datasets which exhibit more variation and more complex structures [30–32]. In another approach, van Tulder and de Bruijne [47] train a convolutional Restricted Boltzmann Machine (RBM) in both a supervised and unsupervised manner by combining a generative and a discriminative learning objective. The generative learning helps describing the large training data by optimizing a reconstruction criteria while the discriminative learning optimizes the separation of the classes in the learned feature space. An RBM is trained in a supervised manner by adding a set of label nodes to the visible layer. A convolutional RBM is similar to a classic RBM although it uses the weight sharing of CNNs, *i.e.*, convolution filters applied across the image. The convolutional RBM therefore benefits from the reduction of parameters and from a simple filter banks design for the analysis of lung CT texture images. In [45], Kallenberg et al. develop a multiscale approach, which also combines supervised and unsupervised learning for breast density segmentation and mammographic risk scoring. The analysis of mammographic images aims at detecting breast tissue structures similar to a texture recognition approach. A three layer (with pooling) CNN is first trained layer by layer in an unsupervised manner as a convolutional sparse auto-encoder. In this generative approach, the network learns simple texture features which are detected at multiple areas in the image from unlabeled training data. The pretrained weights of the last layer are then fine-tuned with multinomial logistic regression in a supervised manner. The proposed approach is applied to patches extracted from the input images at multiple scales, which are trained independently in the generative part and combined in the last convolution layer.

The state-of-the-art methods in texture analysis [31,32] are of high complexity, not trained end-to-end [31] and have not been applied to the classification of biomedical texture images which raise specific needs and challenges (*e.g.*, small training sets, high variability, and large images). In the following sections, we introduce a low complexity CNN architecture trained end-to-end which explores the trainable filter bank approach suitable for texture analysis and develop a method based on this architecture to recognize texture tissue images.

4.4 END-TO-END TEXTURE CNN: PROPOSED SOLUTION

This section introduces our Texture CNN (T-CNN) architecture developed in [30] and used in a dynamic texture classification approach in [48]. As explained previously, CNNs are excellent tools to extract repeated texture patterns from images. The computational complexity of deep learning is constantly increasing with the development of the computational power of computers and GPUs. However, low complexity methods are sometimes required, for example, in portable devices or applications that require real-time analysis (*e.g.*, some applications in biomedical imaging). Therefore we develop a method to exploit the power of CNNs in learning and classifying texture descriptors at a reduced computational cost. We experiment on commonly used texture datasets and compare our results to the state-of-the-art. Finally, we introduce a method to combine a texture and shape analysis within the same CNN architecture.

4.4.1 Method

CNNs naturally seek and extract dense features by the weight sharing and local connectivity of the convolution layers. These layers can be compared to filter banks widely used in texture analysis which extract the response to a set of handcrafted filters. While these patterns are predesigned in filter banks methods, the power of CNNs is to learn appropriate features through backpropagation and gradient descent optimization. We develop a simple network architecture that explores this attribute. As explained in [30, 31], the global spatial information is of minor importance in texture analysis as opposed to the necessity of analyzing the global shape for an object recognition scheme. In texture analysis, we seek measurements of the transition between the pixel values and repeated patterns that characterize the textures. To that end, we introduce a dense orderless pooling of texture descriptors within the CNN architecture. We incorporate an energy layer as described below and use multiple configurations with varying numbers of convolution layers. Each feature map of the last convolution layer is simply pooled by calculating the average of its ReLU activations. It results in one single value per feature map, similar to an energy response to a filter bank. The architecture of the T-CNN3 includes three convolution layers C1, C2 and C3 as shown in Fig. 4.2 (additional details on the layers are given in Table 4.1). Our network is derived from AlexNet [3]. The feature map sizes are given for an image cropped to the size 227 × 227 as an example as this is the AlexNet input size although arbitrary input sizes can be used. The forward and backward propagation of the energy layer E3 are similar to an average pooling layer over the entire feature map, *i.e.*, with a kernel size equal to the size of the feature map.

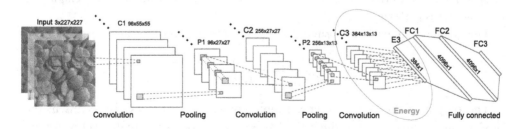

Figure 4.2 T-CNN architecture with three convolution layers (T-CNN3).

Table 4.1 Details of the T-CNN3 architecture [30], with three channel color inputs and N classes. See Section 4.2.4.1 for an explanation of the main building blocks

Layer type	Output size	Kernel, pad, stride	Parameters
Crop	$3 \times 227 \times 227$	–	0
Convolution (C1)	$96 \times 55 \times 55$	11, 0, 4	34,944
ReLU	$96 \times 55 \times 55$	–	0
Pooling (P1)	$96 \times 27 \times 27$	3, 0, 2	0
LRN	$96 \times 27 \times 27$	–	0
Convolution (C2)	$256 \times 27 \times 27$	5, 2, 1	614,656
ReLU	$256 \times 27 \times 27$	–	0
Pooling (P2)	$256 \times 13 \times 13$	3, 0, 2	0
LRN	$256 \times 13 \times 13$	–	0
Convolution (C3)	$384 \times 13 \times 13$	3, 1, 1	885,120
ReLU	$384 \times 13 \times 13$	–	0
Energy	384	–	0
Fully-con. (FC1)	4096	–	1,576,960
ReLU	4096	–	0
Dropout	4096	–	0
Fully-con. (FC2)	4096	–	16,781,312
ReLU	4096	–	0
Dropout	4096	–	0
Fully-con. (FC3)	N	–	$4096 \times N + N$
Softmax	N	–	0

The forward computation of the energy layer is given by

$$e_j = \frac{1}{W \times H} \sum_{x_1=0}^{W} \sum_{x_2=0}^{H} f_j(x_1, x_2), \qquad (4.14)$$

where $e \in \mathbb{R}^N$ is the vector output, $f_j(x_1, x_2)$ is the jth input feature map at spatial location (x_1, x_2). j enumerates the N input feature maps and W, H are respectively the number of columns and rows of the input feature maps. Note that we refer to it as *energy* in reference to the energy response to a filter bank even though an averaging of the responses is performed.

The gradients are backpropagated through the energy layer by computing:

$$\frac{\partial e_j}{\partial f_j(x_1, x_2)} = \frac{1}{W \times H}. \qquad (4.15)$$

The output of the energy layer is simply connected to the FC layers. As in the AlexNet architecture, we use three FC layers which process the high–level reasoning and final classification based on the texture patterns detected and pooled in the energy layer. Similar to other CNNs used in a classification task, the number of neurons in the last FC layer is equal to the number of classes. Except for the layers that are removed and the energy layer, all the other layers are replicated from AlexNet. We experiment with five configurations T-CNN1 to T-CNN5 with one to five convolution layers, respectively. The energy is pooled from the output of the last convolution layer.

The complexity of the proposed T-CNN approach is largely reduced when compared to a classic AlexNet or VGG network. Indeed, convolution layers are removed from the T-CNN architectures (except T-CNN5, which has the same number of convolution layers as the original AlexNet) and the number of parameters between the energy layer and the fully connected layer FC1 is largely reduced as compared to the classic full connection of the last convolution layer. The number of trainable parameters (weights and biases) of the various networks used in the experiments are indicated in Tables 4.1, 4.2, and 4.6.

4.4.2 Experiments

4.4.2.1 Details of the network

The T-CNN architectures are implemented and trained with Caffe [49] and an example can be downloaded from github.[1] The same layers as AlexNet are used in all our architectures for comparison except that the depth is reduced and we use the energy layer (see T-CNN3 in Table 4.1). More details on the implementation are provided in our original paper [30].

4.4.2.2 Datasets

We use a total of ten datasets; seven are texture datasets, the other three are object datasets as described below.

ImageNet 2012 [11] is an image recognition dataset with 1000 classes. The training set contains 1,281,167 images and the testing set 50,000 images (*i.e.*, 50 images per class) of size 256×256 (cropped to 227×227 by the first layer of the networks).

We create three subsets from ImageNet to evaluate the domain transferability of features learned on texture and object image datasets to the classification of texture images. For each subset, we select 28 classes from ImageNet and keep the same training and testing splits as in the full set. Each subset contains 1400 test images and approximately 36,000 training images (35,513, 35,064, 36,318 for ImageNet-T, -S1, and -S2, respectively). The full list of classes selected for each subset is provided in [30].

ImageNet-T is a subset which retains texture classes such as "stone wall," "tile roof" and "velvet." Based on visual examination, 28 classes with high texture content and single texture per image are selected.

---

[1] An implementation to train and test our T-CNN3 network is provided at https://github.com/v-andrearczyk/caffe-TCNN.

Table 4.2 Classification accuracy (%) of various networks trained from scratch and fine-tuned (pretrained on ImageNet). The number of trainable parameters (in millions) is indicated in brackets for 1000 classes. The state-of-the-art results are as reported by the authors in the original papers

	Kylberg	CUReT	DTD	kth-tips-2b	ImNet-T	ImNet-S1	ImNet-S2	ImNet
From scratch								
T-CNN1 (20.8)	89.5 ±1.0	97.0 ±1.0	20.6 ±1.4	45.7 ±1.2	42.7	34.9	42.1	13.2
T-CNN2 (22.1)	**99.2** ±0.3	98.2 ±0.6	24.6 ±1.0	47.3 ±2.0	62.9	59.6	70.2	39.7
T-CNN3 (23.4)	**99.2** ±0.2	98.1 ±1.0	**27.8** ±1.2	**48.7** ±1.3	**71.1**	**69.4**	**78.6**	51.2
T-CNN4 (24.7)	98.8 ±0.2	97.8 ±0.9	25.4 ±1.3	47.2 ±1.4	**71.1**	**69.4**	76.9	28.6
T-CNN5 (25.1)	98.1 ±0.4	97.1 ±1.2	19.1 ±1.8	45.9 ±1.5	65.8	54.7	72.1	24.6
AlexNet (60.9)	98.9 ±0.3	**98.7** ±0.6	22.7 ±1.3	47.6 ±1.4	66.3	65.7	73.1	**57.1**
Fine-tune								
T-CNN1 (20.8)	96.7 ±0.2	99.0 ±0.3	33.2 ±1.1	61.4 ±3.1	51.2	46.2	53.5	–
T-CNN3 (23.4)	**99.4** ±0.2	**99.5** ±0.4	55.8 ±0.8	**73.2** ±2.2	81.2	82.1	87.8	–
AlexNet (60.9)	**99.4** ±0.1	99.4 ±0.4	**56.3** ±1	71.5 ±1.4	**83.2**	**85.4**	**90.8**	–
state-of-the-art	99.7 [52]	99.8 ±0.1 [54]	75.5 ±0.8 [33]	83.3 ±1.4 [33]	–	–	–	–

ImageNet-S1 is another subset with chosen object-like classes such as "chihuahua," "ambulance" and "hammer." Based on visual examination, classes of object images with remarkable global shape are selected.

In the last subset *ImageNet-S2*, 28 classes are chosen randomly from the 1000 classes.

kth-tips-2b [50] contains 11 classes and 432 texture images per class. Each class is made of four samples (*i.e.*, 108 images per sample). Each sample is used once for training while the remaining three samples are used for testing. The images are resized to 227×227 with nearest neighbor interpolation as for the resizing of the following datasets. The results for kth-tips-2b are given as the average classification and the standard deviation over the four splits.

Kylberg [51] is a texture database containing 28 classes of 160 images each of size 576×576. We replicate the validation setup from [52]. We randomly choose one of 12 available orientations for each image and split the images into four nonoverlapping subimages which results in 17,920 images of size 288×288 which we resize to 256×256. We use a ten-folded cross-validation and report the average accuracy and standard deviation. The cross-validation folds are created once and kept fixed throughout the experiments for a fair comparison of the methods.

CUReT [53] is a texture database which contains 61 classes. We reproduce the setup of [54] in which 92 images are used per class, 46 for training, the other 46 for testing. We resize the 200×200 images to 227×227. The experiment is repeated 20 times and we report the average accuracy and standard deviation.

DTD [54] consists of 47 classes which contain 120 images each. The images are of various sizes and even though using multiple input sizes is possible with our T-CNN, we resize all the images to 227×227 for comparison with AlexNet, which requires fixed input images. The dataset includes 10 available annotated splits with 40 training images, 40 validation images, and 40 testing images for each class. We report the average accuracy over 10 splits and the standard deviation.

Macroscopic [55] and *Microscopic forest species databases* [56] are used (in a separate section) to test our approach on larger texture images (respectively 3264×2448 and 1024×768). The Macroscopic dataset consists of 41 classes with over 50 images per class. 50% of the set is used for training, the other 50% for testing. The Microscopic dataset contains 120 classes of 20 images each. In each class, 70% of the images are used for training, the rest for testing. We resize the images of both datasets to 640×640. These experimental setups are reproduced from [35] except that we average the results over ten trials instead of three, for more stable results, and report the standard deviation.

4.4.3 Results and discussions

4.4.3.1 Networks from scratch and pretrained

The results of the various T-CNN architectures (*i.e.*, one to five convolution layers) trained from scratch and fine-tuned (see Section 4.4.3.3) are shown in Table 4.2. Our

T-CNN3 approach outperforms AlexNet on most texture datasets: Kylberg, DTD, kth-tips-2b, and ImageNet-T from scratch and Kylberg, CUReT, and kth-tips-2b in the fine-tuned experiments; while containing nearly three times fewer trainable parameters than AlexNet. Trained from scratch, T-CNN3 also performs well on the object-like and random subsets of ImageNet, *i.e.*, ImageNet-S1 and ImageNet-S2. We believe that our network can learn from the texture regions present in these images. The shallower T-CNN3 with simpler features and fewer parameters is also faster and easier to train on the smaller subdatasets than AlexNet, which explains that T-CNN3 outperforms AlexNet on these datasets from scratch. However, when using pretrained networks, the original AlexNet outperforms the T-CNN architectures on nontexture images (*i.e.*, ImageNet-S1 and ImageNet-S2) as expected as it is designed for object recognition and can learn more complex and larger features from the large training set than the T-CNNs.

Note that our method does not compete with the state-of-the-art as seen in Table 4.2 because the latter use much more complex and deeper architectures. For this reason the comparison to AlexNet is more meaningful.

4.4.3.2 Networks depth analysis

As shown in Table 4.2, the best T-CNN architecture is with three convolution layers (T-CNN3) with T-CNN4 and T-CNN2 close behind. The energy layer measures the response to a set of patterns which are the results of the combination of multiple simple filters. Generally, the more convolution layers we use, the more complex the features sought in the last convolution layer are. In our experiments, using five layers degrades the T-CNN's performance as the fifth layer detects complex object-like features (*e.g.*, entire object or part of an object). Such large and complex features, unlike (simpler) local texture features, are likely to be very sparsely found in the input image. Averaging the response to such features over the entire feature map seems inappropriate as it results in a massive loss of information (see Section 2.3.2 of Chapter 2). Intuitively, a maximum pooling of such deep features could achieve better results than an average pooling, especially in an object recognition task. To evaluate this idea, we experiment with a maximum pooling layer instead of the energy (average) layer. This maximum pooling measures whether a certain feature was found in the input image, disregarding its location and number of occurrences. This maximum pooling layer is computed as follows:

$$e_j = \max(f_j(x_1, x_2)), \qquad (4.16)$$

where x_1 and x_2 span the input feature map f_j. The results in Table 4.3 confirm this idea as the shallow T-CNNs are more accurate with an average energy layer while deeper ones are more accurate with a maximum pooling.

On the other extreme, features sought in the first layer are too simple (mainly edges and colors) to extract meaningful and discriminative information from the images. We

Table 4.3 Accuracy (%) of various network depths with average and maximum pooling of the energy layer

method	ImageNet-T		ImageNet-S1	
	average	max	average	max
T-CNN1	42.7	41.3	34.9	24.1
T-CNN3	71.1	71.0	69.4	70.6
T-CNN5	65.8	67.4	54.7	67.6

first notice that the T-CNN1 obtains better results on texture datasets (ImageNet-T) and the random selection of classes (ImageNet-S2) than on the object-like set (ImageNet-S1). The very simple texture features extracted in the first convolution layer are not able to describe the complex object shapes, which exhibit the most discriminative information. Secondly, even in texture datasets, the accuracy of the T-CNN1 is significantly lower than T-CNN3 as these simpler (and fewer) features cannot represent the more complex patterns found in the texture images. As shown in Table 4.2, T-CNN3 performs significantly better than T-CNN1 both from scratch and fine-tuned.

Note that this depth analysis does not generalize to all network architectures and datasets as a significantly deeper approach with a large number of parameters can implement very complex functions and has shown excellent results in texture classification with different texture descriptors (*e.g.*, FV-CNN [31]). Also, average pooling is successfully used in several recent deep architectures after the last convolution layer, replacing fully-connected layers to enforce the correspondence between the feature maps and the classes probabilities [19]. The analysis of such architectures and descriptors is out of the scope of this chapter.

4.4.3.3 Domain transferability and visualization

Domain transferability is a key concept of deep learning which enables obtaining high accuracy on many tasks with relatively little training data. As it is not possible to obtain labeled datasets like ImageNet for all recognition tasks, in particular for biomedical images, CNNs are often pretrained on this large dataset (source task is object recognition). The pretrained parameters are then fine-tuned on another dataset with a target task (in our case texture recognition), which can be close or far from the object recognition source task. The features learned on ImageNet are transferred to the target task and the weights of the last FC layer are fine-tuned to remodel the combination of the detected features. As described in Section 1.3.3 of Chapter 1, it is important to distinguish between the affine invariances required for general computer vision tasks (*e.g.*, ImageNet) and some biomedical image and texture datasets which only require invariances to rigid motions. For instance, the CNNs trained on ImageNet learned invariances to scale which may have a negative impact on datasets such as CUReT, Kylberg, forest species datasets and tissue images (Section 4.5) with fixed viewpoints and where scale can be

Table 4.4 Classification results (accuracy %) on the kth-tips-2b dataset using networks pretrained on different databases

	ImageNet	ImageNet-T	ImageNet-S1	ImageNet-S2
T-CNN1	**61.4** ±3.1	54.3 ±2.7	52.9 ±2.7	53.1 ±3.6
T-CNN3	**73.2** ±2.2	61.8 ±3.5	56.3 ±3.9	59.0 ±3.0
AlexNet	**71.5** ±1.4	56.9 ±3.9	55.5 ±3.9	58.2 ±2.3

a discriminative property. Yet the networks largely benefit from the pretraining as the invariance to scale is generally learned independently from the recognition of patterns at different scales; *i.e.*, two neurons respond to the same pattern at different scales and are combined in a deeper layer so that one deeper neuron responds to both scales. This is a (simplified) hierarchical property of CNNs in which deep representations are obtained from a composition of simpler and shallower ones. The weights can therefore be easily remodeled through fine-tuning (with sufficient training data) to drop the unnecessary invariances and learn important discriminations from the target datasets.

The accuracy of several networks pretrained on multiple datasets and fine-tuned on a texture dataset are reported in Table 4.4. As expected, a network pretrained on a texture dataset (ImageNet-T) achieves better results on another texture dataset than the same network pretrained on a nontexture dataset (ImageNet-S1 and ImageNet-S2). This is explained as the tasks of object recognition and texture recognition are relatively distant. Yet, we also notice that the amount of pretraining data predominantly influences the accuracy of the fine-tuned network in the target task. Indeed, all the networks pretrained on the entire ImageNet significantly outperform the other ones. Note that the CNNs trained on an object detection task such as ImageNet sometimes learn to recognize parts on the image that we would not expect such as the grass or trees in the background which can be highly textured. When such background is redundant within classes, the networks can learn this discriminant information even though this is not the salient object that is primarily sought. Thus the network is able to learn texture-like patterns from nonsegmented textures (*e.g.*, cheetahs and background textures) and transfers relatively well to the texture recognition task. These observations confirm that domain transferability greatly helps the training (*i.e.*, fine-tuning) on texture datasets and suggest that a very large labeled texture dataset could bring a significant contribution to CNNs applied to texture analysis.

We also want to visualize the features learned by T-CNNs from scratch and fine-tuned. For that, we apply the neuron activation maximization method via gradient ascent optimization [23] (see Section 4.2.4.3). Examples of image patterns that activate the neurons in the third convolution layer (C3) are shown in Fig. 4.3 for T-CNN3 trained on a texture dataset both from scratch and pretrained on ImageNet. Similar examples for the last fully-connected layer (FC3) are shown in Fig. 4.4. We notice that the patterns which activate the neurons of the intermediate layer (C3) are more complex

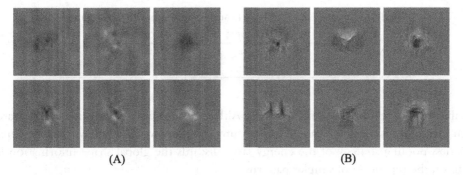

Figure 4.3 Examples of activation maximization of neurons in the third convolution layer. The T-CNN3 networks are trained on kth-tips-2b (a) from scratch and (b) fine-tuned (pretrained on ImageNet). Figures were obtained with the DeepVis toolbox [57].

in fine-tuned networks than those trained from scratch. Note that in our fine-tuning approach, as in most applications, most of the learning is carried by the last FC layer while the weights in intermediate layers vary only slightly from the pretrained weights (*i.e.*, learning controlled by the learning rates). Therefore the features in Fig. 4.4 are similar to the pretrained ones before fine-tuning which explains the complexity. Complexity here refers to the shape of the patterns in the input image and their level of abstraction. For instance, the neurons in the third convolution layer of the fine-tuned network respond to complex patterns such as a head or an object as shown in Fig. 4.3B. The features learned from scratch are simpler texture patterns. However, the features that activate the neurons of the last FC layer (Fig. 4.4) are of similar low complexity for both the networks from scratch and fine-tuned. These neurons mainly detect simple repeated patterns, even though the last neurons respond to patterns which are the

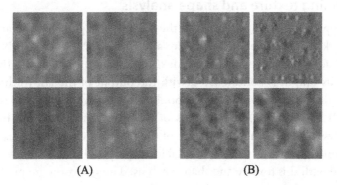

Figure 4.4 Examples of activation maximization of neurons in the last fully-connected layer (FC3). The T-CNN3 networks are trained on kth-tips-2b (a) from scratch and (b) fine-tuned (pretrained on ImageNet). Figures were obtained with the DeepVis toolbox [57].

Table 4.5 Accuracy (%) of the T-CNN3 and comparison with the literature

	Macroscopic	Microscopic
T-CNN3	**97.2** ±0.40	97.0 ±0.61
[35]	95.77 ±0.27	**97.32** ±0.21

combination of features in previous layers. Although the complexity generally increases with the depth, it is not the case in this texture CNN approach. It shows that the dense orderless pooling strategy of the energy layer discards the global shape information to focus on the repetition of simpler patterns.

4.4.3.4 Results on larger images

The previous experiments were conducted on classic texture datasets with medium size images which fit most CNN input dimensions. In this section, our method is evaluated on larger texture images (forest species databases). Our T-CNN approach, due to the average pooling strategy, enables using the same model architecture with various input sizes and transfer features between datasets of varying image sizes. A fully convolutional approach [16] is not necessary here as the energy layer pools a single value per feature map regardless the input size. We can therefore transfer the weights pretrained on Im-ageNet and fine-tune our T-CNN on the forest species images of size 640×640. In Table 4.5, we compare the results obtained with T-CNN3 to the state-of-the-art. We do not compare to the standard AlexNet as it requires fixed input sizes (227×227). Our method slightly outperforms the state-of-the-art [35] on the Macroscopic forest species dataset (+1.4%) and obtains very similar results on the Microscopic one (−0.3%). Note that the method in [35] is of higher computational complexity and uses different hyper-parameters for both datasets.

4.4.4 Combining texture and shape analysis

We also developed a method to combine the texture and global shape analysis within a single network that we refer to as Texture and Shape CNN (TS-CNN). An energy layer is simply added to a classic CNN (AlexNet) at the output of the third convolution layer and its output is concatenated with the output of the last (fifth) convolution layer, skipping the other convolution and pooling layers. In this way, the features of the early layers (C1, C2 and C3) and weights of the FC layers are shared by the texture and shape analysis, keeping the complexity of the network close to the classic AlexNet. More details on this architecture are available in [30]. Table 4.6 shows the improvements obtained with this texture and shape analysis. The networks are pretrained on the ImageNet dataset and fine-tuned on kth-tips-2b.

In the first experiment, the voting scores (outputs of the softmax layer) of T-CNN3 and AlexNet are summed to provide an averaged classification of the two networks. This

Table 4.6 Classification results (accuracy %) on kth-tips-2b using AlexNet and T-CNN3 separately and combined as well as the state-of-the-art method with a medium depth CNN (VGG-M). The number of trainable parameters (in millions) is indicated in brackets (for 1000 classes)

Shape	**AlexNet** (60.9)	71.5 ±1.4
	VGG-M FC-CNN [31]	71.0 ±2.3
Texture	**T-CNN3** (23.4)	73.2 ±2.2
	VGG-M FV-CNN [31]	73.3 ±4.7
Texture and Shape	**sum scores AlexNet T-CNN3** (84.3)	73.4 ±1.4
	TS-CNN3 (62.5)	**74.0** ±1.1
	VGG-M FV+FV-CNN [31]	73.9 ±4.9

simple score summing method obtains 73.4%, an increase of 0.2% as compared to the T-CNN3 and 1.9% as compared to AlexNet. In our second experiment, the network combining texture and shape analysis (TS-CNN) obtains the best results with 74.0%. These experiments show the complementary of the texture and shape analysis and the possibility to share the same simple features in early layers to conduct both analyses within the same network.

Our method does not compete with the state-of-the-art [31] due to a large difference in network depth. However, comparing approaches of similar depth, *i.e.*, our approach and FV-CNN with medium network depth [31] in Table 4.6, we see that our "shape," "texture," and "shape + texture" methods obtain similar results. Also, our algorithm is trained end-to-end in a single CNN whereas [31] uses two CNNs of significantly higher complexity (*i.e.*, more layers and neurons), an encoding and an SVM classifier. Note that we do not report the number of parameters used in [31] in Table 4.6 as this number differs between pretraining and fine-tuning and requires extra processing for FV encoding and SVM classification. As a rough comparison, the VGG-M network has 101.7 million trainable parameters vs. 23.4 million for the T-CNN3 and 62.5 million for the TS-CNN3.

4.5 APPLICATION TO TISSUE IMAGES CLASSIFICATION

The analysis of tissue images is crucial for studying the cells behavior, cellular senescence, detecting abnormal cells, cancers, etc. The automatic analysis of these images helps obtaining more consistent, and less demanding analyses and diagnoses (see Chapter 10). In this section, we focus on two important tasks which are the recognition of malignant lymphomas and the classification of mouse liver tissue based on the age, gender, and diet. The tissue images are obtained from biopsies sectioned and stained with hematoxylin/eosin (H&E). We present an application of the T-CNN on this recognition task as an extension of our paper [58]. As shown in Fig. 4.6, tissues exhibit high texture contents, which make them good candidates for the proposed T-CNN approach. We develop a framework which uses this low complexity network in an ensemble approach and significantly improves the state-of-the-art on several tissue classification benchmarks.

4.5.1 State-of-the-art

We briefly describe the methods used in the literature to which we compare our approach in the experiments in Section 4.5.4. These approaches are based on classic machine learning techniques with handcrafted feature extraction, dimensionality reduction, and classification.

A classification method named WND-CHARM (for Weighted Neighbor Distance - Compound Hierarchy of Algorithms Representing Morphology) is developed in [59] and applied to many tissue images datasets. Its accuracy is reported with a hold-25%-out validation. In this method, a set of features are extracted from the raw image and from a set of transformations of the image. The most discriminant features are classified with a variant of nearest neighbor classifier. CP-CHARM (for Cell Profiler − Compound Hierarchy of Algorithms Representing Morphology) [60] is a more recent approach which extracts a different set of features, reduces its dimension with PCA and uses a LDA classifier. CP-CHARM is tested on the same datasets as WND-CHARM with both hold-25%-out and 10-fold cross-validation.

Texture and color descriptors are evaluated in [61] both separately and jointly with an SVM classifier for the recognition of tissue images. Histograms in multiple color spaces are used for the color features while LBP and cooccurrence matrices are used as texture descriptors. In [62] a set of statistical and transform features are extracted from the images and classified with an ensemble SVM. This approach is applied to the prediction of mouse cells senescence by analyzing liver tissue images.

4.5.2 Method

In this section, we use the T-CNN3 architecture introduced previously. The number of labeled tissue images is often limited due to data privacy and/or the need of experts for labeling the data. On the other hand, the resolution of microscopic tissue images being generally high, the images are large (more than 1000×1000 pixels). Also, the analyzed tissues exhibit highly repetitive texture patterns across the image. Therefore we can split the input images similar to [63] to increase the number of training samples. A sum voting score can then be used at test time to combine the classification of the subimages. Note that the T-CNN3 architecture does not need fixed input sizes and the full size images could be used. However, this voting method enables increasing the training set, taking a collective decision in an ensemble classifier manner and avoiding downsampling, which causes a loss of information. The proposed approach, named collective T-CNN, is illustrated in Fig. 4.5 including the training and testing phases.

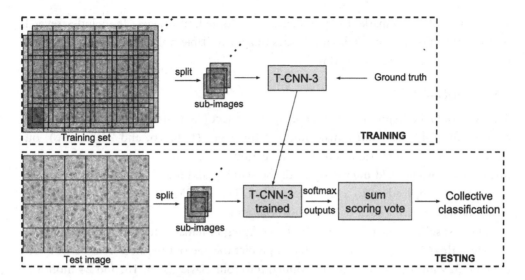

Figure 4.5 Training and testing of our method with subimages and scoring vote.

In the experiments, we use images of size 1388×1040 from the Image Informatics and Computational Biology Unit (IICBU) dataset [64] that we split into 24 nonoverlapping subimages as shown in Fig. 4.5. We resize the resulting images to 227×227 with a nearest neighbor interpolation. In the training phase, we use all the subimages from the training set as independent samples to fine-tune the network pretrained on ImageNet. Each subimage is associated with the label of the original full-size image. In the testing phase, we apply a sum voting score among the 24 subimages to classify each full-size image. The collective score of a given test image is obtained by summing the softmax output vectors of the T-CNN3 for all the subimages. The obtained score vector gives for each class a confidence score (sum of 24 probabilities) to belong to that class. The original full-size image is assigned the class with the largest sum in the vector:

$$c = \arg\max_i \sum_{n=1}^{24} \sigma^{(n)}[i], \tag{4.17}$$

where i spans the classes and $\sigma^{(n)}$ is the softmax output of the nth subimage. This voting algorithm behaves as an ensemble model which takes a collective decision based on the analysis of different nonoverlapping spatial areas of the image.

4.5.3 Datasets

To compare with the state-of-the-art, we reproduce several experiments from the literature. The datasets used in these experiments are part of the IICBU dataset [64], a collection of benchmarks for the analysis of biomedical images. It includes several benchmarks of organelles, cells, and tissues images available online for testing and evaluating automatic analysis algorithms.

4.5.3.1 AGEMAP

The Atlas of Gene Expression in Mouse Aging Project (AGEMAP) [64] contains images of livers from 48 male and female mice of four ages (1, 6, 16, and 24 months), on ad-libitum or caloric restriction diets. The sectioned livers were stained with H&E, and imaged by a bright-field microscope. All the staining and imaging were performed by the same person, leading to a small variability. The AGEMAP dataset is divided into four subdatasets with a total of five validation setups.

The first subdataset and setup is the Liver Aging of female mice on Ad Libitum diet (**LA-female-AL**) [60] in which we try to predict the age of the mice. This benchmark contains 529 images from 11 mice grouped into four classes (1, 6, 16, and 24 months). The validation process is a 10-fold cross-validation repeated 100 times. The accuracy is averaged over the 10 folds for all 100 iterations and the median and standard deviation across the 100 iterations is reported.

The second subdataset/setup is the Liver Aging Across Subjects on Ad Libitum diet (**LA-AS-AL**) reproduced from [62]. It contains four classes (1, 5, 16, and 24 months) and 1027 images from 21 mice. For each mouse, all the images are randomly divided into training (5/6) and test images (1/6). We report the accuracy averaged over 30 runs.

The third subdataset Liver Gender 6 Months on Ad Libitum diet is used in [61,60] in which we try to predict the gender of the mice (binary classification). It contains two classes (male and female) with 265 images from six mice. Note that two different validation setups are used with this subdataset as follows.

The third setup (**LG6M-AL-5p**) is reproduced from [61]. This setup is a hold-95%-out validation, *i.e.*, 5% of the data is used for training, the rest for testing. The reason for this split is to experiment with few training samples to simulate real conditions. We report the Mean Average Precision (MAP) measure averaged over 5000 runs as suggested in [61].

The fourth setup (**LG6M-AL**) is reproduced from [60]. It is the same validation setup as the one used in LA-female-AL with a 10-fold cross-validation repeated 100 times.

The last subdataset and setup derived from the AGEMAP dataset is the Liver Gender 6 Months on Calories Restriction diet (**LG6M-CR**) reproduced from [60]. This subdataset contains two classes (male and female) with a total of 303 images obtained from six mice. The validation setup is the same as LA-female-AL and LG6M-AL with a 10-fold cross-validation repeated 100 times.

4.5.3.2 Lymphoma

The Lymphoma dataset [64] contains images of three types of malignant lymphoma, namely Chronic Lymphocytic Leukemia (CLL), Follicular Lymphoma (FL), and Mantle Cell Lymphoma (MCL). Lymphoma is a cancer affecting the lymph nodes. Its detection and diagnosis can greatly benefit from the automatic analysis of tissue images, which are the result of biopsies stained with H&E. In particular, this dataset allows to experiment methods to automatically distinguish the three types of lymphoma. The samples were prepared by different pathologists at different sites, leading to a large variation in staining and image quality which reflects real acquisition conditions. It also makes the automatic classification a realistic and challenging task.

The Lymphoma dataset contains 374 images grouped into three classes (CLL, FL, and MCL). It is used with two different validation setups as follows.

The first setup (**Lymphoma-5p**) is reproduced from [61]. It is the same validation setup as LG6M-AL-5p with 5% of the data used for training, the rest for testing. The MAP averaged over 5000 runs is reported.

The second setup (**Lymphoma**) is reproduced from [60]. The validation setup is the same as the one used in LA-female-AL, LG6M-AL and LG6M-CR with the repeated 10-fold cross-validations.

To summarize, we use two datasets divided into five different subdatasets for a total of seven experiments, *i.e.*, different validation setups. Example images of the AGEMAP and Lymphoma datasets are shown in Fig. 4.6.

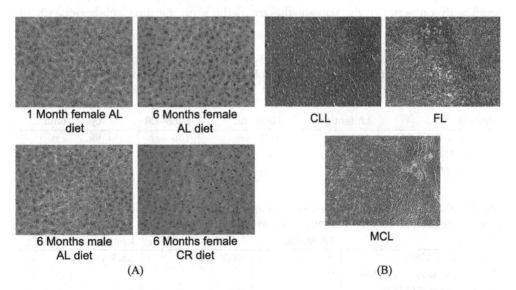

| 1 Month female AL diet | 6 Months female AL diet | CLL | FL |

| 6 Months male AL diet | 6 Months female CR diet | MCL |

(A)   (B)

Figure 4.6 Examples of H&E stained tissue images from the IICBU dataset [64] (a) mice tissue liver images (AGEMAP) and (b) Lymphoma tissue images.

4.5.4 Results and discussions

The results on the 10-fold cross-validation setups are presented in Table 4.7. Note that the results of the WND-CHARM approach are given in [60] and are obtained from a hold-25%-out validation. Yet they are compared to the other results as it was shown in [60] that the average accuracy is similar to a 10-fold cross-validation although the results naturally vary more between the tests (higher standard deviation). Our approach (collective T-CNN) obtains significantly better results with an increase of accuracy of up to 32.7% on the Lymphoma dataset. This large difference in accuracy must be put into perspective as CP-CHARM and WND-CHARM use simple machine learning methods and are generalized image classifiers (not specific to texture analysis). They are not designed for a specific task and perform reasonably well on many various tasks, whereas our approach uses deep learning and is specifically designed for the analysis of texture tissue images.

Our method obtains 100% accuracy on all but one 10-fold cross-validation experiment which demonstrates its accuracy and reliability. To evaluate the benefit of the collective classification (sum scoring vote), we also tested the average accuracy of all subimages. The average per-subimage classification accuracy is 99.7%, 98.2%, 95.2%, and 95.5% for respectively LA-female-AL, LG6M-AL, LG6M-CR, and Lymphoma. We conclude that the collective decision has an important positive impact on the accuracy of the method.

The results on LA-AL-AS, LG6M-AL-5p, and Lymphoma-5p are presented in Table 4.8. They are an updated version of the results provided in [58]. Our approach again significantly outperforms the state-of-the-art methods on the three subdatasets. Of all

Table 4.7 Classification accuracy (%) and standard deviation of our approach (Collective T-CNN) and comparison with the state-of-the-art on the 10-fold cross-validation setups. The WND-CHARM results are obtained from a hold-25%-out validation setup while the other results use a 10-fold cross-validation

Method	LA-fem-AL	LG6M-AL	LG6M-CR	Lymphoma
Collective T-CNN	100 ±0.14	100 ±0.0	100 ±0.0	98.7 ±0.33
CP-CHARM [60]	89 ±0.4	98 ±0.5	99 ±0.1	66 ±0.1
WND-CHARM [59]	93 ±3	98 ±1	99 ±1	79 ±4

Table 4.8 Classification accuracy (%), Mean Average Precision (MAP) (%) and standard deviation of our approach (Collective T-CNN) and comparison with the state-of-the-art on other validation setups

Method	LA-AL-AS	LG6M-AL-5p (MAP)	Lymphoma-5p (MAP)
Collective T-CNN	99.1 ±0.7	98.7 ±3.2	65.1 ±5.7
feature selection + SVM [62]	97.01	–	–
$LBP_{riu} + I1H2H3$ [61]	–	97.3	57.6
$CCOM$ [61]	–	82.5	63.3

Table 4.9 Confusion matrix of our approach on Lymphoma-5p

	True		
	CLL	FL	MCL
CLL	0.56	0.16	0.28
FL	0.1	0.75	0.14
MCL	0.26	0.21	0.53

Table 4.10 Confusion matrix of our approach on LG6M-AL-5p

	True	
	Male	Female
Male	0.93	0.07
Female	0.06	0.94

Table 4.11 Confusion matrix of our approach on LA-AL-AS

	True			
	1 Month	6 Months	16 Months	24 Months
1 Month	0.99	0.01	0	0
6 Months	0	0.99	0.01	0
16 Months	0	0.02	0.98	0
24 Months	0	0	0	1

the experiments, only Lymphoma–5p seems to create problems for our method with 65.1% accuracy. This lower accuracy can be explained by the variation in data acquisition and the low number of training images. The confusion matrices for these datasets are shown in Tables 4.9, 4.10, and 4.11. The confusion matrix of Lymphoma-5p shows that most confusions occur between the CLL and MCL classes. Table 4.11 shows the correlation between the age of the mice and the misclassification as the few confusions occur between images of mice of close age, *e.g.*, 1 month misclassified as 6 months, 16 months misclassified as 6 months, etc. As expected, the similarity between the classes decreases as the difference between the ages increases. The average per-subimage classification accuracy for the LA–AL–AS subdataset is 93.1% vs. 99.1% with the collective decision, which again demonstrates the power of the ensemble approach in the final classification.

Finally, similar observations are made on the visualization of the features learned by the network as in Section 4.4.3. The neurons in the last FC layer respond to simple repeated texture patterns.

4.6 CONCLUSION

This chapter introduced the concepts behind neural networks and convolutional networks, presented a review of the literature of deep learning in texture analysis as well as a novel CNN approach to texture analysis and biomedical texture analysis with a specific application to tissue images classification. Several adaptations of CNN to texture analysis were introduced with a shared key concept of discarding the overall shape analysis by orderless pooling of the features. It was shown that the weight sharing and local connectivity of CNNs are well designed for the analysis of patterns repeated at different locations in texture images. We focused on a low complexity approach (T-CNN) [30] which densely pools the activations of an intermediate convolution layer in a manner similar to a filter bank approach. The proposed method is trained end-to-end with backpropagation and can be combined to a shape analysis within a single network by sharing the features of early layers.

An application to tissue image classification is presented which is based on the analysis of nonoverlapping areas of the images using the T-CNN3 and a scoring vote method. The texture in tissue images, being repetitive across the image, we can treat each image patch as an independent sample at training time and combine the prediction of multiple patches at test time. While labeled training sets for biomedical imaging are generally small, our approach can benefit from the high resolution images by increasing the training samples with this splitting approach and obtain high classification accuracy with the collective classification of subimages. Our collective T-CNN largely improves the state-of-the-art on multiple tissue images benchmarks including microscopic mice liver tissues and lymphoma tissue images [64].

Finally, the proposed approaches can be easily implemented with deeper architectures such as GoogleNet [20] and deep residual networks [21] and we expect an increase of accuracy on larger and more complex datasets.

REFERENCES

[1] I. Goodfellow, Y. Bengio, A. Courville, Deep Learning, MIT Press, 2016, http://www.deeplearningbook.org.
[2] C.M. Bishop, Neural Networks for Pattern Recognition, Oxford University Press, 1995.
[3] A. Krizhevsky, I. Sutskever, G.E. Hinton, Imagenet classification with deep convolutional neural networks, in: Advances in Neural Information Processing Systems, 2012, pp. 1097–1105.
[4] K. Chatfield, K. Simonyan, A. Vedaldi, A. Zisserman, Return of the devil in the details: delving deep into convolutional nets, preprint, arXiv:1405.3531, 2014.
[5] D.E. Rumelhart, G.E. Hinton, R.J. Williams, Learning representations by back-propagating errors, Nature 323 (6088) (1986) 533–536.
[6] Y. Nesterov, A method for unconstrained convex minimization problem with the rate of convergence o $(1/k2)$, Dokl. SSSR 269 (1983) 543–547.
[7] J. Duchi, E. Hazan, Y. Singer, Adaptive subgradient methods for online learning and stochastic optimization, J. Mach. Learn. Res. 12 (Jul) (2011) 2121–2159.

[8] D. Kingma, J. Ba, Adam: a method for stochastic optimization, preprint, arXiv:1412.6980, 2014.

[9] M.D. Zeiler, Adadelta: an adaptive learning rate method, preprint, arXiv:1212.5701, 2012.

[10] X. Glorot, Y. Bengio, Understanding the difficulty of training deep feedforward neural networks, Aistats 9 (2010) 249–256.

[11] O. Russakovsky, J. Deng, H. Su, J. Krause, S. Satheesh, S. Ma, Z. Huang, A. Karpathy, A. Khosla, M. Bernstein, et al., Imagenet large scale visual recognition challenge, Int. J. Comput. Vis. 115 (3) (2015) 211–252.

[12] N. Srivastava, G.E. Hinton, A. Krizhevsky, I. Sutskever, R. Salakhutdinov, Dropout: a simple way to prevent neural networks from overfitting, J. Mach. Learn. Res. 15 (1) (2014) 1929–1958.

[13] D.H. Hubel, T.N. Wiesel, Receptive fields and functional architecture of monkey striate cortex, J. Physiol. 195 (1) (1968) 215–243.

[14] V. Nair, G.E. Hinton, Rectified linear units improve restricted Boltzmann machines, in: Proceedings of the 27th International Conference on Machine Learning (ICML-10), 2010, pp. 807–814.

[15] J. Tobias Springenberg, A. Dosovitskiy, T. Brox, M. Riedmiller, Striving for simplicity: the all convolutional net, preprint, arXiv:1412.6806, 2014.

[16] J. Long, E. Shelhamer, T. Darrell, Fully convolutional networks for semantic segmentation, in: Proceedings of the IEEE Conference on Computer Vision and Pattern Recognition, 2015, pp. 3431–3440.

[17] S. Ioffe, C. Szegedy, Batch normalization: accelerating deep network training by reducing internal covariate shift, preprint, arXiv:1502.03167, 2015.

[18] K. Simonyan, A. Zisserman, Very deep convolutional networks for large-scale image recognition, preprint, arXiv:1409.1556, 2014.

[19] M. Lin, Q. Chen, S. Yan, Network in network, preprint, arXiv:1312.4400, 2013.

[20] C. Szegedy, W. Liu, Y. Jia, P. Sermanet, S. Reed, D. Anguelov, D. Erhan, V. Vanhoucke, A. Rabinovich, Going deeper with convolutions, in: Proceedings of the IEEE Conference on Computer Vision and Pattern Recognition, 2015, pp. 1–9.

[21] K. He, X. Zhang, S. Ren, J. Sun, Deep residual learning for image recognition, preprint, arXiv:1512.03385, 2015.

[22] M.D. Zeiler, R. Fergus, Visualizing and understanding convolutional networks, in: European Conference on Computer Vision, Springer, 2014, pp. 818–833.

[23] D. Erhan, Y. Bengio, A. Courville, P. Vincent, Visualizing Higher-Layer Features of a Deep Network, Technical Report 1341, University of Montreal, 2009.

[24] T. Randen, J.H. Husoy, Filtering for texture classification: a comparative study, IEEE Trans. Pattern Anal. Mach. Intell. 21 (4) (1999) 291–310.

[25] T. Ojala, M. Pietikäinen, D. Harwood, A comparative study of texture measures with classification based on featured distributions, Pattern Recognit. 29 (1) (1996) 51–59.

[26] J. Zhang, M. Marszałek, S. Lazebnik, C. Schmid, Local features and kernels for classification of texture and object categories: a comprehensive study, Int. J. Comput. Vis. 73 (2) (2007) 213–238.

[27] F. Perronnin, C. Dance, Fisher kernels on visual vocabularies for image categorization, in: 2007 IEEE Conference on Computer Vision and Pattern Recognition, IEEE, 2007, pp. 1–8.

[28] H. Jégou, M. Douze, C. Schmid, P. Pérez, Aggregating local descriptors into a compact image representation, in: 2010 IEEE Conference on Computer Vision and Pattern Recognition (CVPR), IEEE, 2010, pp. 3304–3311.

[29] S. Basu, M. Karki, R. DiBiano, S. Mukhopadhyay, S. Ganguly, R. Nemani, S. Gayaka, A theoretical analysis of deep neural networks for texture classification, preprint, arXiv:1605.02699, 2016.

[30] V. Andrearczyk, P.F. Whelan, Using filter banks in convolutional neural networks for texture classification, Pattern Recognit. Lett. 84 (2016) 63–69.

[31] M. Cimpoi, S. Maji, I. Kokkinos, A. Vedaldi, Deep filter banks for texture recognition, description, and segmentation, Int. J. Comput. Vis. 118 (1) (2016) 65–94.

[32] T.-Y. Lin, S. Maji, Visualizing and understanding deep texture representations, preprint, arXiv:1511.05197, 2015.

[33] Y. Song, Q. Li, D. Feng, J.J. Zou, W. Cai, Texture image classification with discriminative neural networks, Comput. Vis. Media 2 (4) (2016) 367–377.

[34] F.H.C. Tivive, A. Bouzerdoum, Texture classification using convolutional neural networks, in: TEN-CON 2006. 2006 IEEE Region 10 Conference, IEEE, 2006, pp. 1–4.

[35] L.G. Hafemann, L.S. Oliveira, P. Cavalin, Forest species recognition using deep convolutional neural networks, in: 2014 22nd International Conference on Pattern Recognition (ICPR), IEEE, 2014, pp. 1103–1107.

[36] L.G. Hafemann, L.S. Oliveira, P.R. Cavalin, R. Sabourin, Transfer learning between texture classification tasks using convolutional neural networks, in: 2015 International Joint Conference on Neural Networks (IJCNN), IEEE, 2015, pp. 1–7.

[37] S. Hochreiter, J. Schmidhuber, Long short-term memory, Neural Comput. 9 (8) (1997) 1735–1780.

[38] W. Byeon, M. Liwicki, T.M. Breuel, Texture classification using 2d LSTM networks, in: 2014 22nd International Conference on Pattern Recognition (ICPR), IEEE, 2014, pp. 1144–1149.

[39] W. Byeon, T.M. Breuel, Supervised texture segmentation using 2D LSTM networks, in: 2014 IEEE International Conference on Image Processing (ICIP), IEEE, 2014, pp. 4373–4377.

[40] R. Girshick, J. Donahue, T. Darrell, J. Malik, Rich feature hierarchies for accurate object detection and semantic segmentation, in: Proceedings of the IEEE Conference on Computer Vision and Pattern Recognition, 2014, pp. 580–587.

[41] T.-Y. Lin, A. RoyChowdhury, S. Maji, Bilinear CNN models for fine-grained visual recognition, in: Proceedings of the IEEE International Conference on Computer Vision, 2015, pp. 1449–1457.

[42] V. Andrearczyk, P.F. Whelan, Texture segmentation with fully convolutional networks, preprint, arXiv:1703.05230, 2017.

[43] L. Liu, P. Fieguth, X. Wang, M. Pietikäinen, D. Hu, Evaluation of LBP and deep texture descriptors with a new robustness benchmark, in: European Conference on Computer Vision, Springer, 2016, pp. 69–86.

[44] D. Marcos, M. Volpi, D. Tuia, Learning rotation invariant convolutional filters for texture classification, preprint, arXiv:1604.06720, 2016.

[45] M. Kallenberg, K. Petersen, M. Nielsen, A.Y. Ng, P. Diao, C. Igel, C.M. Vachon, K. Holland, R.R. Winkel, N. Karssemeijer, et al., Unsupervised deep learning applied to breast density segmentation and mammographic risk scoring, IEEE Trans. Med. Imaging 35 (5) (2016) 1322–1331.

[46] Q. Li, W. Cai, X. Wang, Y. Zhou, D.D. Feng, M. Chen, Medical image classification with convolutional neural network, in: 2014 13th International Conference on Control Automation Robotics & Vision (ICARCV), IEEE, 2014, pp. 844–848.

[47] G. van Tulder, M. de Bruijne, Combining generative and discriminative representation learning for lung ct analysis with convolutional restricted Boltzmann machines, IEEE Trans. Med. Imaging 35 (5) (2016) 1262–1272.

[48] V. Andrearczyk, P.F. Whelan, Convolutional neural network on three orthogonal planes for dynamic texture classification, preprint, arXiv:1703.05530, 2017.

[49] Y. Jia, E. Shelhamer, J. Donahue, S. Karayev, J. Long, R. Girshick, S. Guadarrama, T. Darrell, Caffe: convolutional architecture for fast feature embedding, preprint, arXiv:1408.5093, 2014.

[50] E. Hayman, B. Caputo, M. Fritz, J.-O. Eklundh, On the significance of real-world conditions for material classification, in: European Conference on Computer Vision, Springer, 2004, pp. 253–266.

[51] G. Kylberg, The Kylberg texture dataset v. 1.0, External Report (Blue series), vol. 35, Centre for Image Analysis, Swedish University of Agricultural Sciences and Uppsala University, Uppsala, Sweden, 2011.

[52] G. Kylberg, I.-M. Sintorn, Evaluation of noise robustness for local binary pattern descriptors in texture classification, EURASIP J. Image Video Process. 2013 (1) (2013) 1–20.

[53] K.J. Dana, B. Van Ginneken, S.K. Nayar, J.J. Koenderink, Reflectance and texture of real-world surfaces, ACM Trans. Graph. 18 (1) (1999) 1–34.

[54] M. Cimpoi, S. Maji, I. Kokkinos, S. Mohamed, A. Vedaldi, Describing textures in the wild, in: 2014 IEEE Conference on Computer Vision and Pattern Recognition (CVPR), IEEE, 2014, pp. 3606–3613.

[55] P.L.P. Filho, L.S. Oliveira, S. Nisgoski, A.S. Britto Jr., Forest species recognition using macroscopic images, Mach. Vis. Appl. 25 (4) (2014) 1019–1031.

[56] J. Martins, L.S. Oliveira, S. Nisgoski, R. Sabourin, A database for automatic classification of forest species, Mach. Vis. Appl. 24 (3) (2013) 567–578.

[57] J. Yosinski, J. Clune, A. Nguyen, T. Fuchs, H. Lipson, Understanding neural networks through deep visualization, preprint, arXiv:1506.06579, 2015.

[58] V. Andrearczyk, P.F. Whelan, Deep learning for biomedical texture image analysis, in: Irish Machine Vision & Image Processing Conference Proceedings, IMVIP 2016, 2016.

[59] N. Orlov, L. Shamir, T. Macura, J. Johnston, D. Mark Eckley, I.G. Goldberg, WND-CHARM: multi-purpose image classification using compound image transforms, Pattern Recognit. Lett. 29 (11) (2008) 1684–1693.

[60] V. Uhlmann, S. Singh, A.E. Carpenter, CP-CHARM: segmentation-free image classification made accessible, BMC Bioinform. 17 (1) (2016) 1.

[61] N. Hervé, A. Servais, E. Thervet, J.-C. Olivo-Marin, V. Meas-Yedid, Statistical color texture descriptors for histological images analysis, in: 2011 IEEE International Symposium on Biomedical Imaging: From Nano to Macro, IEEE, 2011, pp. 724–727.

[62] H.-L. Huang, M.-H. Hsu, H.-C. Lee, P. Charoenkwan, S.-J. Ho, S.-Y. Ho, Prediction of mouse senescence from he-stain liver images using an ensemble SVM classifier, in: Asian Conference on Intelligent Information and Database Systems, Springer, 2013, pp. 325–334.

[63] J. Barker, A. Hoogi, A. Depeursinge, D.L. Rubin, Automated classification of brain tumor type in whole-slide digital pathology images using local representative tiles, Med. Image Anal. 30 (2016) 60–71.

[64] L. Shamir, N. Orlov, D. Mark Eckley, T.J. Macura, I.G. Goldberg, IICBU 2008: a proposed benchmark suite for biological image analysis, Med. Biol. Eng. Comput. 46 (9) (2008) 943–947.

CHAPTER 5

Fractals for Biomedical Texture Analysis

Omar S. Al-Kadi
University of Jordan, King Abdullah II School for Information Technology, Amman, Jordan

Abstract

Discriminative visual patterns in tissue texture tends to be fuzzy, and clinicians are usually faced with a certain degree of diagnostic uncertainty. Understanding how biological tissue complexity is manifested at different levels of resolution can contribute toward an effective texture analysis approach. The fractal dimension can quantify the geometrical complexity of tissue structure by the level of surface irregularity. It can characterize texture patterns that naturally exhibit similarities across scales, which are ordered in terms of their statistical distribution. In order to accommodate for tissue heterogeneity and rotation-invariance, the lacunarity parameter provides an additional significant feature of fractal analysis; complementing the attractiveness of the fractal dimension's self-similarity and surface roughness estimation. This chapter presents biomedical imaging applications of fractal analysis to computed tomography, ultrasound, and microscopy modalities, and is more concerned with characterizing tumor tissue, which is known to be highly stochastic texture. Examples of fractal analysis to two main types of tumor textures, namely, fine texture as in lung and liver tumor tissue, and coarse texture represented by brain tumor histopathological tissue would be given. Extending the concept of fractals to metric measurements, that might closely approximate the real magnitude of irregular tumor texture, is significant for understanding tumor behavior, and thus improved management of disease.

Keywords

Fractal dimension, Lacunarity, Tumor texture analysis, Tissue characterization, Pattern classification, Computed tomography, Ultrasound, Histopathology

5.1 INTRODUCTION

Texture analysis continues to attract the attention of the biomedical imaging community as an important means for understanding and quantifying disease. Many researchers have proposed numerous methods that can quantify the textural properties of an image. In fact, the problem of features definition for texture analysis was an increasing research domain for many years [1,2]. Recent studies have shown the ability of texture analysis algorithms to extract diagnostically meaningful information from medical images that were obtained with various imaging modalities, such as microscopy, Computed Tomography (CT), Single Photon Emission Computed Tomography (SPECT), Positron Emission Tomography (PET), UltraSound (US), and Magnetic Resonance Imaging (MRI). However, many developed texture analysis methods assume that texture images are acquired at a single scale and same angle of view. This limitation renders these

Biomedical Texture Analysis
DOI: 10.1016/B978-0-12-812133-7.00005-3

methods not suitable for applications where textures occur with different scales, orientations, or translations. This is especially true when it comes to characterizing biological tissue.

Tissue characteristics may vary according to its location and morphology, and has different properties at microscopic and macroscopic scales. A more challenging example would be characterizing diseased tissue, such as tumors. Tumors have different shapes and happen to occur at different levels, positions, and depths. Also tumor texture tends to be heterogeneous due to the underlying complex nature of tumor behavior. The multiple features and added textural complexity would reduce the efficiency of the texture analysis method. On the other hand an increased interest in utilizing fractals for tissue texture analysis has been shown recently [3]. Fractal geometry is considered a powerful tool to describe the irregular or fragmented shape of natural features. It can characterize real-world physical systems, such as geophysics, biology, or fluid mechanics, using spatial or time-domain statistical scaling laws (power-law behavior) [4]. In this regard, the shape complexity which cannot be described in terms of Euclidean geometry, can be easily represented via its fractal geometry.

The application of fractal geometry has profound relevance in biomedical texture analysis, which is supported by the fact that self-similarity of tissue features cannot be defined at a single or finite scale. A pattern in tissue texture can be defined by other subpatterns, which occur repeatedly according to a statistical placement rule. The resemblance between the smaller subpatterns and the whole tissue pattern is considered approximate, and often exhibits certain similarities at different spatial scales [3]. The subpattern could be subjected to other natural deformations, such as translation and rotation at various scales, meaning that spatially complex patterns can be hard to define by simple texture features. In addition, variability in appearance is inevitable as the various medical imaging modalities tend to physically characterize tissue with diverse types of textures (*e.g.*, Fig. 5.1). Therefore, this requires an effective image analysis technique, which can efficiently relate with underlying physiology in an attempt to provide an improved tissue characterization. Fractal geometry has the ability to characterize the complexity of images texture composition. A practical approach would be to quantify the surface roughness by estimating its fractal geometry, and hence heterogeneity in tissue texture for signs of abnormality. This chapter is more concerned with characterizing tumor tissue, which is known to be highly stochastic. Fractal analysis allows for developing techniques that better understand how tumors affect tissue texture images; giving a precise estimation of irregularity of complex texture structures across image scales. Also, this paves the way for improved feature extraction quality, more reproducible and standard analysis of the experimental data, significantly reduces the cost of experimental studies, and makes it possible to conduct them with more cases.

In this chapter, Section 5.2 relates the concept of texture analysis to tissue characterization, and why use fractals for understanding tissue texture properties. Section 5.3

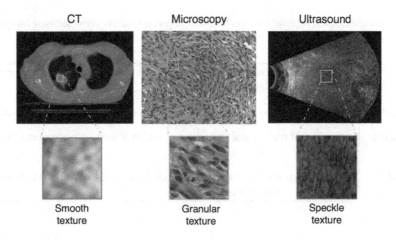

Figure 5.1 Tissue texture patterns can vary according to the utilized imaging system.

gives a background on fractal geometry and the main characteristics sought in defining a fractal pattern. This is followed by the main fractal methods used for quantifying texture properties (presented in Section 5.4). The different types of fractals and issues related to improved fractal parameter estimation is discussed in Sections 5.5 and 5.6, respectively. Another aspect for further characterizing tissue texture heterogeneity and rotation-invariance, and its usefulness in predicting tumor response to treatment is presented in Section 5.7. Practical applications of fractal analysis in three different biomedical imaging modalities, namely computed tomography, ultrasound and microscopy for characterizing texture properties (*i.e.*, tissue characteristics) is demonstrated in Section 5.8. Challenges related to fractal analysis in tissue characterization is discussed in Section 5.9. Finally, the chapter concludes with a summary on how fractal analysis can be used to improve the quality of the extracted texture features.

5.2 TISSUE TEXTURE

Typically, we can simply conceive of texture, which is the characteristic appearance of an area or surface, as arrangement of elementary elements which jointly compose the surface pattern. For any pixel in an image $f(x)$, where x corresponds to the vector of spatial coordinates as $x = (x_1, x_2, \ldots, x_D)$ and f is the amplitude representing the intensity at any x. The unique arrangement of pixels in a finite set of x and f characterize the texture, which will give the sense of the surface. This similarly applies when characterizing the surface of 3D objects, where voxels would jointly form the texture volume. Understanding the conspicuous visual features in an image gives the characteristic appearance of texture surface. Texture appearance can be decomposed into several perceptual properties such as regularity, directionality, contrast, size, and coarse-

Table 5.1 Main histological textural features for the four grade I meningioma subtype images shown in Fig. 5.20

Subtype	Characteristics
Fibroblastic	Spindle-shaped cells resembling fibroblasts in appearance, with abundant amounts of pericellular collagen
Meningothelial	Broad sheets or lobules of fairly uniform cells with round or oval nuclei
Psammomatous	A variant of transitional meningiomas with abundant psammoma bodies and many cystic spaces
Transitional	Contains whorls, few psammoma bodies, and cells having some fibroblastic features (*i.e.*, spindle-shaped cells)

ness to perform automatic texture segmentation and classification. Texture analysis was shown to be useful in approaching a number of problems, including texture segmentation [5–8], texture recognition [9–11], shape-from-texture [12–14], as well as texture synthesis [15].

Studying tissue texture appearance can play an important role in tumor grading and staging. The grade of a histopathological brain tumor refers to the cell shape when observed under a digital microscope. Classification is done by grades – from low (grade I) to high (grade IV). Cells from higher grade tumors are more abnormal looking and generally grow faster than cells from lower grade tumors, while distinctive features in lower grade tumors are subtle and relatively harder to discern. Table 5.1 shows an example of the main histopathological texture features referring to four different grade I meningioma subtype images (see Fig. 5.20). In respect to lung and liver tumor staging, the American Joint Committee on Cancer TNM[1] staging system can be used (see Table 5.2). In brief, the T stands for tumor size and invasiveness and is staged from T1 to T4. Tumors with size less than 3 cm are staged as T1, while for greater sizes (*e.g.*, if the tumor is grown to the main bronchus in lungs or blood vessel in the liver) it is T2. T3 means the tumor has reached the chest wall or the hepatic vein, or at least one tumor is more than 5 cm across. T4 tumors have invaded the nearby organs of the lung or liver, or because it involves the lining of the lung or the thin layer covering the liver. The N letter stands for the degree of lymph node involvement and is represented from N1 to N3, and M represents the presence or absence of metastases, which is staged as 1 for presence and 0 for otherwise. For example, if there was a stage two diagnosed tumor represented as T2N1M0, then it can be interpreted as a lung or liver tumor having a size greater than 3 cm, with first degree lymph node involvement and no metastasis.

---

[1] In the TNM system, *T* stands for the original (primary) tumor, *N* stands for nodes, and *M* stands for metastasis (whether the cancer has spread to distant parts of the body).

Table 5.2 Lung and liver tumor staging based on TNM system

Stage	Size	Lymph	Metastasis
Stage IA	T1	N0	M0
Stage IB	T2	N0	M0
Stage IIA	T1	N1	M0
Stage IIB	T2	N1	M0
	T3	N0	M0
Stage IIIA	T1	N2	M0
	T2	N2	M0
	T3	N1	M0
	T3	N2	M0
Stage IIIB	Any T	N3	M0
	T4	Any N	M0
Stage IV	Any T	Any N	M1

However, characterizing patterns in tissue texture is challenging. Texture therein is inherently heterogeneous and includes both global and local process. Tissue texture exists on a small and varying spatial extent, resulting in multiple local areas within the texture pattern (revisit Section 1.2.3.1 of Chapter 1). It is difficult to consider tissue as an aggregation of nonoverlapping local regions in isolation of its global appearance. Regions tend to intertwine together especially when tissue property begins to change (*e.g.*, disease progression or regression), leaving certain regions with unclear borders. Besides, tissue texture rarely shows a clear periodic pattern, which can be dealt with as a global spatial stochastic process alone. Characterizing biomedical texture analysis is not a straightforward task, given the complex disease behavior and associated noise with the used modality [16,17]. The combination of both global and local textural features complement the task of tissue characterization and overcomes possible deficiencies associated with capturing the various texture spatial variations. Furthermore, the presence of multiple textural properties within the same examined pattern increases the complexity of the texture pattern, and hence reduces the discriminative ability between the different tissue characteristics. To untangle texture complexity, this calls for understanding the texture characteristics beforehand. The concept of fractal is one of the fundamental paradigms of extracting meaningful information in texture analysis. It can be used to measure the complexity of an object through self-similarity across scales in a multiresolution representation.

Texture with fractal characteristics has spatial structure that is invariant with scale. Fractal analysis facilitates the investigation of tissue textural characteristics at higher or deeper levels of resolution. It can also break down the global statistical complexity of texture for a better separability between the different regions. Moreover, its high sensitivity to local features facilitates the processes of preattentive or subtle texture discrimination [18]. In fact, many of our internal organs and structures display fractal properties. This fractal phenomenon can often be detected in various biological tissue structures ranging from subcellular (macro structures) to tissue (micro structures). Examples of the former include changes in size and spatial distribution of cancer cell nuclei, while for the tissue-level it appears in the cerebral cortex folding of the brain [19], or in the development of lung lobes by splitting into smaller lobules (smaller magnifications). All natural fractals of this kind, as well as some mathematical self-similar ones, are stochastic, or random; they thus scale in a statistical sense. Understanding image texture irregularity, hierarchy, and self-similarity by means of fractals can provide meaningful insights on tissue characterization.

5.3 BASIC CONCEPTS OF FRACTAL GEOMETRY

Being curious by nature and the quest to know everything was always been one of the human characteristics. The desire to learn or know about something, enables to see new worlds and possibilities which are normally not visible. This applies in analogy to fractal geometry, in a sense of being curious to know about the final shape of the fractal surface. The more the characteristics of the fractal object are probed, the infinite the details and length becomes, leading curiously to explore deeper levels of magnification. The main characteristics shaping the fractal pattern is discussed next.

5.3.1 Self-similarity

A fractal, also known as evolving symmetry, is a geometrical structure having an irregular or fragmented shape at different scales of magnification. The term fractal stems from the Latin adjective *fractus* meaning "broken," where it was first popularized by Mandelbrot as a useful tool for modeling different natural phenomena [4]. Since then, the field of *fractal geometry* emerged where structural information can be extracted with a higher degree of preciseness. It can give a more detailed description of spatially nonuniform phenomena in nature such as clouds, mountains, and coastlines [4]. The structure of the fractal object is repeated iteratively at progressively smaller scales to produce self-similar or irregular shapes and surfaces that cannot be represented by classical Euclidean geometry. Possessing the property of self-similarity means that component fragments of the object resembles the object as a whole; see Fig. 5.3. This similarity is a consequence of the fact that fractal objects do not have a characteristic scale, but look alike or approximately similar on the different scales, and can in theory continue indefinitely.

5.3.2 Ordered hierarchy

The hierarchical order across many scales is another important aspect of fractals. Fractal structures are self-similar at different levels of hierarchy, which is associated with different scales and sizes. Classical Euclidean geometry does not consider difference in size as fundamental, but as an added local property. However, fractals can be seen as static mathematical objects where scaling is an intrinsic aspect of their description. This leads for a better understanding of mathematical models where size becomes a meaningful attribute of a shape. Such that self-similar objects remain invariant under change of scale, *i.e.*, it has scaling symmetry. This ordered hierarchical property comes handy when dealing with uneven distribution in image intensity. Since intensity variation could give different texture densities, it would implicitly increase the likelihood of incorporating different patterns whose densities have low interscale variation, or subdividing a pattern into several patterns due to high intrascale variation. This may lead for the generation of false patterns within the true pattern. The variation in fractal pattern density gives a hierarchy of fractals that can be described by a multifractal distribution, *i.e.*, the two notions of scales and levels in fractal geometry are not equivalent, especially in biology. Scales appear through quantities varying in magnitude; they can have a dimensionality. This case is not seen as problematic in biology, since scales are mostly inherited from physically defined quantities [20]. By contrast, levels of structural shapes appear through geometric changes between objects that are organized in a hierarchical manner (*e.g.*, molecules, cells, organs, etc.), and characterized by more than one dimension or scaling rule.

5.3.3 Degree of irregularity

The third property relates to the local and global details of *irregularity* of a form. Here, irregularity means forms which are continuous but not smooth; focusing on measuring the surface roughness as illustrated in Fig. 5.2. It describes how accurately a fractal appears to fill a line, a plane, or space, whereas the more contorted a form is, the rougher the surface and vice versa. This leads to the definition of the *Fractal Dimension* (FD), which is a fundamental characteristic of a fractal. It serves as a mathematical parameter that measures the complexity of the shape, based on irregularity related to scale dependency and self-similarity. The FD values can quantify the details of the structure with varying scale. Therefore fractal objects do not have absolute or integral dimensions, but instead have fractional dimensions.

Unlike Euclidean or topological dimensions – such as circles or spheres, fractal objects do not have an absolute measure as it changes at every scale of observation. Namely, dimensional quantities such as length, area and volume cannot be measured directly since fractal objects do not necessarily have characteristic scales. The geometry can only be realized using a recursion of the iterative map (see Fig. 5.3). The repetitive pattern exhibited by the fractal structure has certain textural properties, *e.g.*, curve perimeter,

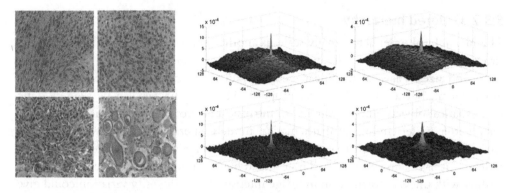

Figure 5.2 (Left) Meningioma fibroblastic, meningothelial, psammomatous, and transitional histopathological subtypes, respectively, and (right) corresponding surface roughness plots – as normalized auto-covariance functions – showing different degrees of texture irregularity.

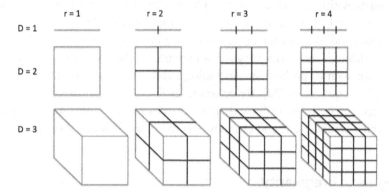

Figure 5.3 Example of self-similarity for a line (1D), square (2D), and cube (3D) by reducing its corresponding size (length, area, or volume) up to the fourth scale ($r = 4$), where r represents the scaling factor describing coarse to fine structures.

which behaves as if the dimensions of the structure (*i.e.*, fractal dimensions) are greater than the spatial or topological dimensions D_T, where D_T corresponds to the number of independent variables needed to describe the object's shape. Fractal geometry tends to characterize object geometry details as if it would fall "in-between" dimensions, *i.e.*, if a mathematical set describing a point would have a 0-dimension, then the estimated fractal dimension would lie between 0 and 1 for 1D fractal lines; 1 and 2 for 2D fractal surfaces; or between 2 and 3 for 3D fractal volumes. In general, given a texture function f in an n-dimensional Euclidean space $\mathbb{R}^n$, the FD of f can be described as $(n-1) \leq FD \leq n$. Thus a curve with fractal dimension very near to 1, *e.g.*, 1.1, behaves quite like an ordinary line, but a curve with fractal dimension 1.9 would be closer to a surface. Similarly, a surface with fractal dimension of 2.1 fills space very much like an

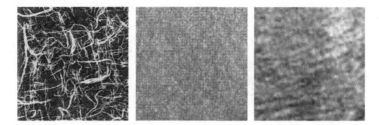

Figure 5.4 (Left–right) The first two images represent synthetic textures with a fractal dimension of 1.2 and 1.9, respectively, while the last is a CT lung tumor tissue texture image having a fractal dimension value of 2.1.

ordinary surface, but one with a fractal dimension of 2.9 would fill the space rather like a volume (see Fig. 5.4). In other words, most object structures would have a fractional rather than an integral dimension, thus the Euclidean geometry can be considered as being a special case of fractal geometry, which is the rule and not the exception.

The complexity of a given object can be illustrated as a FD by considering the well-known Sierpinski triangle[2] as an example: It has the basic fractal properties described before as being self-similar and recursive. Assuming the initial triangle is equilateral with sides one-unit long. The typical deterministic uniform fractal is recursively subdivided by using the midpoints of each vertex to create four equilateral smaller triangles and removing the middle one. The initial (larger) triangle is identical to a two-fold magnification of each of the three small remaining triangles on which the process is iterated. Hence the scaling factor at each stage is two. FD denotes the power to which 2 (scale changes by 2 with each iteration i) must be raised to produce 3 (number of smaller self-similar triangles δ at i, where δ_{i+1} is $3 \times \delta_i$), i.e., $2^{FD} = 3$. Applying the natural logarithm (base e), the dimension of the Sierpinski curve is thus $FD = \log(3)/\log(2)$, or roughly 1.585. Thereby, the Sierpinski triangle shown in Fig. 5.5, can be self-similar on infinite scales as

$$FD = \frac{\log(3)}{\log(2)} = \frac{\log(9)}{\log(4)} = \frac{\log(3^k)}{\log(2^k)} \approx 1.585$$

Thereby theoretically, in the Euclidean n-space, the local size of a bounded set of numbers $\mathbb{S}$ can be considered statistically self-similar if $\mathbb{S}$ is the union of N_r nonoverlapping subsets with respect to a scaling factor r, each of which is of the form $r(\mathbb{S}_n)$ where the N_r and $\mathbb{S}_n$ sets are congruent in the spatial distribution to $\mathbb{S}$. Hence, the *Hausdorff–Besicovitch* dimension – which can be considered as a generalization of the Euclidean space – of a bounded set $\mathbb{S}$ is a real number used to characterize the geometric complexity of $\mathbb{S}$. The fractal dimension FD of strictly self-similar objects can be derived

---

[2] A special mathematical set discovered by Waclaw Sierpinski in the early 1915.

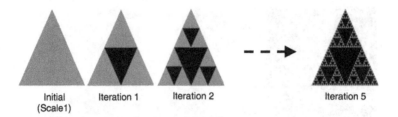

Figure 5.5 Sierpinski triangle fragmented up to six scales.

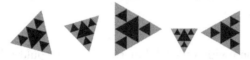

Figure 5.6 Rotated Sierpinski triangle shape in 15°, 45°, 90°, 180°, 270°, and with different magnification scales.

mathematically and is given by [4]:

$$FD = \frac{\log(N_r)}{\log(1/r)} \tag{5.1}$$

where a plot of the quantity on a log–log graph (having base e) versus scale then gives a straight line, whose slope is FD.

Also as the FD characterizes object's self-similarity irrespective of scale, rotated copies of the mother shape at smaller scales might be subjected to arbitrary rotations; see Fig. 5.6. The FD is not invariant to rotations of the fractal object. Section 5.7 introduces an additional parameter that is invariant to rotation at a given scale.

5.4 METHODS FOR COMPUTING FRACTAL DIMENSIONS

The application of fractal geometry in image texture analysis is more concerned with the evaluation of the fractal dimension (referred to as FD). The calculation methods of FD can be categorized into *power-law* and *size-measurement* fitting methods [21]. The former recursively measures the area of a surface or the length of a curve at different scales. The latter aims to fit a curve or a surface to a function known a *priori* to have a fractal behavior. The majority of the methods in the literature are based on the power-law relationship; however, they do not generally give the same results, especially if the object is multifractal (*i.e.*, a single fractal dimension is not enough to describe the texture characteristics). Many methods were developed to compute this dimension; each of which has its own theoretical foundations related to its assumptions about the nature of the object. This fact often leads to obtaining different dimensions by different methods for the same feature. These variations are attributed to the Hausdorff–Besicovitch dimension D_h not estimated in this form in most methods. Eq. (5.1) corresponds to D_h as the relation between the affine transformation of an object (N) to the associated reduction ratio (r) on a logarithmic scale. Namely, a fractal object respecting Eq. (5.1) would consist of N patterns with a reduced size of a factor r. Moreover, the fractal dimension of nondeterministic (*i.e.*, random and quasifractal) objects cannot be derived analytically. Therefore a number of methods have been proposed to empirically estimate the fractal (monofractal) dimension of natural objects. These methods differ in the way the quantity N_r in Eq. (5.1) is approximated, yet they are similar in the sense of exploiting the statistical relationship between the measured quantities of an object and step sizes to derive the estimates of FD. The "quantity" of an object is expressed in terms of length, area, or number of boxes required to overlay the object. "Step size" refers to the scale or resolution of measuring units used. Although the applied algorithms differ, they employ the common procedures summarized in the following three steps: a) measure the quantities of the object using various step sizes; b) plot the measured quantities versus the step sizes on a log–log-plot and fit a least squares regression line through the data points; c) estimate FD using the slope of the regression line.

Broadly, there are a number of approaches for computing the FD. The most well-known methods in the literature are grouped into three main classes: differential box-counting, fractional Brownian motion, and area measurement methods. The image texture surface is approached by different perspectives, where the first method relies on a finite scale grid, the second characterizes the object irregularity via fractional Brownian functions, and the third class measures the areas and the respective perimeters of the different curves.

5.4.1 Differential Box-Counting (DBC) method

This method proposed by Sarkar and Chaudhuri [22,23] can be thought of as a variant of the classical box-counting approach [24]. Unlike the classical method which requires signal binarization (see Section 3.5 of Chapter 3) and the computation of the FD is box size sensitive [25], the DBC method has the advantage of being able to work on gray-scale images, and thus the binarization step is avoided. It is a recursive method that involves covering the contour lines with a grid of n-dimensional boxes having a side-length r and counting the number of nonempty boxes $N(r)$. An overlaying grid approximates each contour line and then the number of boxes or cells intersections is counted by means of the difference between the minimum and the maximum gray levels in the n-dimensional box. The smaller the size of the boxes becomes, the larger the length estimate as finer details are captured. For a one-dimensional curve having a length L, the relation can be expressed as [4,8]: $N(r) \approx \frac{L}{r}$. Then the generalized form of the slope of the natural logarithmic plot of the number of boxes against their size would represent the FD:

$$N(r) \propto \frac{1}{r^{FD}} \qquad (5.2)$$

$$FD = \lim_{r \to 0} -\frac{\log N(r)}{\log(r)} \qquad (5.3)$$

The DBC estimation process is repeated for all pixels in $f(x)$ and a parametric FD image $(\mathcal{H})$ is generated as in Eq. (5.4), where b represents the slope of the linear regression line of Eq. (5.3):

$$\mathcal{H}\{f\}(x) = \begin{pmatrix} b_{11} & b_{12} & \cdots & \cdots & b_{1M_2} \\ b_{21} & b_{22} & \cdots & \cdots & b_{2M_2} \\ \vdots & \vdots & \ddots & & \vdots \\ \vdots & \vdots & & \ddots & \vdots \\ b_{M_11} & b_{M_12} & \cdots & \cdots & b_{M_1M_2} \end{pmatrix} \qquad (5.4)$$

5.4.2 Fractional Brownian motion methods

The fractional Brownian motion (fBm) is a nonstationary model known for its capability to describe random phenomena [26]. It is a generalization of the Brownian motion where the increments of the process are normally distributed but not independent [4]. The correlation of a random process $B(t)$ for any time $t \geq 0$ and $h > 0$ can be represented as the expected value $E(B)$ of the product of nonoverlapping increments of the fBm process:

$$E\Big((B(t) - B(0)) \cdot (B(t+h) - B(t)) \Big) = ((t+h)^{2H} - t^{2H} - h^{2H})/2, \qquad (5.5)$$

where the real number H is called the Hurst index falls in the range $(0, 1)$, and describes the roughness of the motion, with a higher value leading to a smoother motion [27]. The value of H determines the process of the fBm, such that $H > \frac{1}{2}$ indicates a positive correlation between the increments. A Brownian motion is achieved for the special case $H = \frac{1}{2}$, and the increments are negatively correlated when $H < \frac{1}{2}$. For a texture image, H is practically estimated by plotting the mean absolute difference of pixel pairs in all (horizontal, vertical, and diagonal) directions as a function of scale on a log–log scale (see Fig. 5.7), where H will represent the slope of the curve that is used to estimate the FD as: $FD = 3 - H$. The two algorithms that consider an image to be a 2D fractal Brownian function are described in Sections 5.4.2.1 and 5.4.2.2.

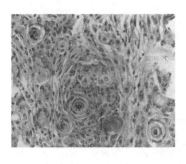
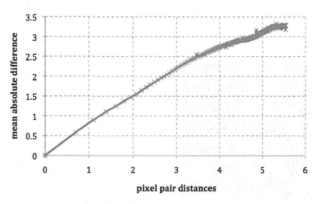

Figure 5.7 Estimating the Hurst index (*H*) from the natural log–log plot of mean absolute difference versus pixel pair distance for a meningioma (transitional subtype) histopathological texture image.

5.4.2.1 Power spectrum

Another method for computing the FD is based on the power spectrum dependence of fractional Brownian motion [28,23]. The slope is estimated from the relation between the Fourier power spectral density versus the frequency for computing FD. It can be shown that given a step size r, $P_f(f) \propto f^{-(2H+1)}$, where $P_f(f)$ is the power spectrum of a fractional Brownian function representing the image itself f, and $H = 2 - FD_p$. The fractal dimension of the surface (FD_p) is obtained from the slope of the regression line of the log–log plot of $P_f(f)$ versus f. The FD of the surface is computed as $FD = FD_p + 1$. The Fourier-based method is ideal for isotropic self-affine surface analysis but generally runs slowly [29,3].

5.4.2.2 Variogram

This method tends to quantify the autocorrelation structure of the texture surface [30, 31,23]. It assumes a fractional Brownian surface and estimates FD based on the mean squared elevation difference (*i.e.*, variance) versus the distance, such that the expectation value $E[(f(x_p) - f(x_q))^2 \propto (\|x_p - x_q\|)^{2H}]$, where $f(x_p)$ and $f(x_p)$ are elevations at points p and q, $\|x_p - x_q\|$ is the distance between p and q. The FD is estimated by plotting $\log\{E[\ldots]\}$ versus $\|x_p - x_q\|$ and the FD would be estimated from the slope $2H$; see Fig. 5.8. The variogram method is relatively simpler to calculate and assumes a weaker model of statistical stationarity as compared to the power spectrum. However, it yields unstable results for higher dimension surfaces [32].

5.4.3 Area-based methods

The area $A(r)$ of a texture surface at scale r is estimated using different structuring elements, such as triangle, erosion, dilation, etc., at various r. The most applied algorithms for this class of FD estimation are described in Sections 5.4.3.1 to 5.4.3.4.

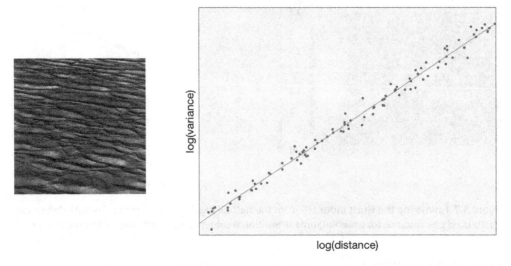

Figure 5.8 Sample texture image (D37) from the Brodatz album and corresponding natural log–log plot of the variance versus distance for a variogram.

5.4.3.1 Triangular prism

The method compares the surface areas of four-sided triangular prisms with the pixels area (step size squared) in log–log form [33]. A relation is derived between the total area of the tops of prisms versus the side length of analysis windows (*i.e.*, corner points of squares) such that $s(r) \propto r^{2-FD}$, where $s(r)$ is the area, r is the side length of the analysis window; see Fig. 5.9. FD is estimated from the slope $(2 - FD)$ of the log–log plot of $s(r)$ versus r. Generally, the method is less computationally intensive than other methods such as the variogram or Fourier power spectrum methods [33].

5.4.3.2 Isarithm

The basic idea of the method is approximating the complexity of the surface based on the complexity of contour lines [34], which represent horizontal cross-sections of the texture surface. The FD of each isarithm can be estimated with the *walking divider* method as $L(r) \propto r^{1-FD_{contour}}$, where $L(r)$ is the length of the contour line (*i.e.*, number of boundary pixels), r is the step size, and $FD_{contour}$ is the fractal dimension of a contour line. The method operates by iteratively walking the divider along the curve and then counting the number of steps required to cover the curve for each case. For each contour line, we plot $\log L(r)$ versus $\log(r)$ estimating the slope $(1 - FD_{contour})$ of the linear regression line. This process is repeated at different resolutions, and then FD would represent the average of all $FD_{contour} + 1$. However, the computed FD will vary depending on the direction that the data is measure in (horizontal, vertically, or diagonally). Also the selected step size may affect the stability of the results [35].

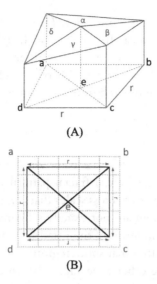

Figure 5.9 Triangular prism method [33,23] showing (A) a 3D view of elevation values at the corners of squares (*a*, *b*, *c* and *d*), (B) top view of the corner pixels (*a*, *b*, *c*, and *d*) and the center point (*e*) with step size (*r*) = 4. The method operates by interpolating a center value (*e*), dividing the square into four triangles (*abe*, *bce*, *cde*, and *dae* as in Fig. 5.9(B)), and then computing the top surface areas *s*(*r*) of the prisms which result from raising the triangles to their given elevations (α, β, γ, and δ in Fig. 5.9(A)). By repeating this calculation for geometrically increasing square sizes (*r*), the relationship between the total upper surface area of the prisms (*i.e.*, the sum of areas α, β, γ, and δ in Fig. 5.9(A)) and the spacing of the squares (*i.e.*, step size *r*) can be established, and used to estimate FD.

5.4.3.3 Robust estimator

Also based on the walking-divider method, a relation is set between the length of the profile and the step size such that $L(r) \propto r^{1-FD_{profile}}$, where $L(r)$ is the length of the profile, r is the step size, and $FD_{profile}$ is the corresponding fractal dimension [36]. The method computes for each pixel the average of $FD_{profile}$ in both north–south and east–west directions. Then the FD of the surface is obtained by combining the FD values of each pixel using a weighted average and adding 1. Although proposed for providing stability in computed FD, it yields a lower FD value as compared to the triangular prism and variogram methods [36,23].

5.4.3.4 Epsilon-blanket

The iterative method computes the FD of a scale-varying curve (or surface) based on the corresponding area (or volume) [37]. The set of points that lie at a certain distance ε from the curve are used in forming a "blanket" having a strip of width of 2ε covering the curve, and for different values of ε. This blanket is defined by two surfaces, an upper surface, and a lower surface (defined by dilatation and erosion of the image). The FD for a curve can be calculated using the relation $L(\varepsilon) \propto \varepsilon^{1-FD}$, where $L(\varepsilon)$ is the curve length defined from the blanket area $A(\varepsilon)$ as $L(\varepsilon) = A(\varepsilon)/2\varepsilon$. For the case of a surface, the blanket is formed from a set of points in the three-dimensional space, which are also ε far from the surface and having a width of 2ε. The surface area is defined as $A(\varepsilon) = V(\varepsilon)/2\varepsilon$, and the FD is estimated from the relation $A(\varepsilon) = \varepsilon^{2-FD}$.

5.5 TYPES OF FRACTALS

Fractals can be divided into three categories based on their self-similarity property (see Section 5.3.1): as having a mathematical fractal pattern (*i.e.*, *deterministic*), which are well specified and manifest similarity across scales that can be modeled exactly; a natural fractal pattern (*i.e.*, *random*) which are not exactly self-similar, but manifest similarities across scales which are ordered in terms of their statistical distribution; or a combination of both (*i.e.*, *quasifractal*) suggesting that some regions within the fractal pattern exhibit stronger self-similarity than the others; see Fig. 5.10. In addition, from the notion of normalized distributions (measures) and the associate scaling properties, fractals can also be subdivided as uniform and nonuniform. The former are known as "monofractals" or simply fractals, whereas nonuniform fractals are known as "multifractals" as they tend to scale with multiple scaling rules (*i.e.*, different regions within the texture pattern are characterized with different fractal dimensions) [38,39]. As the name specifies, a deterministic or idealized–forms fractal is generated by a deterministic rule, such as a given iterative map $\mathbb{I} \rightarrow \mathbb{I}$ (*i.e.*, a map from some set $\mathbb{I}$ to itself), which is obtained by composing

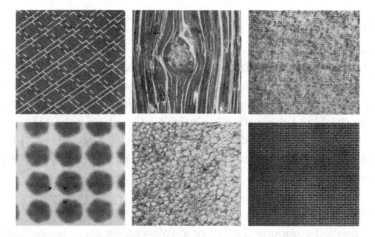

Figure 5.10 Sample texture images from the Brodatz album representing (column 1) deterministic D47 and D48, (column 2) quasifractal D72 and D112, and (column 3) random D19 and D34 fractal structures, respectively. The self-similarity of the texture patterns in column 1 appear more prominent as compared to column 2 where some areas do not exhibit pronounced self-similarity; whereas the texture patterns are closer to be considered as random in column 3.

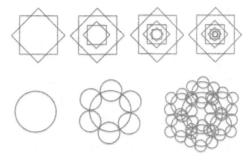

Figure 5.11 Fractal recursions: (first row) octagon-shape deterministic fractal representing an example of disjointed-fractal iteration, and (second row) interlaced-fractal iteration of a circle where leaf circle deterministic fractals are generated from each circle perimeter.

another map $\mathbb{X} : \mathbb{I} \to \mathbb{I}$ with itself a certain number of times. However, quasifractal and random fractals are generated by stochastic rules. They are further generated by recursive processes either in a disjoint or interlaced manner; see Fig. 5.11. Random fractals, whether uniform or otherwise, are always statistically self-similar: the normalized distributions of the gray level values show the same statistical properties across many scales. In other words, such random fractals are not strictly geometrical self-similar, but exhibit self-similarity of an approximate pattern with approximate scaling. Random fractals can represent natural phenomena such as coastlines, land surfaces, roughness, cloud boundaries, etc. Particularly in biomedical image analysis, tissue texture is considered to be a random fractal as it does not exhibit exactly the same structure at all scales, but with an approximate pattern appearing at different scales. This statistical self-similarity mainly depends on the underlying physiology of the pattern itself (describing the chaotic phenomena) as well as on the spatial resolution of the imaging modality. How far can the random fractal pattern recur at progressively smaller scales needs to be investigated, as this depends mainly on the imaging modality spatial resolution and scale. This variation in the fractal pattern details is mainly attributed to the fundamental texture elements, called "textons" (see Section 1.2.4 of Chapter 1), which form the perceived textural pattern. What dictates the appearance of a pattern is its texton properties, which come in different: sizes (height, area, or number); colors (hue and saturation); roughness (fine or coarse); orientations (angular displacement); and various geometric attributes (shapes).

5.6 FD PARAMETER ESTIMATION OPTIMIZATION

Since most biological tissues are considered random and/or quasifractals, they scale in a statistical fashion. The resemblance between patterns at different scales in this case is approximately self-similar. In theory, an examined fractal structure should have invariant self-similar fragmented and irregular shapes at all scales of measurement reaching to

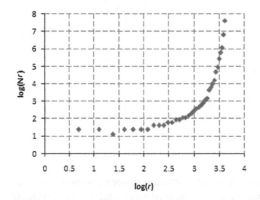

Figure 5.12 Scaling factor (*r*) vs. required number of boxes to cover each image pixel (*N_r*) on a natural log–log scale for CT lung tumor tissue [40].

infinity. Yet, in biological tissue texture this could only be true for a finite number of scales, depending on the resolution and depth of the examined structure, which means there is a finite scaling range, such that the estimation of the slope of the linear regression line is optimal. Whereas below or above this scale, the relation between the fractal object and scaling factor would give a biased estimation. In this manner, irrelevant points are excluded from the analysis to avoid regression using larger points than the best scale. This feature appears especially useful for the analysis of tissue images as it ensures that random Gaussian noise in the image will not be taken into account in the estimation process. Fig. 5.12 shows that the FD estimation has to be made at a well-defined scale $(2 \leq r \leq 9)$.

5.7 LACUNARITY ANALYSIS

5.7.1 Assessing image texture sparsity

In clinical practice, it is very common to encounter texture with a conjecture of patterns, due to the varying tissue characteristics occurring as a natural phenomenon. Also a certain degree of randomness is added if the analyzed tissue is abnormal. These mixtures of texton's properties draws the complex relationship of the tissue texture, making the texture surface to appear heterogeneous. This complexity may reduce the efficiency of the FD computation methods, where textures which appear visually different may get characterized with an identical FD value. Moreover, other factors such as the size and distribution of a textural feature and its spatial relations to other features may also play an important part in differentiating one type of texture from another. This leads to the fact that using FD alone may not be sufficient in characterizing image textures. Poor texture separability can be dealt with by adding another dimension, which can complement how a fractal fills the space. *Lacunarity analysis* is a technique introduced to deal

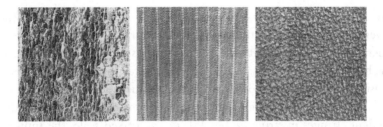

Figure 5.13 (Left–right) Texture images from the Brodatz album (D12, D85, and D92) showing an equal FD of 1.8, while having different lacunarity ($\mathcal{L}$) values of 0.031, 0.050, and 0.025, respectively.

with fractal objects of the same dimension with different textural appearances [4]; see Fig. 5.13. The lacunarity parameter describes the spatial complexity of a texture pattern, on different scales, based on the spatial distribution of gaps of a specific size. Namely, lacunarity measures the sparsity of the fractal texture, providing additional information on the uniformity or irregularity in the way space is filled.

The idea of homogeneity can be evoked for understanding how lacunarity characterizes texture heterogeneity. For example, Fig. 5.14 illustrates simple texture patterns consisting of alternating ones (bright color) and zeros (dark color), where ones represent the gaps in texture. The pattern in the first two images of Fig. 5.14 is translated uniformly by steps of 1 and 3, respectively. Since the gap sizes in the two images tend to not change, *i.e.*, uniformly distributed, this indicates that the texture is homogeneous, and thus has a low lacunarity. Whereas the last two images in Fig. 5.14 presents an example of heterogeneous pattern, with gaps having a varying size; giving a higher lacunarity as the gap size distribution varies. It is worth noting here that texture homogeneity refers to a smooth or periodic structure defined in terms of the spatial relationships between the bright pixels (*i.e.*, gaps in Fig. 5.14) and their surrounding neighbors (dark pixels in Fig. 5.14) for a given fractal pattern. The gaps define how the main fractal pattern in a texture appear smooth or rough/periodic or irregular. Specifically, how controlled the sparsity (or density) and periodicity (or irregularity) of the gaps in the

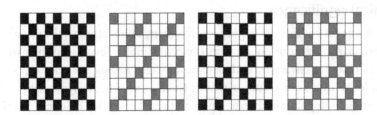

Figure 5.14 Lacunarity versus texture homogeneity: (left–right) homogeneous pattern with a gap size of one, homogeneous pattern with a gap size of 3; heterogeneous pattern with a varying gap size of 1 and 3, heterogeneous pattern with gap size changing randomly, respectively.

fractal pattern plays a fundamental role in defining texture homogeneity (or hetero-geneity). Hence, lacunarity quantifies the fractal pattern deviation from homogeneity. It is considered a scale-dependent parameter because sets that are uniform at a coarse scale might be heterogeneous at a finer scale, and vice versa [41–43]. Lacunarity can thus be considered a scale-dependent measure of textural heterogeneity. Also it acts as a translation-invariant feature which can quantify aspects of fractal patterns that exhibit scale-dependent changes in structure [44,45], *i.e.*, at a fixed scale, the similarity of the parts from different regions of a pattern to each other is assessed. Patterns with high la-cunarity are considered heterogeneous and not translational-invariant if a wide range of gap sizes exist, and vice versa [42]. Given a fractal texture pattern $f_{FD}(x)$, the lacunarity ($\mathcal{L}$) can be defined in terms of the ratio of the variance over the mean value as in (5.6), where M_1 and M_2 represent the size of $f_{FD}(x)$ [46]:

$$\mathcal{L} = \frac{\frac{1}{M_1 M_2} \sum_{x_1=0}^{M_1-1} \sum_{x_2=0}^{M_2-1} f_{FD}(x)^2}{(\frac{1}{M_1 M_2} \sum_{x_1=0}^{M_1-1} \sum_{x_2=0}^{M_2-1} f_{FD}(x))^2} - 1 \tag{5.6}$$

5.7.2 Rotation-invariance

Rotation-invariance is one of the important aspects in texture analysis. A common lim-itation is to assume that texture is always acquired from the same viewpoint. Texture tend to occur with different scales, orientations, or translations. Therefore scale and ori-entation invariant texture analysis approaches are essential for texture characterization. A fractal texture pattern is said to have rotation invariance if its value does not change when arbitrary rotations are applied at a specific scale.

In order to investigate how robust the lacunarity parameter to rotations, textures in Fig. 5.10 were first transformed pixel-by-pixel via local windows to generate corre-sponding FD parametric images as shown in Fig. 5.15. Then the surface roughness is further analyzed by investigating their rotation-invariance using lacunarity; see Fig. 5.16. The majority of the analyzed FD images showed invariance to rotation with a small de-viation in textures D72 and D112.

5.7.3 Clinical significance

Feature-invariance is an important property for many applications in biomedical texture analysis. Particularly, rotations do not occur only at a global level, as in the patient's body rotation in an imaging system or with acquired images having variations in viewpoints, but also at a local level as in tissue structures (such as vessels, necrotic tissue, or clusters of cells in histopathology), which can appear with local random orientations. This is es-pecially true for tissue characterization as tumor texture is chaotic and happen to occur at different depths and perspectives depending on the relative position of the imaging device and the tumor surface and metastasis. As a scale-dependent measure for spatial

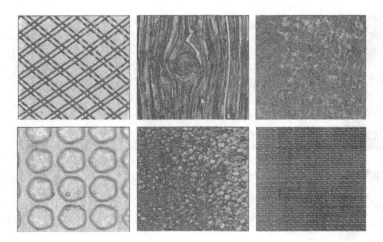

Figure 5.15 Corresponding FD parametric surface images computed via local windows for the textures in Fig. 5.10.

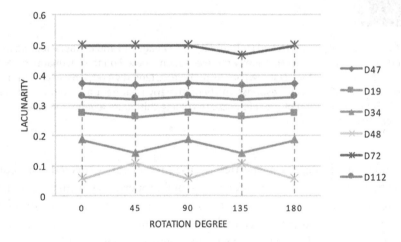

Figure 5.16 Rotation-invariance of lacunarity features for FD images of Fig. 5.15.

heterogeneity, a lacunarity parameter has been applied to characterize diseases, assisting in improved diagnosis [49]. The lacunarity serves as both a measure of heterogeneity, as well as the degree of invariance to change of the fractal pattern. Whereas the fractal dimension (geometric complexity) indicates how much space is filled by the pattern, lacunarity describes the distribution of the sizes of the gaps or lacunae surrounding the pattern within the image. This parameter, which has the capacity to recognize different fractal structures owing the same fractal dimension, was employed to quantitatively assessing tumor response to chemotherapy treatment; see Fig. 5.17. Tumors tend to progress in a chaotic manner as compared to normal tissue. Therefore if we relate tex-

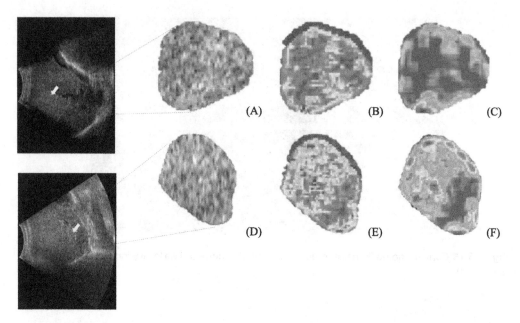

Figure 5.17 Heterogeneity maps: (A and D) Tumor B-mode images, (B and E) fractal texture maps and (C and F) corresponding tissue heterogeneity representation – based on the Lacunarity parameter as defined in Subsection 5.7.1 – for a (1st row) nonresponsive vs. (2nd row) responsive case, respectively. Red regions in (C) and (F) indicate response to chemotherapy treatment according to RECIST criteria [26,47,48]. (For interpretation of the references to color in this figure legend, the reader is referred to the web version of this chapter.)

ture with underlying biological properties, necrotic regions are expected – an indication of response to treatment – to have different textural properties as compared to active or aggressive regions within the tumor tissue. However, these textural variations within the tumor tissue overlap together, causing for discriminative features for the different regions to be obscured. Having the capacity to quantify tumor tissue heterogeneity can give interesting indications on the response to chemotherapy treatment [26].

5.8 TUMOR TISSUE CHARACTERIZATION

In many clinical applications, tumor tissue characterization is an important task which depends strongly on the signatures used to characterize a given region of the image. The fractal signatures based on estimating the similarity of tissue structures across scales – which other texture measures do not characterize it as such – enable to describe the scale-to-scale propagation of heterogeneity of the tumor texture. The concept of image surface roughness (variation of pixels distribution in a texture) and how it relates to fractal analysis is essential for quantifying tumor texture complexity.

5.8.1 Surface roughness

Formally, surface roughness is a measure of surface texture. It is quantified by the vertical deviations of a real surface from its geometric ideal form (*i.e.*, perfect condition without geometric distortion). If these deviations are large, the surface is rough; if they are small, the surface is smooth. The roughness varies between fine and coarse depending on the characteristics of the surface. Rough or fragmented features that cannot be described in terms of Euclidean geometry can be better represented by fractal geometry, in analogy of using *fuzzy logic* to describe fuzzy subsets of a crisp set. The degree of roughness or fragmentation can be accurately fine-tuned via considering its fractal properties over a range of scales: the larger the fractal dimension is the rougher the surface and vice versa; see Fig. 5.18.

Tumor, defined at tissue-level, is characterized by dense vascular networks having an abnormal structure that results in an increased resistance to blood flow; as the vast majority of these blood vessels are not functioning. It is also known for its abnormal and uncontrolled cell growth; enabling angiogenesis, cell invasion, and metastasis [50, 51]. This chaotic behavior reflects on the tissue texture visual appearance in the form of spatial intensity variations, which can also be described as an increase in surface roughness. Indeed, the surface roughness is also a reflection of abnormal activities and tumor cell morphologies in the background (*i.e.*, cellular-level). Therein, texture is coarse, and the distinctive features defined in the morphology of cell nuclei is different to normal tissue. Cell nuclei play the main role in assessing tumor malignancy, as changes in its surface, volume, shape, density, as well as structure homogeneity or heterogeneity defines the characteristics of the tumor texture.

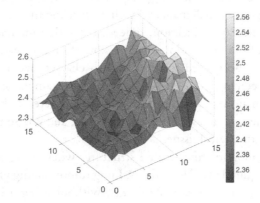

Figure 5.18 Fractal dimension surface illustrating the irregularity of the D112 quasifractal texture image of Fig. 5.10.

5.8.2 Tumor fractal analysis

Although it may appear random, tumor texture, similar to normal tissue texture, tends to exhibit statistical self-similar behaviors at different scales of observation [52]. In fact, the self-similarity is considered approximate and a varying degree of heterogeneity is associated with the distinctive pathological features; adding complexity to the tumor's recursive morphology. This is especially the case if the tumor texture itself is analyzed for the purpose of staging or subtype classification, where differences in texture are inconspicuous to an untrained eye. A way to quantify the tumor surface complexity, which cannot be described in terms of the integral dimension, is by using its fractals geometry. Fractal analysis of a tumor texture can be viewed as complex surface whose elevation is proportional to the metabolic activity of the tumor, *i.e.*, the aggressive the tumor the more random the tissue texture becomes, and thus the rougher the texture surface. Unlike the geometry of idealized forms, the resemblance between the successive smaller patterns in tumor texture tend to have a random probability distribution of surface structure that can be analyzed statistically. The surface of structures of tumor cells are known to be rough [40]. These convoluted structures display fractal properties which vary during the different stages of the cancer cell's growth. Using the FD as a quantitative measure for analyzing texture, *i.e.*, fractal analysis, can be useful for predicting a precise estimation of the texture dimension, and thus surface roughness, in analogy of how fuzzy logic handles the concept of partial truth. As the truth value in fuzzy logic may range between completely true and completely false, the surface dimensionality would vary between dimensions. Another important aspect in characterizing tumor texture is morphology, where it can give important signs on tumor behavior. In a sense that malignant tumors grow rapidly, invade adjacent tissue and spread by metastasis; while benign tumors are slow and grow with the formation of a capsule. Between these extremes there may be tumors of intermediate behavior, and thus having varying surface roughness. Here, fractal geometry could also serve as a useful criterion for diagnosing varying tumor malignancies by characterizing tumor shape irregularities, and thus surface roughness quantification. Indeed, estimating the fractal dimension can reflect important cue of the progression or regression of tumor treatment [26,53,47].

Fractal analyses have been used to study and to characterize a wide range of signals in biology and medicine [54]. Herein, we give an example of the application of fractal analysis to two main types of tumor texture encountered in biomedical imaging modalities, namely, fine texture as in lung tumor tissue, and coarse texture represented by brain tumor histopathological tissue.

Fig. 5.19 is an example of fine texture structure, where lung tumor contrast enhanced images were acquired via CT modality. Lung tumor staging predication accuracy was improved from conventional CT alone, *i.e.*, without using a PET scan, through identifying malignant aggressiveness of lung tumors by examining vascularized tumor regions that exhibit strong fractal characteristics [40,55]. Besides the side effect of edge

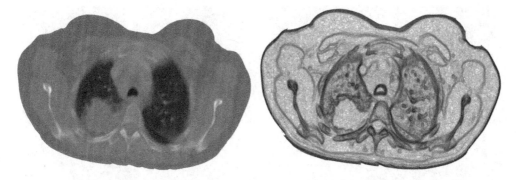

Figure 5.19 CT lung image (tumor region shown in lower part of right lung) and corresponding fractal dimension parametric image using the fractional Brownian motion algorithm.

enhancement, the computation of FD for each pixel in the image can assist in better tumor delineation and characterization of surface roughness. Fractals show several types and degrees of self-similarity and detail that may not be easily visualized. The overlapping of the various descriptive tissue features could obscure tissue patterns deemed to be essential in tumor characterization. Although the textural pattern in Fig. 5.19 does show some well-defined fractal characteristics, the surface roughness can be used in improving lung tumor staging accuracy from CT images [56,40].

Fractal analysis can also be used to characterize histopathological tissue, which has different textural properties compared to CT. Histopathological texture usually has a macro or coarse structure where the cell nuclei color, shape, and orientation defines the general texture structure. Overall, fractals in histopathological tissue texture show several types and degrees of self-similarity and detail that are imperceptible at lower scales of magnification. For example, meningiomas brain tumors exhibit a wide range of fractal patterns and a single subtype may show a combination of patterns [57]. This includes arrangement of the meningioma subtype fibroblastic and transitional cells in swirls or concentric whorls in different shapes and scales, while the large round nuclei in the meningothelial and the many cystic spaces in the psammomatous bodies contribute to the coarseness of the texture pattern (see Fig. 5.20). Each subtype shows a conjecture of texture patterns, characterized in the cell nuclei shape, arrangement, density, and color. This variation in tissue characteristics adds complexity to texture visual appearance, and may not always be resolvable into easily grasped units of detail and scale. The parametric FD images computed via local windows can quantify the surface roughness of the different patterns (mixture of different fractals), assisting in highlighting the aforementioned textural properties which distinct one subtype from another. FD was employed for the purpose of developing an automated system for improving histopathological meningioma brain tumors discrimination [58–60,57].

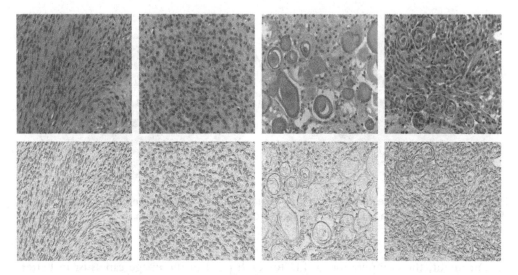

Figure 5.20 Histopathological tissue characterization: (first row) meningioma images representing fibroblastic, meningothelial, psammomatous, and transitional subtypes; (second row) corresponding fractal dimension parametric images using the fBm algorithm.

Also, local FD computed for each pixel in an image can be used as well for texture segregation. Patterns having common tissue variations are located and grouped together, forming textural boundaries that can visually separate the different patterns. Fig. 5.21 shows an example of the relevance of FD in tumor segregation from the surround ultrasound liver tissue. This can be used to localize tumors that tend to have high metabolic activity, which reflects in the degree of randomness of the tumor texture surface [61,26, 47].

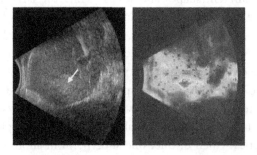

Figure 5.21 Clinical ultrasound B-mode image showing a liver tumor (indicated by an arrow), and corresponding FD-based parametric image showing tumor.

5.9 CONSIDERATIONS FOR FRACTAL ANALYSIS

5.9.1 Choosing a suitable method for estimating the FD

The selection of the suitable fractal algorithm plays a crucial role in shaping the correctness of the estimated fractal dimensions; otherwise, it would be difficult to assess whether the variation in the estimated FD values are attributed to the algorithm and/or parameter settings, or due to the inherit spatial variations of the tissue texture itself. An improvement in results could be achieved if the bias in the estimated values could be reduced. This suggests performing beforehand a comparative evaluation of the different FD computation methods to select the most suitable algorithm for the analyzed texture (revisit Section 5.4).

5.9.2 Multivariate fractal analysis

Using fractal techniques for analyzing biomedical texture images applied only to a single modality might not deliver an improved texture description. Since pathology is a bottom-up approach, where changes begin to occur at the cellular level before appearing at tissue level, which requires a multimodal fractal analysis to investigate disease at its different levels. A better approach would be to relate the spatial changes and variation in tissue to initial cellular level morphological behavior. It would be advantageous to develop a "multivariate fractal method," which should enable us to analyze all features extracted from the different modalities together (*e.g.*, combining microscopy, CT, and MRI).

5.9.3 Performing fractal or multifractal analysis

Some types of rough surfaces might vary considerably in space in a random manner. In this situation, assuming a single (monofractal) dimension to characterize both local and global scales properties might not suffice. Nevertheless, extracting a spectrum or a set of fractal dimensions can better adapt to the curvature of the surface, and hence covering different scaling properties in different texture regions. Recent contributions suggested that the structures underlying most tissue textures are most likely multifractals, and has shown interesting results in some imaging modalities like CT, MR, US, and microscopy [44,62–66]. This implies that fractal analysis of tissue texture images without checking their dimensionalities may be limiting. Furthermore, investigating whether multifractal models could provide better texture characterization is needed, taking into consideration the numerical differences between the different methods.

5.10 CONCLUSION

The problem of tumor diagnosis, treatment, and prevention continues to come to the forefront as a challenge. There is a compelling need for deriving meaningful features that could complement clinicians' diagnostic capability to discriminate higher order statistical textural information. Tissue characterization is a key step in many medical imaging based procedures. Many researchers have proposed numerous methods to solve specific image texture analysis problems; however, different tissue has different properties and there is still a lot we do not know about tissue texture heterogeneities. The inherent complexity associated with tissue characterization and tumor spatial heterogeneity renders the understanding of texture irregularity in medical images challenging.

In many clinical applications, image tissue characterization is an important task, which depends strongly on the signatures used to describe a pattern in an image. Fractal analysis is considered an effective mean of capturing textural descriptors for tissue characterization. This is mainly attributed to the discontinuous, complex, and fragmented shape of tissue texture in medical images. All tissue textures follow statistical placement rules and involve some roughness and a lot of irregularity. Surface roughness or fragmented features that cannot be described in terms of Euclidean geometry can be better represented by its fractal geometry. The quality of the extracted texture features can be improved by expressing the surface geometric complexity, which can be summarized as:

a) The fractal dimension parameter, serving as a degree of the geometric irregularity of image texture, provides the unique ability to measure self-similarity of patterns across scales. The self-similarity accounts for either the pattern geometry or the statistical distribution of gray levels.

b) Tissue texture might be acquired at different viewpoints; thus the FD's scale symmetry alone might not suffice. Another significant parameter of fractal analysis is the Lacunarity, which measures how sparse a fractal texture is. Its translation and rotation-invariance property provides a multiscaled measure of texture's heterogeneity. Therefore combining both parameters allows for a global and local approach in capturing the various texture heterogeneities. In a sense, the FD characterizes texture surface roughness, while lacunarity further looks deep into the details between the fractal texture structures.

REFERENCES

[1] R.M. Haralick, K. Shanmuga, I. Dinstein, Textural features for image classification, IEEE Trans. Syst. Man Cybern. SMC3 (6) (1973) 610–621.
[2] W.K. Pratt, O.D. Faugeras, A. Gagalowicz, Visual-discrimination of stochastic texture fields, IEEE Trans. Syst. Man Cybern. 8 (11) (1978) 796–804.
[3] R. Lopes, N. Betrouni, Fractal and multifractal analysis: a review, Med. Image Anal. 13 (4) (2009).
[4] B.B. Mandelbrot, Fractal Geometry of Nature, Freeman, 1982.
[5] O.S. Al-Kadi, Supervised texture segmentation: a comparative study, in: IEEE Jordan Conference on Applied Electrical Engineering and Computing Technologies, 2011, pp. 1–5.
[6] J. Garding, Surface orientation and curvature from differential texture distortion, in: Proceedings of IEEE International Conference on Computer Vision, 2002, pp. 733–739.

[7] J. Malik, S. Belongie, T. Leung, J. Shi, Contour and texture analysis for image segmentation, Int. J. Comput. Vis. 43 (1) (2001) 7–27.

[8] M.J. Turner, P.R. Andrews, J.M. Blackledge, Fractal Geometry in Digital Imaging, Academic Press, 1998.

[9] O.S. Al-Kadi, Combined statistical and model based texture features for improved image classification, in: IET International Conference on Advances in Medical, Signal and Information Processing – MEDSIP, 2008, pp. 1–4.

[10] T. Randen, J.H. Husoy, Filtering for texture classification: a comparative study, IEEE Trans. Pattern Anal. Mach. Intell. 21 (4) (1999) 291–310.

[11] M. Varma, A. Zisserman, Texture classification: are filter banks necessary?, in: IEEE Computer Society Conference on Computer Vision and Pattern Recognition, vol. 2, 2003, pp. 691–698.

[12] J. Garding, Shape from texture for smooth curved surfaces in perspective projection, J. Math. Imaging Vis. 2 (4) (1992) 327–350.

[13] A. Lobay, D.A. Forsyth, Shape from texture without boundaries, Int. J. Comput. Vis. 67 (1) (2006) 71–91.

[14] J. Malik, R. Rosenholtz, Computing local surface orientation and shape from texture for curved surfaces, Int. J. Comput. Vis. 23 (2) (1997) 149–168.

[15] A. Lobay, D.A. Forsyth, Recovering shape and irradiance maps from rich dense texton fields, in: Proceedings of the IEEE Computer Society Conference on Computer Vision and Pattern Recognition, vol. 1, 2004.

[16] O.S. Al-Kadi, Assessment of texture measures susceptibility to noise in conventional and contrast enhanced computed tomography lung tumour images, Comput. Med. Imaging Graph. 34 (6) (2010) 494–503.

[17] O.S. Al-Kadi, D. Watson, Susceptibility of texture measures to noise: an application to lung tumor CT images, in: IEEE International Conference on BioInformatics and BioEngineering, 2008, pp. 1–4.

[18] I. Pantic, S. Dacic, P. Brkic, I. Lavrnja, T. Jovanovic, S. Pantic, S. Pekovic, Discriminatory ability of fractal and grey level co-occurrence matrix methods in structural analysis of hippocampus layers, J. Theor. Biol. 370 (2015) 151–156.

[19] A. Di Ieva, O.S. Al-Kadi, Computational Fractal-Based Analysis of Brain Tumor Microvascular Networks, Springer New York, New York, NY, 2016, pp. 393–411.

[20] G. Longo, M. Montevil, A. Pocheville, From bottom-up approaches to levels of organization and extended critical transitions, Front. Physiol. 3 (2012).

[21] M.G. Turner, R.H. Gardner, Quantitative Methods in Landscape Ecology, Springer-Verlag, 1991.

[22] N. Sarkar, B.B. Chaudhuri, An efficient differential box-counting approach to compute fractal dimension of image, IEEE Trans. Syst. Man Cybern. 24 (1) (1994) 115–120.

[23] W. Sun, G. Xu, P. Gong, S. Liang, Fractal analysis of remotely sensed images: a review of methods and applications, Int. J. Remote Sens. 27 (22) (2006) 4963–4990.

[24] R.F. Voss, Fractals in Nature: From Characterization to Simulation, Springer, 1988, pp. 21–70.

[25] S.A. Pruess, Some remarks on the numerical estimation of fractal dimension, in: C.C. Barton, P.R. La Pointe (Eds.), Fractals in the Earth Sciences, Springer US, 1995, pp. 65–75.

[26] O.S. Al-Kadi, D.Y.F. Chung, R.C. Carlisle, C.C. Coussios, J.A. Noble, Quantification of ultrasonic texture intra-heterogeneity via volumetric stochastic modeling for tissue characterization, Med. Image Anal. 21 (1) (2015) 59–71.

[27] B.B. Mandelbrot, J.W. van Ness, Fractional Brownian motions, fractional noises and applications, SIAM Rev. 10 (4) (1968) 422–437.

[28] A.P. Pentland, Fractal-based description of natural scenes, IEEE Trans. Pattern Anal. Mach. Intell. 6 (6) (1984) 661–674.

[29] P. Heinz-Otto, S. Dietmar, The Science of Fractal Images, Springer-Verlag, 1988.

[30] D.M. Mark, P.B. Aronson, Scale-dependent fractal dimensions of topographic surfaces – an empirical-investigation, with applications in geomorphology and computer mapping, J. Int. Assoc. Math. Geol. 16 (7) (1984) 671–683.

[31] P. Soille, J.F. Rivest, On the validity of fractal dimension measurements in image analysis, J. Vis. Commun. Image Represent. 7 (3) (1996) 217–229.

[32] N. Lam, H. Qiu, D. Quattochi, C. Emerson, An evaluation of fractal methods for characterizing image complexity, Cartogr. Geogr. Inf. Sci. 29 (1) (2002) 25–35.

[33] K.C. Clarke, Computation of the fractal dimension of topographic surfaces using the triangular prism surface-area method, Comput. Geosci. 12 (5) (1986) 713–722.

[34] M. Shelberg, N. Lam, H. Moellering, Measuring the fractal dimension of surfaces, in: Proceedings of the Sixth International Symposium on Computer-Assisted Cartography, 1983, pp. 319–328.

[35] N. Lam, L. De Cola, Fractals in Geography, Blackburn Press, 2002.

[36] K.C. Clarke, D.M. Schweizer, Measuring the fractal dimension of natural surfaces using a robust fractal estimator, Cartogr. Geogr. Inf. Syst. 18 (1) (1991) 37–47.

[37] S. Peleg, J. Naor, R. Hartley, D. Avnir, Multiple resolution texture analysis and classification, IEEE Trans. Pattern Anal. Mach. Intell. 6 (4) (1984) 518–523.

[38] A.C.J. Luo, Regularity and Complexity in Dynamical Systems, Springer-Verlag, New York, 2012.

[39] B. Pesquet-Popescu, J.L. Vehel, Stochastic fractal models for image processing, IEEE Signal Process. Mag. 19 (5) (sep 2002) 48–62.

[40] O.S. Al-Kadi, D. Watson, Texture analysis of aggressive and non-aggressive lung tumor CE CT images, IEEE Trans. Biomed. Eng. 55 (7) (2008) 1822–1830.

[41] C. Allain, M. Cloitre, Characterizing the lacunarity of random and deterministic fractal sets, Phys. Rev. A 44 (6) (1991) 3552–3558.

[42] R.E. Plotnick, R.H. Gardner, R.V. Oneill, Lacunarity indexes as measures of landscape texture, Landsc. Ecol. 8 (3) (1993) 201–211.

[43] A. Roy, E. Perfect, W.M. Dunne, L.D. McKay, A technique for revealing scale-dependent patterns in fracture spacing data, J. Geophys. Res., Solid Earth 119 (7) (2014) 5979–5986.

[44] Y. Gefen, Y. Meir, B.B. Mandelbrot, A. Aharony, Geometric implementation of hypercubic lattices with noninteger dimensionality by use of low lacunarity fractal lattices, Phys. Rev. Lett. 50 (3) (1983) 145–148.

[45] B. Lin, Z.R. Yang, A suggested lacunarity expression for Sierpinski carpets, J. Phys. A, Math. Gen. 19 (2) (1986) L49–L52.

[46] M. Petrou, P. Gacia Sevilla, Image Processing: Dealing with Texture, Wiley, 2006.

[47] O.S. Al-Kadi, D. Van De Ville, A. Depeursinge, Multidimensional Texture Analysis for Improved Prediction of Ultrasound Liver Tumor Response to Chemotherapy Treatment, Springer, 2016, pp. 619–626.

[48] E.A. Eisenhauer, P. Therasse, J. Bogaerts, L.H. Schwartz, D. Sargent, R. Ford, J. Dancey, S. Arbuck, S. Gwyther, M. Mooney, L. Rubinstein, L. Shankar, L. Dodd, R. Kaplan, D. Lacombe, J. Verweij, New response evaluation criteria in solid tumours: revised RECIST guideline (version 1.1), Eur. J. Cancer 45 (2) (2009) 228–247.

[49] B. Nielsen, F. Albregtsen, H.E. Danielsen, Fractal Signature Vectors and Lacunarity Class Distance Matrices to Extract New Adaptive Texture Features from Cell Nuclei, Birkhäuser Basel, Basel, 2002, pp. 55–65.

[50] P.R. Band, Early Detection and Localization of Lung Tumors in High Risk Groups, Springer, 1982.

[51] C.D.M. Fletcher, Diagnostic Histopathology of Tumors, Churchill Livingstone Elsevier, 2007.

[52] A. Asher, P. Burger, D. Croteau, T. Mikkelsen, T. Omert, N. Paleologos, J. Barnholtz-Sloan, A Primer of Brain Tumors: A Patient's Reference Manual, American Brain Tumor Association, 1998.

[53] O.S. Al-Kadi, D.Y.F. Chung, C.C. Coussios, J.A. Noble, Heterogeneous tissue characterization using ultrasound: a comparison of fractal analysis backscatter models on liver tumors, Ultrasound Med. Biol. 42 (7) (2016) 1612–1626.

[54] G.A. Losa, D. Merlini, T. Nonnenmacher, E. Weibel, Fractals in Biology and Medicine, vol. 3, 3rd edition, Springer, Basel AG, 2002.

[55] J.V. Raja, M. Khan, V.K. Ramachandra, O. Al-Kadi, Texture analysis of CT images in the characterization of oral cancers involving buccal mucosa, Dentomaxillofacial Radiol. 41 (6) (2012) 475–480.

[56] O.S. Al-Kadi, Tumour Grading and Discrimination based on Class Assignment and Quantitative Texture Analysis Techniques, PhD Thesis, University of Sussex, Informatics Department, 2009.

[57] O.S. Al-Kadi, A. Di Ieva, Histological Fractal-Based Classification of Brain Tumors, Springer, 2016, pp. 371–391.

[58] O.S. Al-Kadi, A fractal dimension based optimal wavelet packet analysis technique for classification of meningioma brain tumours, in: IEEE International Conference on Image Processing, 2009, pp. 4125–4128.

[59] O.S. Al-Kadi, Texture measures combination for improved meningioma classification of histopathological images, Pattern Recognit. 43 (6) (2010) 2043–2053.

[60] O.S. Al-Kadi, A multiresolution clinical decision support system based on fractal model design for classification of histological brain tumours, Comput. Med. Imaging Graph. 41 (2015) 67–79.

[61] O.S. Al-Kadi, Multiscale Nakagami parametric imaging for improved liver tumor localization, in: IEEE International Conference on Image Processing, 2016, pp. 3384–3388.

[62] C.M. Breki, A. Dimitrakopoulou-Strauss, J. Hassel, T. Theoharis, C. Sachpekidis, L.Y. Pan, A. Provata, Fractal and multifractal analysis of PET/CT images of metastatic melanoma before and after treatment with ipilimumab, EJNMMI Res. 6 (2016).

[63] Y. Ding, W.O.C. Ward, J.M. Duan, D.P. Auer, P. Gowland, L. Bai, Retinal vasculature classification using novel multifractal features, Phys. Med. Biol. 60 (21) (2015) 8365–8379.

[64] R. Lopes, P. Dubois, N. Makni, W. Szurhaj, S. Maouche, N. Betrouni, Classification of brain SPECT imaging using 3D local multifractal spectrum for epilepsy detection, Int. J. Comput. Assisted Radiol. Surg. 3 (3) (2008) 341–346.

[65] H.J. Ni, L.P. Zhou, X.B. Ning, L. Wang, Alzheimer's Disease Neuroimaging Initiative, Exploring multifractal-based features for mild Alzheimer's disease classification, Magn. Reson. Med. 76 (1) (2016) 259–269.

[66] T. Stosic, B.D. Stosic, Multifractal analysis of human retinal vessels, IEEE Trans. Med. Imaging 25 (8) (2006) 1101–1107.

CHAPTER 6

Handling of Feature Space Complexity for Texture Analysis in Medical Images

Yang Song, Weidong Cai

The University of Sydney, School of Information Technologies, Sydney, Australia

Abstract

In this chapter, we focus on the issue of feature space complexity when performing texture classification in medical images. In particular, to obtain accurate image classification, we would expect clear feature space separation between images of different classes. This means that ideally there should be low intraclass variation and interclass ambiguity in the feature space. However, such expectation is rarely met in real applications and the classification performance is thus affected. While the majority of existing studies focus on designing more representative and discriminative feature descriptors, we suggest that designing a classifier model that explicitly addresses the feature space complexity is an alternative direction in research. We provide a comprehensive review of such classification methods, including ensemble classification, subcategorization, and sparse representation. We also describe in detail a specific design on subcategory-based ensemble classification, with application in tissue pattern classification of Interstitial Lung Diseases (ILD) in High-Resolution Computed Tomography (HRCT) images. Some experimental results are presented as well.

Keywords

Ensemble classification, Sparse representation, Subcategorization, Lung disease, CT imaging

6.1 INTRODUCTION

Automated classification of image textures is a major research topic in medical image analysis. It can be the fundamental technique in a large variety of applications, such as lesion detection, disease categorization, and image retrieval. Section 1.3.1 of Chapter 1 describes the operators and aggregation functions for computing the texture features of medical images. The commonly used texture analysis methods can be considered as particular cases of texture operators and aggregation functions. Given a set of medical images, the multidimensional texture features extracted from all images form a multidimensional feature space of the images. The classification task can be achieved by separating the feature space of different classes. One of the main difficulties in this task is the large intraclass variation and interclass ambiguity. In other words, images of the same class can exhibit different texture patterns, while images of different classes can display similar patterns. As illustrated in Fig. 6.1, the feature space can be complicated with scattering regions within the same class and overlapping areas between different

Biomedical Texture Analysis
DOI: 10.1016/B978-0-12-812133-7.00006-5

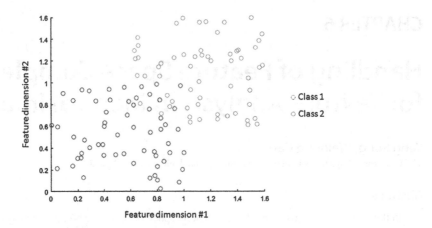

Figure 6.1 Illustration of a 2D feature space. The blue circles represent the data instances of one class, and the red circles indicate another class. It can be seen that the data of one class have varying feature (intraclass variation), and data of different classes can overlap in the feature space (interclass ambiguity). (For interpretation of the references to color in this figure legend, the reader is referred to the web version of this chapter.)

classes, even with domain–customized feature design. Such issues inevitably cause difficulties to the classifier. How to best represent the texture features and design a classifier to cope with the intraclass variation and interclass ambiguity is thus a challenging issue.

In this chapter, we focus on the classification method design for texture analysis. In Section 6.2, we first review some applications of texture analysis in the medical imaging domain, including lesion detection, disease categorization, and image retrieval. We then provide a comprehensive review of classification methods used in medical imaging in Section 6.3. Next, we focus on the application in Interstitial Lung Disease (ILD) tissue pattern classification for lung disease categorization, and we present two subcategory-based ensemble classification models in Section 6.4. The experimental study and results are described in Section 6.5. Finally, Section 6.6 concludes the chapter.

6.2 APPLICATIONS OF TEXTURE ANALYSIS

6.2.1 Lesion detection

An example of lesion detection is to detect lung tumors and abnormal lymph nodes from Positron Emission Tomography – Computed Tomography (PET-CT) thoracic images. Lesions typically exhibit higher uptake, *e.g.*, the ^{18}F-*fluoro-Deoxy-Glucose* (FDG) uptake than the normal tissues and can be seen as dark areas in PET images (Fig. 6.2). However, the main challenges to automated lesion detection are low contrast between normal anatomical structures and lesions, and intersubject variations in tumoral FDG uptake

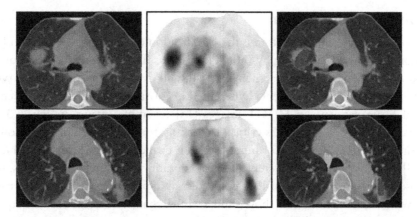

Figure 6.2 Examples of primary lung tumors and abnormal lymph nodes. Each row shows one example, using a transaxial slice view for easier visualization. The left column shows the CT slices, the middle column the PET slices, and the right column shows the lung tumors (red) and abnormal lymph nodes (orange). (For interpretation of the references to color in this figure legend, the reader is referred to the web version of this chapter.)

[1]. On CT, lesions usually appear similar to the soft tissues, which is problematic when lesions are located adjacent to the chest wall or mediastinal structures. For PET, although lesions are typically more FDG-avid than surrounding structures, some lesions have only mild FDG uptake; and there can be elevated FDG uptake in normal structures. Such low contrast implies that detecting lesions based on the within-subject information can be complicated. Furthermore, adding supervised information from other subjects might not be very useful as well, due to the inter-subject variation. Different subjects can display different ranges of FDG uptake, in both the normal anatomical structures and lesions. The separation criteria would thus vary between subjects.

Most lesion detection methods on PET-CT images are based on thresholding. The Standard Uptake Value (SUV), which is a semiquantitative measure of FDG uptake, is widely used to determine the threshold. Traditionally fixed SUV values have been used as the threshold [2,3]. More recently, adaptive threshold values have been proposed to better accommodate the subject- or region-level characteristics, such as the contrast-based threshold [4–7], and the iterative threshold [8,9]. However, the derived thresholds might not represent a suitable level of spatial scope, and this can affect the detection performance. Apart from the unsupervised SUV-based algorithms, other studies incorporate more complex features and apply classification techniques to detect lesions [10–12]. To avoid the intersubject variation, within-subject information are used as references for patch-wise labeling in a multiatlas model [13,1].

6.2.2 Disease categorization

An example of disease categorization is the classification of ILD tissue patterns in High-Resolution Computed Tomography (HRCT) images. ILD represents a group of more than 150 disorders of the lung parenchyma [14], and determining the specific type of disorder is important for treatment. Different ILDs normally exhibit different combinations of tissue patterns on HRCT images [15] (example images in Fig. 6.3), and differentiating the tissue patterns is critical to identify the actual type of ILD. However, interpreting the HRCT images for lung diseases is challenging even for trained radiologists [14]. Patients also have different physical conditions and medical histories, hence even those with the same type of ILD could display quite different tissue patterns and patients with different ILDs can show similar visual characteristics. As a consequence, automated classification is difficult due to the feature space complexity.

To achieve effective classification of tissue patterns, currently the main research focus is on visual feature extraction. Texture features are typically used, including second-order statistics such as the Gray-Level Co-occurrence Matrices (GLCM) and Gtay-Level Run Length Matrices (GLRLM) [17–19]. It is also common to use filter-based features, *e.g.*, Gaussian and wavelets [20–25], to highlight specific image features such as edges. More recent descriptors in the general imaging domain have also been incorporated

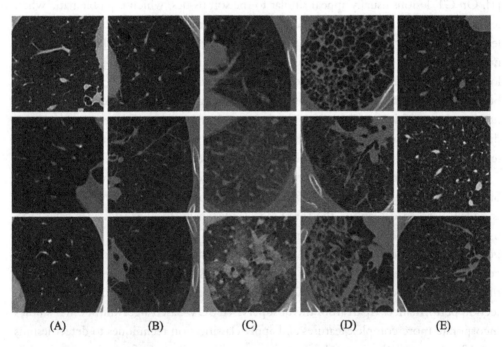

(A)   (B)   (C)   (D)   (E)

Figure 6.3 Three example images of each tissue category. From left to right: normal, emphysema, ground glass, fibrosis, and micronodules [16].

for lung imaging. Some examples include the Scale-Invariant Feature Transform (SIFT) [26,27], and Local Binary Patterns (LBP) [21,23]. Such feature descriptors are comprehensive in extracting small image details, and yet they provide multiresolution and histogram quantization properties that are especially useful for accommodating feature variations. Features that are customized for ILD images have also been designed, such as the locally-oriented Riesz components [22], Rotation-invariant Gabor-Local Binary Patterns (RGLBP) texture descriptor and Multicoordinate Histogram of Oriented Gradients (MCHOG) gradient descriptor [28]. Recently automated feature learning based on techniques such as the Restricted Boltzmann Machine (RMB) and Convolutional Neural Networks (CNN) has also been investigated in ILD classification [29–33]. Some approaches have shown promising results in slice-level disease categorization, without the need to delineate the region of interest required for image patch classification [33]. More detailed information of the various features can be found in Chapters 3 and 4.

6.2.3 Image retrieval

In the past three decades, the volume of medical image data has rapidly expanded. The vast amount of data opens an opportunity for case-based reasoning or evidence-based medicine support [35,36]. When a physician makes a diagnosis based on a patient scan, the physician may choose to browse through similar images in the database as a reference set to help reach the correct diagnosis. Fig. 6.4 gives an example of retrieving CT images of similar nodule types. The ability to find images with similar content is thus important to facilitate diagnosis.

Content-Based Image Retrieval (CBIR) methods support retrieval of similar images based on visual content or image features at a perceptual level. The majority of image retrieval studies have focused on feature extractions [37–44], designed for specific anatomical structures. Texture feature is one of the main feature types used in this application area [37,39,40,42,45–47]. Similarity between images is then obtained by comparing the visual features.

The retrieval performance is however often hindered by visual variations between images of similar categories and visual similarities between images of different categories [34]. In other words, images with similar diagnosis may show different patterns of anatomical structures; on the other hand, the irrelevant cases may show visually similar structures. Thus it is important to design a descriptive and discriminative feature descriptor so that only images with similar diagnosis or semantics will be retrieved. Further encoding of texture features into visual words has been proposed as one way to enhance the discriminative power of features to fill the semantic gap [42,48,49,34]. For example, images are first represented as a mixture of latent semantic topics in an image pair context, then the semantic correlation between image pairs is modeled based on canonical correlation analysis to generate the similarity scores [34]. Deep learning-based

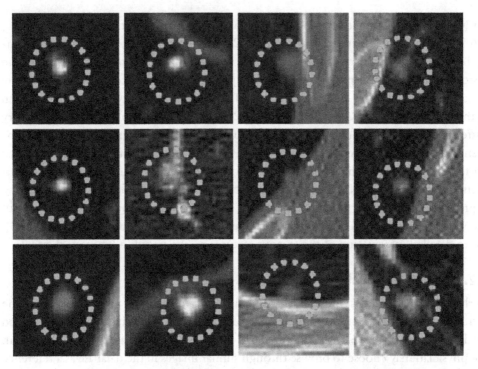

Figure 6.4 Example transaxial CT images [34] with typical nodules of four types. For left to right: well-circumscribed, vascularized, juxta-pleural, and pleural-tail. The nodules are circled in red. (For interpretation of the references to color in this figure legend, the reader is referred to the web version of this chapter.)

approach can potentially be used in this area as well. The readers can refer to Chapter 4 for more in-depth understanding of deep learning in the biomedical imaging domain.

6.3 REVIEW OF CLASSIFICATION METHODS

While currently most studies have focused on the feature design, as reviewed in the previous section, the design of classifiers is less studied and usually a standard classifier is used. For example, the most commonly used classifiers include k-nearest neighbor (kNN) [20,21,50,17,25], Support Vector Machine (SVM) [51,22,52,18,24,19,53], Linear Discriminant Analysis (LDA) [23,26], and Bayesian classifiers [54–56]. Due to the inevitable intraclass feature variation and interclass ambiguity, the classification performance would be enhanced by explicitly handling the feature space complexity in the classifier design. In this section, we review three types of classifiers that have this capability built in the classification model.

6.3.1 Ensemble classification

Classification based on an ensemble of classifiers has been quite popular in medical image analysis. The basic principle of ensemble classification is that by integrating multiple base classifiers, better classification performance would be obtained than using the individual base classifiers [57]. Most of the existing ensemble classification methods can be categorized as the bagging [58–64] or boosting [65–68] models. A major difference between bagging and boosting is that the base classifiers in bagging are normally trained independently while in boosting the successive base classifiers are influenced by the prior ones. In both models a single learning algorithm is used as the base classifier, including SVMs [64], decision or regression tree [66,58–63], logistic regression [65], and sparse representation [67,68]. Boosting algorithms can also be the building blocks in a tree structured classifier [69–71]. The choice of base classifier is motivated by the particular imaging application, for which the base classifiers need to provide diverse and good classification performance to enable more accurate classification by the ensemble. The outputs from base classifiers are then fused via some variants of weighted averaging. However, the weighting method is normally a simple pooling technique or uses a standard boosting algorithm, and might not adapt well to the specific data to be classified.

Generation of training subsets for the base classifiers is typically performed randomly or following a certain distribution. Random selection of feature subspace is also conducted with the random forest [72] models [58–63] to introduce further diversity in the base classifiers. The concept of representing multimodal classes with multiple pseudoclasses and classifying them with separation hyperplanes between the pseudoclasses is proposed [73]. Such a concept can be useful to classification problems with large intraclass variation. However, without a fusion component, the effectiveness of classification would be highly dependent on the individual separation hyperplanes. The undesirable effect of an averaging technique is illustrated in Fig. 2.6 of Chapter 2.

We can also consider the k-Nearest Neighbor (kNN) classifier as an ensemble model. With kNN, each data sample serves as a base classifier and majority voting from these base classifiers leads to the fused classification output. kNN is non-parametric and can naturally handle a large number of classes. Its effectiveness is, however, affected by the data distribution in the feature space. To improve the classification performance, discriminative learning has been incorporated. For example, in the SVM-KNN method [74], SVM-based classification is applied in the localized feature space determined by the nearest neighbors of the test data. SVM can also be applied first to identify data samples that are close to the decision boundary, based on which cluster centers are derived and kNN is used to classify the data [75]. Learning-based distance metric is also a popular trend, such as the Large Margin Nearest Neighbor (LMNN) method [76] that learns a Mahalanobis distance metric so that data from the same class would be more similar than those from different classes. Such a distance metric, however, is monolithic and could still be sensitive to the feature space complexity.

6.3.2 Subcategorization

As a means to tackle the feature space complexity, recently subcategorization classification models [77–81] have been proposed. Subcategorization models can be generally grouped into two types. In the first type [77–79], the clustering objective is integrated into the discriminative classifier, so that the classification is optimized based on the separation between the subcategories. The second type [80,81] clusters each class into subcategories, creates an individual classifier for each subcategory, and then fuses the subcategory-level classification outputs to obtain the final result. The advantage of the first type is that the clustering and classification objectives can be effectively integrated, but the multistage design of the second type is clearer and each stage can be customized in a modular manner.

Subcategorization models can be considered a type of ensemble learning since reference images are divided into subsets and a classification decision is made based on the subsets collectively. One of the main distinctions of subcategorization models is that the reference images of a certain class are usually divided into clusters based on some optimization criteria, not randomly. In addition, while most of the existing models have been built based on discriminative classifiers such as LDA and SVMs, subcategorization has recently been extended to the sparse representation classifier as well [82,83]. Notable performance has been reported in the field of lung image classification.

6.3.3 Sparse representation

Sparse representation can be considered similar to kNN that a test image is classified based on similar images in the reference set. The major difference from kNN is that sparse representation identifies similar images by sparse approximation of the test image from linear combination of reference images, while kNN uses direct distance computation. Sparse representation does not require a parametric model to characterize the feature space separation (*e.g.*, SVMs) and can thus be particularly effective to handle the feature space complexity. Classification using sparse representation has recently been applied in many medical imaging applications [84–87,28,88,89,1,68,82,67]. Sparse representation has also been applied in multiatlas models and the approximation coefficient is used as the weight vector to fuse the multiple atlases [90–92,1]. The weights derived in this way would be adaptive to the test data and normally provide more desirable results than using predefined weighting schemes.

The effectiveness of sparse representation is largely affected by the quality of the reference data [93]. In particular, if the reference set has large intraclass variation and interclass ambiguity, a diverse set of reference images could be selected during the sparse approximation and this could result in better approximation from the wrong class than the correct class. One way to address this issue is to adapt the reference set to the test image. For example, the reference dictionary can be rescaled to increase the feature distance between the test image and the wrong class [28]. The Locality-constrained Linear Coding (LLC) [94] has become widely popular in general computer vision. The essential idea is to encourage higher weights to be assigned to reference images that are more similar to the test image. The locality information in the feature space is thus exploited and the sparse approximation can be efficiently solved analytically. LLC has been directly applied for medical imaging and good performance has been demonstrated [95–97]. Another way is to partition the reference set into subsets and use sparse presentation as the base classifiers in a boosting [67,68] or subcategorization [82,83] model. The subsets are expected to have lower intraclass variation and interclass ambiguity, hence the sparse approximation at the subset-level could be more representative of the test image. The subcategorization method [82,83] is expected to provide higher classification performance than the boosting-based methods [67,68], with its clustering-based reference partition and learning-based large margin fusion of base classifiers. Here, large margin refers to the discriminative learning formulation, which aims to maximize the separation margin between difference classes. This is the concept used in the SVMs model. However, the large margin fusion is relatively complex and could be sensitive to the selection of training set.

6.4 SUBCATEGORY-BASED ENSEMBLE CLASSIFICATION

In this section, we present two subcategory-based ensemble classification methods, which are evaluated on the classification of HRCT image patches of five ILD tissue patterns. The first method, namely the Large Margin Local Estimate (LMLE) method [83], consists of 1) clustering the reference dictionary of each class into subcategories; 2) obtaining a local sparse representation of the test image from each reference subcategory; and 3) computing the class-level similarity by aggregating the local estimates with a large-margin supervised learning method. The second method, namely the Locality-constrained Subcluster Representation ensemble (LSRE) method [16], works by 1) partitioning the images into subclusters; 2) deriving the sparse representations of the test image from each subcluster; and 3) fusing the subcluster-level representation with various weights to obtain the image classification.

While both methods are based on subcategorization and sparse representation models, the essential differences between the two methods are as follows. First, in LMLE the subcategories are generated from the reference (training) images and subcategorization is performed for each image class separately. This means that all images in one subcategory would belong to the same class. On the other hand, in LSRE the subcategories are generated from all images regardless of the image class. Therefore a subcategory in

LSRE would contain a mixture of image classes and to avoid confusion, we name this subcategory as a subcluster. Second, the way of fusing the subcategory-level sparse representation is different. In LMLE, a discriminative learning model based on large margin distance metric is designed to derive the class label based on the aggregated distance between the test image and the various subcategory-level sparse representations. In LSRE, each subcluster-level sparse representation is treated as a base classifier and fusing of all subcluster-level representations is constructed based on statistically computed weights. In addition, our evaluation results show that LMLE provides more accurate classification than LSRE with its discriminative learning model, but LSRE is more computationally efficient with the analytical weight computation.

6.4.1 Large Margin Local Estimate (LMLE)

Our LMLE model is designed as follows. First, the reference dictionary of each class is clustered into subcategories using a weighted clustering method. Then, using each subcategory as a reference dictionary, a local estimate is generated for the test image with sparse representation. Next, the similarity between the test image and each class is computed by fusing the local estimates with a learning-based large margin aggregation method. Finally, the test image is classified according to its similarities with the various classes. Our hypothesis is that by embedding the sparse representation classifier in a subcategorization construct, our classifier can effectively address the issues of intraclass variation and interclass ambiguity. Fig. 6.5 shows the overall flow of the method.

We define the classification objective as finding the class label $\mathcal{L}(f)$ of a test image f, with $\mathcal{L}(f) \in \{1, ..., L\}$ and L is the number of classes. In a supervised setting, L sets of reference images $\{\boldsymbol{R}_l : l = 1, ..., L\}$ are given. Each reference set $\boldsymbol{R}_l$ is a matrix containing N_l feature vectors of reference images from class l. For notation simplicity, we denote both the reference image and its H-dimensional feature vector by $\boldsymbol{r}_l^i$ with $i = 1, ..., N_l$. Also for clear distinction, we use f to specifically refer to the test image and its feature vector, while $\boldsymbol{r}$ indicates the reference image. Assume feature vectors are precomputed for all images. The classification problem is to derive $\mathcal{L}(f)$ based on $\{\boldsymbol{R}_l\}$.

6.4.1.1 Reference subcategorization

In the first step the reference set $\boldsymbol{R}_l$ is subcategorized into K_l clusters/subcategories. We denote one subcategory as $\boldsymbol{R}_l^k \subset \boldsymbol{R}_l : k = 1, ..., K_l$. With subcategorization, each reference image $\boldsymbol{r}_l^i \in \boldsymbol{R}_l$ is assigned to one of the subcategories. Our subcategorization method is designed based on the k-means clustering technique.

Our design objective is that reference images with similar features and similar separations from the other classes should be clustered into the same subcategory. Therefore with two reference images $\boldsymbol{r}_l^i$ and $\boldsymbol{r}_l^j$, our objective can be defined based on two distance terms: (i) distance between feature vectors: $\|\boldsymbol{r}_l^i - \boldsymbol{r}_l^j\|$; and (ii) distance between feature separations: $\|\boldsymbol{d}_l^i - \boldsymbol{d}_l^j\|$. The feature separation $\boldsymbol{d}_l^i$ is an $L-1$ dimensional vector, in which

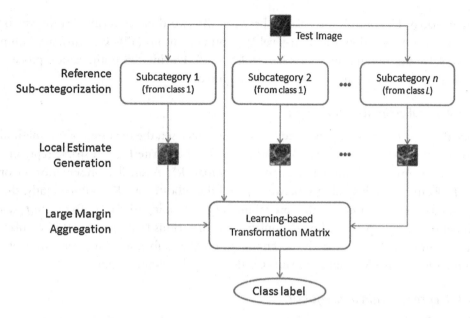

Figure 6.5 Overview of our LMLE model, which consists of reference sub-categorization (Section 6.4.1.1), local estimation generation (Section 6.4.1.2), and large margin aggregation (Section 6.4.1.3).

each element $d_l^i(l')$ is the mean Euclidean distance between r_l^i and the reference set $R_{l'}$, with $l' = 1, ..., L$ and $l' \neq l$.

The actual subcategorization objective is to minimize the within-subcategory distances, which is defined based on the two distance terms:

$$\underset{\{R_l^k\}_k}{\operatorname{argmin}} \sum_{k=1}^{K_l} \sum_{r_l^i \in R_l^k} (r_l^i - \mu_l^k)^T W_l (r_l^i - \mu_l^k)$$
$$+ (d_l^i - \theta_l^k)^T U_l (d_l^i - \theta_l^k), \tag{6.1}$$

where μ_l^k and θ_l^k represent the mean r_l^i and d_l^i of the subcategory R_l^k, and W_l and U_l are diagonal matrices containing weight factors. This formulation is equivalent to

$$\underset{\{R_l^k\}_k}{\operatorname{argmin}} \sum_{k=1}^{K_l} \sum_{r_l^i \in R_l^k} (x_l^i - \phi_l^k)^T Q_l (x_l^i - \phi_l^k), \tag{6.2}$$

where $x_l^i \in \mathbb{R}^{H+L-1}$ is the concatenation of r_l^i and d_l^i, $\phi_l^k \in \mathbb{R}^{H+L-1}$ is the concatenation of μ_l^k and θ_l^k, and $Q_l \in \mathbb{R}^{(H+L-1) \times (H+L-1)}$ represents the diagonal matrix combining W_l and U_l. The weight matrix W_l represents the variable weights and it is class-specific. The weight matrix U_l controls the balance between the two distance terms (feature vector and separation). We adopt the Two-level Weighting k-means (TW-k-means) algorithm [98] to solve this minimization problem in Eq. (6.2). During this optimization process, the weights W_l and U_l are automatically computed.

6.4.1.2 Local estimate generation

Once the subcategories are generated for each reference set, the next step of our method is to compute the local estimates. We define the local estimate $f_{l,k}$ as a sparse representation of an image f from a reference subcategory R_l^k. A small representation error $\|f - f_{l,k}\|^2$ means high similarity between f and the subcategory R_l^k. Subsequently, this high similarity would imply a high probability of f belonging to class l. To obtain $f_{l,k}$, a number of existing sparse coding formulations with various ℓ_1 [99] or ℓ_0 [100] regularizations can be used. In our study, we choose to use the ℓ_0 formulation for its simplicity, and the Orthogonal Matching Pursuit (OMP) [100] algorithm is applied.

6.4.1.3 Large margin aggregation

With the local estimates, the final step is to classify the test image f. This can be done by aggregating the similarity $S(f, l, R_l)$ between the test image f and each reference set R_l based on the local estimates $\{f_{l,k} : k = 1, ..., K_l\}$. The similarity value gives a measure of the probability of f belonging to class l. If we can obtain a large similarity from the correct class and small similarities from the wrong classes, the test image would be classified correctly. However, with interclass ambiguity, local estimates from the wrong class could appear closer to the test image than local estimates from the correct class, and such confusions would affect the classification result.

To overcome this issue, we designed a large margin aggregation algorithm with a learning-based transformation matrix. Our hypothesis is that by transforming the feature space of the test image and local estimates, the test image would become closer to the local estimates of the correct class and more different from those of the wrong classes.

Specifically, we first define the similarity between the test image f and reference set R_l as

$$S(f, l, R_l) = \frac{1}{\{1 + D(f, l, R_l)\}}, \tag{6.3}$$

where $D(f, l, R_l)$ is the distance between f and R_l. The distance is computed as the accumulated transformed Euclidean distance between f and M local estimates that are the closest to f:

$$D(f, l, R_l) = \sum_{m=1}^{M} \|F_l f - F_l f_{l,m}\|^2. \tag{6.4}$$

Here, $F_l \in \mathbb{R}^{H \times H}$ is the learned transformation matrix and is specific to class l. The selection of M closest local estimates $\{f_{l,m} : m = 1, ..., M\}$ is based on the Euclidean distance between the transformed vectors $F_l f$ and $F_l f_{l,k}$. Assume that the class label of f is l. We expect that ideally $D(f, l, R_l) < D(f, l, \{R_{l'} : l' = 1, ..., L, l' \neq l\})$, which is

$$\sum_{m=1}^{M} \|F_l f - F_l f_{l,m}\|^2 < \sum_{m'=1}^{M} \|F_l f - F_l f_{l',m'}\|^2, \tag{6.5}$$

where m' is the index of the M closest local estimates $f_{l',m'}$ from all the wrong classes $l' \neq l$.

The unknown variable in Eqs. (6.4) and (6.5) is the transformation matrix F_l. To learn F_l in a supervised manner, we first gather a set of I training samples from the reference set R_l, which are denoted as $\{R_l^i : i = 1, ..., I\}$. For a certain training sample R_l^i, the closest local estimates are obtained based on the representation error. We then denote the mth closest local estimates from the correct class as $R_{l,m}^i$, and the m'th closest local estimate from the wrong class as $R_{l',m'}^i$.

Next, our goals of training are defined as: (i) to minimize the total distance between the transformed feature $F_l R_l^i$ and M closest local estimates $\{F_l R_{l,m}^i\}$ from the correct class; and (ii) to impose a large margin constraint so that the transformed local estimate from the wrong classes is one unit further away than that from the correct class, for all pairs of $F_l R_{l,m}^i$ and $F_l R_{l',m'}^i$. Formally, the training objective is formulated as

$$\operatorname*{argmin}_{F_l} \sum_{i=1}^{I} \sum_{m=1}^{M} \|F_l(R_l^i - R_{l,m}^i)\|^2 + $$
$$\sum_{i=1}^{I} \sum_{m=1}^{M} \sum_{m'=1}^{M} [1 + \|F_l(R_l^i - R_{l,m}^i)\|^2 - \|F_l(R_l^i - R_{l',m'}^i)\|^2]_+, \tag{6.6}$$

where $[z]_+ = \max(0, z)$ is the standard hinge loss. With this optimization objective the transformed feature $F_l R_l^i$ is expected to become more similar to $\{F_l R_{l,m}^i : m = 1, ..., M\}$ than $\{F_l R_{l',m'}^i : m' = 1, ..., M\}$. Subsequently, the transformed feature and local estimates would lead to accurate classification of class l. Eq. (6.6) is solved following the semidefinite programming model [76] to obtain the transformation matrix F_l for each class.

With the learned transformation matrices, the test image f is classified based on its similarities with the reference sets $\{R_l\}$. This classification process is defined as

$$\mathcal{L}(f) = \operatorname*{argmax}_{l} \{ \max_{l'=1,...,L} S(f, l', R_l) \}, \tag{6.7}$$

with $l = 1, ..., L$.

6.4.2 Locally-constrained subcluster representation ensemble

The LSRE method contains three components as shown in Fig. 6.6. First, the images are partitioned hierarchically into subclusters based on spectral clustering with a sparse representation-based affinity matrix. Next, basis representations of the test image are obtained by deriving the sparse representation of the test image using each subcluster as the reference dictionary. Finally, the basis representations are fused based on approximation- and distribution-based weights and the class label is determined accordingly. By using subcategories/subclusters, we expect to discover localized regions in the feature space and capture the intraclass variation and interclass ambiguity explicitly. Subsequently, we expect the ensemble of the basis representations would provide more accurate classification than using a single classifier obtained from the entire reference set.

The classification method is formally defined as follows. Given a set of N images (i.e., HRCT image patches), each with an H-dimensional feature vector $\boldsymbol{x}_i \in \mathbb{R}^H$ and $\mathcal{X} = \{x_i : i = 1, ..., N\}$, our aim is to determine the class label of each image in a multi-class classification setting.

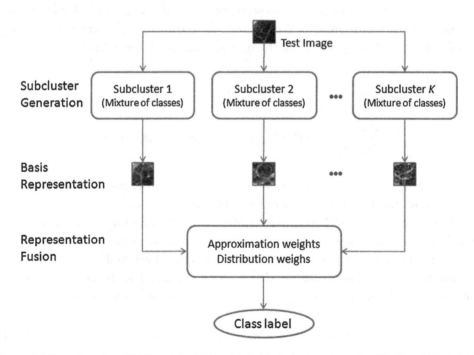

Figure 6.6 Overview of our LSRE model, which consists of subcluster generation (Section 6.4.2.1), basis representation (Section 6.4.2.2), and representation fusion (Section 6.4.2.3).

6.4.2.1 Subcluster generation

The first step of our LSRE method is to divide the full image set $\mathcal{X}$ into a union of K subclusters $\{S_k : k = 1, ..., K\}$. We expect to minimize the within-subcluster feature variation during this process, and we design a hierarchical spectral clustering method for this aim.

With spectral clustering, assume an affinity matrix $A \in \mathbb{R}^{N \times N}$ is defined, in which the (i, j)th element A_{ij} indicates the similarity between images x_i and x_j. $\mathcal{X}$ is then divided into clusters by applying the normalized cuts algorithm on A [101]. The unknown variable here is the affinity matrix A. Similar to the Sparse Subspace Clustering (SSC) method [102], we design a sparse representation-based method to derive the affinity matrix A. Different from SSC, we use LLC as the sparse coding algorithm, which is highly efficient with an analytical solution. The computational cost of our clustering method thus becomes much lower than SSC. Specifically, LLC is applied to get a sparse representation of each feature vector x_i in $\mathcal{X}$ using all other vectors as the reference data. The sparse coefficient z_i then indicates the degree of similarity between the feature vector x_i and the other data. By concatenating the coefficient vectors derived for all images $\{x_i : i = 1, ..., N\}$, we obtain the coefficient matrix $Z \in \mathbb{R}^{N \times N}$. The affinity matrix A is then computed as $(|Z| + |Z^T|)/2$, to have symmetric measure of similarities between image pairs.

To generate K subclusters, we apply this spectral clustering approach in a hierarchical manner. At the first level, the affinity matrix A is computed for the entire image set $\mathcal{X}$ with the number of clusters the same as the number of classes in the dataset. At the subsequent level $t > 1$, a cluster from the previous level, indexed by k_{t-1}, is subclustered by computing the affinity matrix for the images within this cluster. The number of subclusters is set to $\lfloor N_{k_{t-1}}/(\eta C_1/t) \rfloor$, where $N_{k_{t-1}}$ denotes the number of images in cluster k_{t-1} and η is a scaling constant. With T levels of hierarchy, a total of K subclusters $\{S_k : k = 1, ..., K\}$ are finally generated at the last level T.

6.4.2.2 Basis representation

With the subclusters generated, the second step is to obtain the basis representations of the test image x based on the subclusters $\{S_k : k = 1, ..., K\}$. The basis representation helps to approximate x by using each subcluster S_k as a reference dictionary, and can be considered as a base classifier for x.

Formally, given a test image x, we derive the sparse representation of x from each subcluster S_k. This is done by first constructing the reference dictionary matrix $\bar{S}_{x,k} \in \mathbb{R}^{H \times N_{x,k}}$, concatenating the $N_{x,k}$ feature vectors of images in S_k that are from different subjects as x. Images from the same subject as x are excluded to ensure complete separation between the test and reference data. With this reference dictionary, LLC is then used to derive the sparse coefficient vector $p_{x,k} \in \mathbb{R}^{N_{x,k}}$ as the basis representation of x from the subcluster S_k. The approximation output of x is $\bar{S}_{x,k}p_{x,k}$ from the subcluster S_k.

We can consider the basis representation $p_{x,k}$ as a base classifier, since x can be classified by finding the class that is assigned the highest total weight in the basis representation:

$$\underset{l}{\mathrm{argmax}} \ p_{x,k}^T I_{x,k,l} \qquad (6.8)$$

where l denotes the class label. $I_{x,k,l} \in \mathbb{R}^{N_{x,k}}$ is a vector indicating the indices of reference images of class l. An element 1 in $I_{x,k,l}$ denotes that the reference image of that index is of class l, and an element 0 indicates otherwise.

6.4.2.3 Representation fusion

After the basis representations $\{p_{x,k} : k = 1, ..., K\}$ are obtained, the final step of our method is to fuse the basis representations to classify the test image x. An intuitive way is to generate a weighted fusion of the outputs from the base classifiers as the final classification result. The weight variables thus become the key component in such an algorithm. Our idea is that by approximating the test image x from the basis representations, the approximation coefficient can be used as the weights. We also restrict the approximation to be sparse so that only the top related basis representations would contribute to the fusion. On the other hand, while some basis representations might produce very good approximation of x, the corresponding labels derived using Eq. (6.8) might actually be incorrect. To address this issue, we note that subclusters containing a balanced mixture of image classes are usually less discriminative or reliable than subclusters with images of mainly a single class. We thus design a distribution-based weight vector to represent this reliability measure.

To generate the approximation-based weights, we formulate the following objective function to approximate x:

$$\min_{w_x} \|x - V_x w_x\|^2 + \lambda \|d_x \odot w_x\|^2 \qquad (6.9)$$
$$s.t. \ \mathbf{1}^T w_x = 1, \ \|w_x\|_0 \leq C$$

Here, the reference dictionary $V_x \in \mathbb{R}^{H \times K}$ is constructed by concatenating all approximations of x from the K basis representations: $V_x = \{\bar{S}_{x,k}p_{x,k} : k = 1, ..., K\}$. The vector $d_x \in \mathbb{R}^K$ contains the Euclidean distances between the each approximation in V_x and x. This $\|d_x \odot w_x\|^2$ term encourages to assign lower weights to basis representations that are more different from x. The coefficient $w_x \in \mathbb{R}^K$ is obtained analytically using LLC, and it represents the weights of fusion. The objective function helps to encourage a good approximation of x while penalizing higher weights for basis representations that are more different from x. Therefore we expect to obtain higher weights for basis representations that are more related to x, and only the top C similar approximations of x would contain nonzero weights.

The distribution-based weight, $u_{x,l}$, is defined to estimate the reliability of the basis representations $\{p_{x,k} : k = 1, ..., K\}$ in classifying x as class l. To design this weight vector, first we note that a subcluster S_k normally contains images from a variety of classes. If images in S_k are mostly from a certain class l, the classification output using $p_{x,k}$ would be quite reliable without the influence from images of other classes. On the other hand, if images in S_k are equally distributed among all the L number of classes, S_k would represent a localized region in the feature space that different classes are indistinguishable. The reliability of classification using such a basis representation would then be lower. To this end, we define the weight as

$$u_{x,k,l} = (\mathbf{1}^T I_{x,k,l}/N_{x,k}) \log(1 + \sigma\{\mathbf{1}^T I_{x,k,l}\}_{l=1}^L) \tag{6.10}$$

Here, $(\mathbf{1}^T I_{x,k,l}/N_{x,k})$ computes the percentage of class l images in $\bar{S}_{x,k}$, and $\log(\cdot)$ gives a log representation of the standard deviation σ of numbers of images belonging to the various classes. A larger percentage or standard deviation would lead to a higher reliability of the basis representation $p_{x,k}$.

Finally, the classification probabilities are computed based on the fusion weights. We define the classification probability of x belonging to class l as

$$P_{x,l} = \sum_{k=1}^K p_{x,k}^T I_{x,k,l} w_{x,k} u_{x,k,l} \tag{6.11}$$

Here, $p_{x,k}^T I_{x,k,l}$ is the classification probability when the basis representation $p_{x,k}$ is used as a base classifier. The weighted fusion of K probabilities is then computed based on the approximation- and distribution-based weights $w_{x,k}$ and $u_{x,k,l}$. The test image x is classified to the class with the highest probability: $\mathrm{argmax}_l P_{x,l}$.

6.5 EXPERIMENTS

6.5.1 Dataset and implementation

We used the ILD database [15] in this study. The database contains 113 HRCT images, and altogether 2062 2D Regions-Of-Interest (ROI) manually annotated with 17 ILD tissue class. The annotation was performed by two radiologists with 15 and 20 years of experience. Following the setup in [15,24,28,68], we selected the ROIs belonging to five major ILD tissue classes: NorMal (NM), EMphysema (EM), Ground Glass (GG), FiBrosis (FB), and MicroNodules (MN). We divided the axial slices into a grid of half-overlapping image patches with 31×31 pixels. The image patches with centroids inside the annotated ROIs were included in our experimentation. Our dataset thus comprised a total of 23131 image patches from 93 HRCT images/subjects, with 6438 NM,

Table 6.1 Number of image patches and subjects of the five tissue classes in the ILD database

Class	NM	EM	GG	FB	MN
# image patches	6438	1474	2974	4396	7849
# subjects	12	5	35	35	16

1474 EM, 2974 GG, 4396 FB, and 7849 MN image patches. Table 6.1 summarizes the dataset. Note that some subjects contain more than one tissue type.

Each image patch was thus an "image" to be classified to one of the five ILD tissue classes ($Y = 5$). The Texture-Intensity-Gradient (TIG) feature vector [28] was used to describe each image. The feature vector is 176-dimensional ($H = 176$) containing three types of information: rotation–invariant local binary patterns based on Gabor-filtered images, intensity histogram, and histogram of oriented gradients with multiple coordinates. This feature vector was specifically designed for the ILD classification problem and showed good performance previously [28,68,82]. To use a consistent test setup with our previous studies [68,82] for convenient performance comparison, we divided the dataset sequentially into four subsets of similar numbers of subjects and a leave-one-subject-out testing scheme was then performed for each subset. Note that due to the small number of subjects of the EM class, three of the five subjects were duplicated so that each subset contained two EM subjects.

To apply the LMLE method, we found empirically that if using the entire reference set as the reference dictionary (without subcategorization), good sparse representations could be obtained by combining a minimum of 10 reference images. We thus decided that the number of reference images in a subcategory should be large enough that at least 10 of them could be sparsely selected. Consequently, we chose to set the number of subcategories $K_l = \lfloor N_l/40 \rfloor$ in Eq. (6.2), so that a subcategory would contain around 40 images. We preferred not to use a number larger than 40 since we also required a sufficient number of subcategories to fuse the local estimates. For large margin aggregation, various settings of the number of closest local estimates M were experimented, and $M = 5$ was found to provide the best classification results.

For LSRE, we experimented with various possible parameter settings on each subset. The set of values that provided good classification for all subsets was then used as the best parameters to evaluate our method performance. Our design choice mainly involved the ranges of possible parameter settings. In particular, for λ in Eq. (6.9), we experimented with $\lambda = 10^{-4}$ to 10^{-1} and set λ to 10^{-2}. The sparsity constants C was related to sparse approximation of image features, and we found that with the standard sparse representation classifier, a sparsity constant of 10 provided the best performance. The number of levels of subclustering T (Section 6.4.2.1) is set to 6, and the scaling factor η to determine the number of clusters is set to 20.

It is worth mentioning that the selection of the five tissue classes was motivated by the fact that they were the most common tissue patterns in ILDs [15] and the existing studies for this database focused on these five tissue classes. We adopted the same aim of study so that we could compare with the state-of-the-art directly. The remaining twelve tissue classes include consolidation (12 subjects), bronchial wall thickening (1 subject), reticulation (10 subjects), macronodules (5 subjects), cysts (1 subject), peripheral micronodules (5 subjects), bronchiectasis (5 subjects), air trapping (1 subject), early fibrosis (1 subject), increased attenuation (2 subjects), tuberculosis (1 subject), and pcp (2 subjects). It would be interesting to see how our method would extend to more classes, especially consolidation and reticulation.

6.5.2 Results of patch classification

Tables 6.2 and 6.3 show the confusion matrices of the classification results using LMLE and LSRE, respectively. Most of the tissue classes obtained higher than 80% classification rates with both methods. Also, in both methods, a large number of EM images were misclassified as NM, while only a small amount of EM images were misclassified as the other three classes. This can be explained by the large interclass ambiguity, *i.e.*, visual similarity, between the EM and NM images, as shown in Fig. 6.3. The NM images exhibited high similarity with both EM and MN images, hence the misclassification of NM images was mainly among the EM and MN images. Another observation is that GG, FB, and MN images were rarely misclassified as EM. This could be explained by the small number of EM images compared to the other classes. The small number

Table 6.2 Confusion matrix of ILD classification using LMLE

Ground Truth	Prediction				
	NM	EM	GG	FB	MN
NM	**0.861**	0.045	0.028	0.013	0.053
EM	0.129	**0.801**	0.005	0.065	0.000
GG	0.076	0.000	**0.830**	0.043	0.052
FB	0.005	0.014	0.059	**0.874**	0.048
MN	0.030	0.000	0.058	0.036	**0.877**

Table 6.3 Confusion matrix of ILD classification using LSRE

Ground Truth	Prediction				
	NM	EM	GG	FB	MN
NM	**0.885**	0.045	0.010	0.007	0.054
EM	0.182	**0.796**	0.022	0.000	0.000
GG	0.069	0.000	**0.800**	0.068	0.064
FB	0.007	0.028	0.059	**0.854**	0.053
MN	0.034	0.000	0.046	0.048	**0.872**

Table 6.4 Classification recall, precision, and F-score of each tissue type, and the average results, using LMLE

	NM	EM	GG	FB	MN	Avg
Recall (%)	86.1	80.1	83.0	87.4	87.7	86.1
Precision (%)	89.3	73.8	76.3	86.9	90.0	86.4
F-score (%)	87.7	76.8	79.5	87.2	88.8	86.2

Table 6.5 Classification recall, precision, and F-score of each tissue type, and the average results, using LSRE

	NM	EM	GG	FB	MN	Avg
Recall (%)	88.5	79.6	80.0	85.4	87.2	85.8
Precision (%)	89.1	70.0	79.0	85.2	89.3	85.9
F-score (%)	88.7	74.5	79.5	85.3	88.3	85.8

implies a lower probability of EM images selected for basis representation and a lower distribution-based weight of the EM class, and hence a lower probability of labeling the other classes as EM. Tables 6.4 and 6.5 summarize the classification recall, precision and F-score of each tissue class using LMLE and LSRE, respectively. Overall, the results show relatively balanced performance among the different tissue classes. Note that while the rates of misclassifying the other classes as EM were low, the precision of EM was low affected by the small number of EM images.

The performances of the LMLE and LSRE models were compared with the other existing methods reported for ILD classification: (i) Localized Features (LF) [24], which used localized features with the SVM classifier; (ii) Patch-Adaptive Sparse Approximation (PASA) [28], which was based on sparse representation with reference adaptation; and (iii) Boosted Multifold Sparse Representation (BMSR) [68], which was based on sparse representation in an AdaBoost construct. The latter two approaches used the same TIG feature vector as in LMLE and LSRE, hence the comparison with them demonstrated the effect of the subcategory-based ensemble classification methods. The comparison with LF reflected the performance difference of the overall framework. We note that the results of LF were obtained directly from the paper [24], which used a slightly different selection of images from our dataset. For a fair comparison, PASA and BMSR were rerun to follow the same leave-one-subject-out test setup as this study; and the parameter settings followed those reported in these studies.

Fig. 6.7 shows the classification recall and precision of our LMLE and LSRE models and the compared approaches. Overall, LMLE and LSRE show comparable results, and they achieved more balanced results among the five tissue types compared to BMSR and PASA. Recall that LMLE uses large margin learning for fusion of base classifiers. The comparable results of LSRE suggests that while LSRE does not involve discriminative learning, it could actually obtain high classification performance with the subcluster-

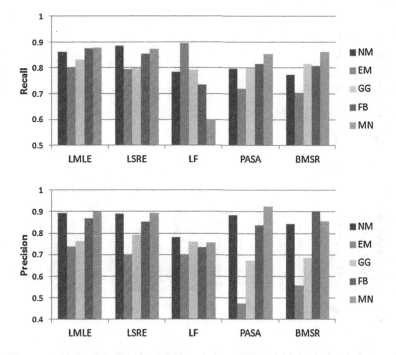

Figure 6.7 Classification recall and precision of the various methods reported for ILD classification.

based representation and fusion. The subclusters generated by LSRE can be used as base classifiers with sparse representation and a relatively uncomplicated fusion algorithm is required. On the other hand, LMLE is based on subcategorization of individual classes separately; and the subcategories would not provide discriminative information between classes. The fusion component, *i.e.*, the large margin learning of transformation matrices, is thus particularly important for classification. It can be seen that LMLE provided highest recall for four tissue types including EM, GG, FB, and MN, and the highest precision for NM and EM, among all five compared approaches.

BMSR is also an ensemble classifier can be considered closely related to LSRE. However, its subclusters are created with random partition and the number of subclusters is much smaller (about 3% of that in LSRE); its base classifier is the ℓ_0 regularized sparse representation; and the fusion of base classifiers is based on a boosting algorithm. The improvement of LSRE over BMSR thus demonstrates the advantage of the overall ensemble method design. PASA is similar to the standard sparse representation algorithm, but involves adaptation of the reference data to the test images. The advantage over PASA implies that our ensemble models LMLE and LSRE were more effective than using a global sparse representation classifier. Finally, our methods are completely different from LF, which uses a different feature set and the SVM classifier. This comparison could however be biased since we used a different subset of the ILD database

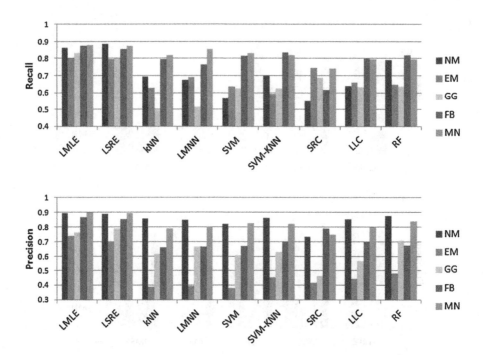

Figure 6.8 Classification recall and precision of our LMLE and LSRE models and the other popular classifiers.

from LF. In particular, the two datasets contained different numbers of images but had very similar distributions of images among the five tissue classes. We would thus like to refer the readers to Fig. 6.8 for comparison between SVM and LSRE, and show the comparison with LF as a general view of our method performance in the area of ILD tissue classification.

We also compared with the standard classifiers that are popular in medical imaging, including the kNN based on Euclidean distance, LMNN, SVM, SVM-KNN, Sparse Representation Classifier (SRC) based on ℓ_0 regularization, the LLC model, and the Random Forest (RF) classifier. For all these approaches, we used the same TIG feature vector and leave-one-subject-out test setup as our LSRE model. The best performing parameters were set for each classifier: three nearest neighbors for kNN and LMNN, polynomial kernel for SVM and SVM-KNN (order of 3 and regularization parameter $C = 1.6$), 50 nearest neighbors for SVM-KNN, 10-sparsity for both SRC and LLC, the balancing parameter for LLC as $1e - 2$, and 150 trees for RF.

As shown in Fig. 6.8, LMLE and LSRE achieved large improvement over the compared approaches. The kNN and LMNN classifiers are nearest neighbor-based methods. The kNN classifier did not work well since an image could appear similar to images of different classes, due to large interclass ambiguity. While LMNN includes a learning-based distance metric, the learning algorithm was monolithic and affected by the large number of contradicting constraints. The large margin learning in LMLE has a mathematically similar formulation to LMNN, but the underlying concepts are different. In particular, when applying LMNN to classify the images, the distances are computed between the reference image and its neighboring images from the same and different classes. In LMLE, the distances are computed between the reference image and its local estimates, and there is no nearest-neighbor relationship between the reference images. The advantage of LMLE over LMNN indicates the benefit of our subcategory-based model and similarity computation based on subcategory-level representations.

SRC and LLC are sparse representation models. SRC did not gain advantage over kNN, mainly due to the feature space complexity causing selection of reference images from the wrong class for sparse approximation. LLC performed better than kNN and SRC, implying that incorporating the locality constraints into the approximation objective helped to accommodate the feature space complexity. The main difference between LSRE and LLC is that we used LLC at the subcluster-level to compute the basis representations and the LLC-based base classifiers are fused to obtain the final classification. The advantage of LSRE over LLC thus implies that our ensemble classifier was more effective than using a single LLC based on the entire reference set.

RF is a popular ensemble classifier based on tree bagging. Different from LSRE, RF creates the subclusters based on random sampling, uses decision tree as the base classifiers, and applies majority voting to obtain the fused classification output. RF also involves the additional random selection of feature subsets for further improvement. LSRE outperformed RF largely, and we suggest the essential cause was that the subclusters in our model were generated by clustering. The base classifiers built on these subclusters adapted to the local distributions in the feature space, and fusing these local classifiers could provide more accurate results than those created based on random subsets.

Among the compared approaches, SVM is the most different classifier from our LMLE and LSRE models. While SVM is typically a highly discriminative classifier, with large intraclass variation and interclass ambiguity, it could become overfitted to the training data. The overfitting problem could be partially addressed by training local classifiers based on the nearest neighbors of the test data as in SVM-KNN, which obtained higher performance over the standard SVM. Classification with SVM-KNN, however, involved a single local classifier. The advantage of LMLE and LSRE over SVM-KNN thus indicates the benefit of using an ensemble of base classifiers.

We would like to mention that the drawback of our LMLE and LSRE methods are the relative complexity. The ensemble classifier design is notably more complicated than kNN and SVM, and the global sparse representation-based classifiers (SRC, LLC, and PASA). On the other hand, BMSR, LMLE, and RF are also ensemble classifiers, which however provided lower classification performance than LMLE and LSRE.

6.6 CONCLUSIONS

Texture classification is a key component in many medical imaging applications, such as lesion detection, disease categorization, and image retrieval. While many different types of texture descriptors have been designed over the years, the discriminative power of feature descriptors is still not sufficient to warrant a highly accurate classification of image textures. This is mainly because medical images typically exhibit varying visual patterns in images of the same classes but similar characteristics in images of different classes. Subsequently, the feature space also exhibits large intraclass variation and interclass ambiguity, and this lowers the discriminative capability of the features. We suggest that to tackle this issue, an ensemble classification method based on subcategorization modeling is more effective than the commonly used classifiers. We applied the classification method to tissue pattern classification of interstitial lung disease, and the results show good improvement compared to the benchmark approaches and the other popular classifiers.

REFERENCES

[1] Y. Song, W. Cai, H. Huang, X. Wang, Y. Zhou, M. Fulham, D. Feng, Lesion detection and characterization with context driven approximation in thoracic FDG PET-CT images of NSCLC studies, IEEE Trans. Med. Imaging 33 (2) (2014) 408–421.

[2] Y. Hashimoto, T. Tsujikawa, C. Kondo, M. Maki, M. Momose, A. Nagai, T. Ohnuki, T. Nishikawa, K. Kusakabe, Accuracy of PET for diagnosis of solid pulmonary lesions with ^{18}F-FDG uptake below the standardized uptake value of 2.5, J. Nucl. Med. 47 (2006) 426–431.

[3] H. Guan, T. Kubota, X. Huang, X.S. Zhou, M. Turk, Automatic hot spot detection and segmentation in whole body FDG-PET images, in: Proc. ICIP, 2006, pp. 85–88.

[4] Y. Song, W. Cai, S. Eberl, M.J. Fulham, D. Feng, Automatic detection of lung tumor and abnormal regional lymph nodes in PET-CT images, J. Nucl. Med. 52 (Supplement 1) (2011) 211.

[5] Y. Song, W. Cai, S. Eberl, M.J. Fulham, D. Feng, Discriminative pathological context detection in thoracic images based on multi-level inference, in: Proc. MICCAI, vol. 6893, 2011, pp. 185–192.

[6] Y. Song, W. Cai, D.D. Feng, Global context inference for adaptive abnormality detection in PET-CT images, in: Proc. ISBI, 2012, pp. 482–485.

[7] H. Cui, X. Wang, D. Feng, Automated localization and segmentation of lung tumor from PET-CT thorax volumes based on image feature analysis, in: Proc. EMBC, 2012, pp. 5384–5387.

[8] H. Ying, F. Zhou, A.F. Shields, O. Muzik, D. Wu, E.I. Heath, A novel computerized approach to enhancing lung tumor detection in whole-body PET images, in: Proc. EMBC, 2004, pp. 1589–1592.

[9] N. Zsoter, P. Bandi, G. Szabo, Z. Toth, R.A. Bundschuh, J. Dinges, L. Papp, PET-CT based automated lung nodule detection, in: Proc. EMBC, 2012, pp. 4974–4977.

[10] G.V. Saradhi, G. Gopalakrishnan, A.S. Roy, R. Mullick, R. Manjeshwar, K. Thielemans, U. Patil, A framework for automated tumor detection in thoracic FDG PET images using texture-based features, in: Proc. ISBI, 2009, pp. 97–100.

[11] M.S. Sharif, M. Abbod, A. Amira, H. Zaidi, Artificial neural network-based system for PET volume segmentation, Int. J. Biomed. Imaging 2010 (2010) 1–11.

[12] C. Lartizien, S. Marache-Francisco, R. Prost, Automatic detection of lung and liver lesions in 3-D positron emission tomography images: a pilot study, IEEE Trans. Nucl. Sci. 59 (1) (2012) 102–112.

[13] Y. Song, W. Cai, H. Huang, X. Wang, S. Eberl, M.J. Fulham, D. Feng, Similarity guided feature labeling for lesion detection, in: Proc. MICCAI, vol. 8149, 2013, pp. 284–291.

[14] W.R. Webb, N.L. Muller, D.P. Naidich, High-Resolution CT of the Lung, Lippincott Williams Wilkins, 2008.

[15] A. Depeursinge, A. Vargas, A. Platon, A. Geissbuhler, P.A. Poletti, H. Muller, Building a reference multimedia database for interstitial lung diseases, Comput. Med. Imaging Graph. 36 (3) (2012) 227–238.

[16] Y. Song, W. Cai, H. Huang, Y. Zhou, Y. Wang, D. Feng, Locality-constrained subcluster representation ensemble for lung image classification, Med. Image Anal. 22 (1) (2015) 102–113.

[17] P.D. Korfiatis, A.N. Karahaliou, A.D. Kazantzi, C. Kalogeropoulou, L.I. Costaridou, Texture-based identification and characterization of interstitial pneumonia patterns in lung multidetector CT, IEEE Trans. Inf. Technol. Biomed. 14 (3) (2010) 675–680.

[18] J. Yao, A. Dwyer, R.M. Summers, D.J. Mollura, Computer-aided diagnosis of pulmonary infections using texture analysis and support vector machine classification, Acad. Radiol. 18 (3) (2011) 306–314.

[19] U. Bagci, J. Yao, A. Wu, J. Caban, T.N. Palmore, A.F. Suffredini, O. Aras, D.J. Mollura, Automatic detection and quantification of tree-in-bud (TIB) opacities from CT scans, IEEE Trans. Biomed. Eng. 59 (6) (2012) 1620–1632.

[20] I.C. Sluimer, M. Prokop, I. Hartmann, B. van Ginneken, Automated classification of hyperlucency, fibrosis, ground glass, solid, and focal lesions in high-resolution CT of the lung, Med. Phys. 33 (7) (2006) 2610–2620.

[21] L. Sorensen, S.B. Shaker, M. de Bruijne, Quantitative analysis of pulmonary emphysema using local binary patterns, IEEE Trans. Med. Imaging 29 (2) (2010) 559–569.

[22] A. Depeursinge, A. Foncubierta-Rodriguez, D.V. de Ville, H. Muller, Lung texture classification using locally-oriented Riesz components, in: Proc. MICCAI, vol. 6893, 2011, pp. 231–238.

[23] C. Jacobs, C.I. Sanchez, S.C. Saur, T. Twellmann, P.A. de Jong, B. van Ginneken, Computer-aided detection of ground glass nodules in thoracic CT images using shape, intensity and context features, in: Proc. MICCAI, vol. 6893, 2011, pp. 207–214.

[24] A. Depeursinge, D.V. de Ville, A. Platon, A. Geissbuhler, P.A. Poletti, H. Muller, Near-affine-invariant texture learning for lung tissue analysis using isotropic wavelet frames, IEEE Trans. Inf. Technol. Biomed. 16 (4) (2012) 665–675.

[25] L. Sorensen, M. Nielsen, P. Lo, H. Ashraf, J.H. Pedersen, M. de Bruijne, Texture-based analysis of COPD: a data-driven approach, IEEE Trans. Med. Imaging 31 (1) (2012) 70–78.

[26] A. Farag, S. Elhabian, J. Graham, A. Farag, R. Falk, Toward precise pulmonary nodule descriptors for nodule type classification, in: Proc. MICCAI, vol. 6363, 2010, pp. 626–633.

[27] Y. Song, W. Cai, Y. Wang, D.D. Feng, Location classification of lung nodules with optimized graph construction, in: Proc. ISBI, 2012, pp. 1439–1442.

[28] Y. Song, W. Cai, Y. Zhou, D. Feng, Feature-based image patch approximation for lung tissue classification, IEEE Trans. Med. Imaging 32 (4) (2013) 797–808.

[29] Q. Li, W. Cai, D. Feng, Lung image patch classification with automatic feature learning, in: Proc. EMBC, 2013, pp. 6079–6082.

[30] Q. Li, W. Cai, X. Wang, Y. Zhou, D. Feng, M. Chen, Medical image classification with convolutional neural network, in: Proc. ICARCV, 2014, pp. 844–848.

[31] G. van Tulder, M. de Bruijne, Combining generative and discriminative representation learning for lung CT analysis with convolutional restricted Boltzmann machine, IEEE Trans. Med. Imaging 35 (5) (2016) 1262–1272.

[32] M. Anthimopoulos, S. Christodoulidis, L. Ebner, A. Christe, S. Mougiakakou, Lung pattern classification for interstitial lung diseases using a deep convolutional neural network, IEEE Trans. Med. Imaging 35 (5) (2016) 1207–1216.

[33] H. Shin, H.R. Roth, M. Gao, L. Lu, Z. Xu, I. Nogues, J. Yao, D. Mollura, R.M. Summers, Deep convolutional neural networks for computer-aided detection: CNN architecture, dataset characteristics and transfer learning, IEEE Trans. Med. Imaging 35 (5) (2016) 1285–1298.

[34] F. Zhang, Y. Song, W. Cai, S. Liu, S. Liu, S. Pujol, R. Kikinis, Y. Xia, M. Fulham, D. Feng, Pairwise latent semantic association for similarity computation in medical imaging, IEEE Trans. Biomed. Eng. 63 (5) (2016) 1058–1069.

[35] H.D. Tagare, C.C. Jaffe, J. Duncan, Medical image database: a content-based retrieval approach, J. Am. Med. Inform. Assoc. 4 (3) (1997) 184–198.

[36] H. Muller, N. Michoux, D. Bandon, A. Geissbuhler, A review of content-based image retrieval systems in medical applications – clinical benefits and future directions, Int. J. Med. Inform. 73 (2004) 1–23.

[37] W. Cai, S. Liu, L. Wen, S. Eberl, M.J. Fulham, D. Feng, 3D neurological image retrieval with localized pathology-centric CMRGlc patterns, in: Proc. ICIP, 2010, pp. 3201–3204.

[38] B. Fischer, A. Brosig, P. Welter, C. Grouls, R.W. Gunther, T.M. Deserno, Content-based image retrieval applied to bone age assessment, in: Proc. SPIE, 2010, p. 762412.

[39] D. Unay, A. Ekin, R.S. Jasinschi, Local structure-based region-of-interest retrieval in brain MR images, IEEE Trans. Inf. Technol. Biomed. 14 (4) (2010) 897–903.

[40] L. Sorensen, M. Loog, P. Lo, H. Ashraf, A. Dirksen, R.P.W. Duin, M.D. Bruijne, Image dissimilarity-based quantification of lung disease from CT, in: Proc. MICCAI, vol. 6361, 2010, pp. 37–44.

[41] X. Qian, H.D. Tagare, R.K. Fulbright, R. Long, S. Antani, Optimal embedding for shape indexing in medical image databases, Med. Image Anal. 14 (2010) 243–254.

[42] U. Avni, H. Greenspan, E. Konen, M. Sharon, J. Goldberger, X-ray image categorization and retrieval on the organ and pathology level, using patch-based visual words, IEEE Trans. Med. Imaging 30 (3) (2011) 733–746.

[43] Y. Song, W. Cai, D.D. Feng, Hierarchical spatial matching for medical image retrieval, in: Proc. ACM MM Workshop, 2011, pp. 1–6.

[44] Y. Song, W. Cai, Y. Zhou, L. Wen, D. Feng, Pathology-centric medical image retrieval with hierarchical contextual spatial descriptor, in: Proc. ISBI, 2013, pp. 202–205.

[45] A. Depeursinge, A. Vargas, A. Platon, A. Geissbuhler, P.A. Poletti, H. Muller, Building a reference multimedia database for interstitial lung diseases, Comput. Med. Imaging Graph. 2011 (2011).

[46] G. Ng, Y. Song, W. Cai, Y. Zhou, S. Liu, D. Feng, Hierarchical and binary spatial descriptors for lung nodule image retrieval, in: Proc. EMBC, 2014, pp. 6463–6466.

[47] F. Zhang, Y. Song, W. Cai, M. Lee, Y. Zhou, H. Huang, S. Shan, M. Fulham, D. Feng, Lung nodule classification with multi-level patch-based context analysis, IEEE Trans. Biomed. Eng. 61 (4) (2014) 1155–1166.

[48] A. Foncubierta-Rodriguez, A.G.S. de Herrera, H. Muller, Medical image retrieval using bag of meaningful visual words: unsupervised visual vocabulary pruning with PLSA, in: Proc. MIIRH, 2013, pp. 75–82.

[49] F. Zhang, Y. Song, W. Cai, A.G. Hauptmann, S. Liu, S. Pujol, R. Kikinis, M. Fulham, D. Feng, M. Chen, Dictionary pruning with visual word significance for medical image retrieval, Neurocomputing 177 (2016) 75–88.

[50] C.S. Mendoza, G.R. Washko, J.C. Ross, A.A. Diaz, D.A. Lynch, J.D. Crapo, E.K. Silverman, B. Acha, C. Serrano, R.S.J. Estepar, Emphysema quantification in a multi-scanner HRCT cohort using local intensity distributions, in: Proc. ISBI, 2012, pp. 474–477.

[51] M.J. Gangeh, L. Sorensen, S.B. Shaker, M.S. Kamel, M. de Bruijne, M. Loog, A texton-based approach for the classification of lung parenchyma in CT images, in: Proc. MICCAI, vol. 6363, 2010, pp. 595–602.

[52] R. Xu, Y. Hirano, R. Tachibana, S. Kido, Classification of diffuse lung disease patterns on high-resolution computed tomography by a bag of words approach, in: Proc. MICCAI, vol. 6893, 2011, pp. 183–190.

[53] Y. Song, W. Cai, J. Kim, D. Feng, A multi-stage discriminative model for tumor and lymph node detection in thoracic images, IEEE Trans. Med. Imaging 31 (5) (2012) 1061–1075.

[54] Y. Xu, M. Sonka, G. McLennan, J. Guo, E.A. Hoffman, MDCT-based 3-D texture classification of emphysema and early smoking related lung pathologies, IEEE Trans. Med. Imaging 25 (4) (2006) 464–475.

[55] O.S. Al-Kadi, D.Y.F. Chung, R.C. Carlisle, C.C. Coussios, J.A. Noble, Quantification of ultrasonic texture intra-heterogeneity via volumetric stochastic modeling for tissue characterization, Med. Image Anal. 21 (1) (2015) 59–71.

[56] M. Dang, J.T. Jysack, T. Wu, T.W. Matthews, S.P. Chandarana, N.T. Brockton, P. Bose, G. Bansal, H. Cheng, J.R. Mitchell, J.C. Dort, MRI texture analysis predicts p53 status in head and neck squamous cell carcinoma, Am. J. Neuroradiol. 36 (1) (2015) 166–170.

[57] L. Rokach, Ensemble-based classifiers, Artif. Intell. Rev. 33 (1–2) (2010) 1–39.

[58] G. Lee, S. Ali, R. Veltri, J.I. Epstein, C. Christudass, A. Madabhushi, Cell orientation entropy (core): predicting biochemical recurrence from prostate cancer tissue microarrays, in: Proc. MICCAI, 2013, pp. 396–403.

[59] P. Chatelain, O. Pauly, L. Peter, S. Ahmadi, A. Plate, K. Botzel, N. Navab, Learning from multiple experts with random forests: application to the segmentation of the midbrain in 3D ultrasound, in: Proc. MICCAI, 2013, pp. 230–237.

[60] A. Criminisi, D. Robertson, E. Konukoglu, J. Shotton, S. Pathak, S. White, K. Siddiqui, Regression forests for efficient anatomy detection and localization in computed tomography scans, Med. Image Anal. 17 (8) (2013) 1293–1303.

[61] M. Yaqub, M.K. Javaid, C. Cooper, J.A. Noble, Investigation of the role of feature selection and weighted voting in random forests for 3-D volumetric segmentation, IEEE Trans. Med. Imaging 33 (2) (2014) 258–271.

[62] D. Allen, L. Lu, J. Yao, J. Liu, E. Turkbey, R.M. Summers, Robust automated lymph node segmentation with random forests, SPIE Med. Imaging (2014) 90343X.

[63] Q. Zhao, K. Okada, K. Rosenbaum, L. Kehoe, D.J. Zand, R. Sze, M. Summar, M.G. Linguraru, Digital facial dysmorphology for genetic screening: hierarchical constrained local model using ICA, Med. Image Anal. 18 (5) (2014) 699–710.

[64] E. Parrado-Hernandez, V. Gomez-Verdejo, M. Martinez-Ramon, J. Shawe-Tylor, P. Alonso, J. Pujol, J.M. Menchon, N. Cardoner, C. Soriano-Mas, Discovering brain regions relevant to obsessive-compulsive disorder identification through bagging and transduction, Med. Image Anal. 18 (3) (2014) 435–448.

[65] C. Jacobs, C.I. Sanchez, S.C. Saur, T. Twellmann, P.A. de Jong, B. van Ginneken, Computer-aided detection of ground glass nodules in thoracic CT images using shape, intensity and context features, in: Proc. MICCAI, 2011, pp. 207–214.

[66] L. Gorelick, O. Veksler, M. Gaed, J.A. Gomez, M. Moussa, G. Bauman, A. Fenster, A.D. Ward, Prostate histopathology: learning tissue component histograms for cancer detection and classification, IEEE Trans. Med. Imaging 32 (10) (2013) 1804–1818.

[67] X. Huang, D.P. Dione, C.B. Compas, X. Papademetris, B.A. Lin, A. Bregasi, A.J. Sinusas, L.H. Staib, J.S. Duncan, Contour tracking in echocardiographic sequences via sparse representation and dictionary learning, Med. Image Anal. 18 (2) (2014) 253–271.

[68] Y. Song, W. Cai, H. Huang, Y. Zhou, Y. Wang, D. Feng, Boosted multifold sparse representation with application to ILD classification, in: Proc. ISBI, 2014, pp. 1023–1026.

[69] Z. Tu, Probabilistic boosting-tree: learning discriminative models for classification, recognition, and clustering, in: Proc. ICCV, 2005, pp. 1589–1596.

[70] C. Lu, Y. Zheng, N. Birkbeck, J. Zhang, T. Kohlberger, C. Tietjen, T. Boettger, J.S. Duncan, S.K. Zhou, Precise segmentation of multiple organs in CT volumes using learning-based approach and information theory, in: Proc. MICCAI, 2012, pp. 462–469.

[71] J. Feulner, S.K. Zhou, M. Hammon, J. Hornegger, D. Comaniciu, Lymph node detection and segmentation in chest CT data using discriminative learning and a spatial prior, Med. Image Anal. 17 (2) (2013) 254–270.

[72] L. Breiman, Random forests, Mach. Learn. 45 (1) (2001) 5–32.

[73] G. Yu, Y. Feng, D.J. Miller, J. Xuan, E.P. Hoffman, R. Clarke, B. Davidon, I. Shih, Y. Wang, Matched gene selection and committee classifier for molecular classification of heterogeneous disease, J. Mach. Learn. Res. 11 (2010) 2141–2167.

[74] H. Zhang, A.C. Berg, M. Maire, J. Malik, SVM-KNN: discriminative nearest neighbor classification for visual category recognition, in: Proc. CVPR, 2006, pp. 2126–2136.

[75] M. Liu, L. Lu, X. Ye, S. Yu, H. Huang, Coarse-to-fine classification via parametric and nonparametric models for computer-aided diagnosis, in: Proc. CIKM, 2011, pp. 2509–2512.

[76] K.Q. Weinberger, L.K. Saul, Distance metric learning for large margin nearest neighbor classification, J. Mach. Learn. Res. 10 (2009) 207–244.

[77] M. Zhu, A. Martinez, Subclass discriminant analysis, IEEE Trans. Pattern Anal. Mach. Intell. 28 (8) (2006) 1274–1286.

[78] N. Gkalelis, V. Mezaris, I. Kompatsiaris, T. Stathaki, Linear subclass support vector machines, IEEE Signal Process. Lett. 19 (9) (2012) 575–578.

[79] M. Hoai, A. Zisserman, Discriminative sub-categorization, in: Proc. CVPR, 2013, pp. 1666–1673.

[80] S. Escalera, D.M.J. Tax, O. Pujol, P. Radeva, R.P.W. Duin, Subclass problem-dependent design for error-correcting output codes, IEEE Trans. Pattern Anal. Mach. Intell. 30 (6) (2008) 1041–1054.

[81] J. Dong, W. Xia, Q. Chen, J. Feng, Z. Huang, S. Yan, Subcategory-aware object classification, in: Proc. CVPR, 2013, pp. 827–834.

[82] Y. Song, W. Cai, H. Huang, Y. Zhou, D. Feng, M. Chen, Large margin aggregation of local estimates for medical image classification, in: Proc. MICCAI, 2014, pp. 196–203.

[83] Y. Song, W. Cai, H. Huang, Y. Zhou, D. Feng, Y. Wang, M. Fulham, M. Chen, Large margin local estimate with applications to medical image classification, IEEE Trans. Med. Imaging 34 (6) (2015) 1362–1377.

[84] M. Liu, L. Lu, X. Ye, S. Yu, M. Salganicoff, Sparse classification for computer aided diagnosis using learned dictionaries, in: Proc. MICCAI, 2011, pp. 41–48.

[85] N. Weiss, D. Rueckert, A. Rao, Multiple sclerosis lesion segmentation using dictionary learning and sparse coding, in: Proc. MICCAI, 2013, pp. 735–742.

[86] Y. Xu, X. Gao, S. Lin, D.W.K. Wong, J. Liu, D. Xu, C. Cheng, C.Y. Cheung, T.Y. Wong, Automatic grading of nuclear cataracts from slit-lamp lens images using group sparsity regression, in: Proc. MICCAI, 2013, pp. 468–475.

[87] T. Tong, R. Wolz, P. Coupe, J.V. Hajnal, D. Rueckert, ANDI, Segmentation of MR images via discriminative dictionary learning and sparse coding: application to hippocampus labeling, NeuroImage 76 (2013) 11–23.

[88] U. Srinivas, H.S. Mousavi, V. Monga, A. Hattel, B. Jayarao, Simultaneous sparsity model for histopathological image representation and classification, IEEE Trans. Med. Imaging 33 (5) (2014) 1163–1179.

[89] L. Wang, F. Shi, Y. Gao, G. Li, J.H. Gilmore, W. Lin, D. Shen, Integration of sparse multi-modality representation and anatomical constraint for isointense infant brain MR image segmentation, NeuroImage 89 (2014) 152–164.

[90] S. Zhang, Y. Zhan, Y. Zhou, M. Uzunbas, D.N. Metaxas, Shape prior modeling using sparse representation and online dictionary learning, in: Proc. MICCAI, 2012, pp. 435–442.

[91] Y. Song, W. Cai, Y. Zhou, D. Feng, Thoracic abnormality detection with data adaptive structure estimation, in: Proc. MICCAI, 2012, pp. 74–81.

[92] S. Liao, Y. Gao, J. Lian, D. Shen, Sparse patch-based label propagation for accurate prostate localization in CT images, IEEE Trans. Med. Imaging 32 (2) (2013) 419–434.

[93] J. Wright, Y. Ma, J. Mairal, G. Sapiro, T.S. Huang, S. Yan, Sparse representation for computer vision and pattern recognition, Proc. IEEE 98 (6) (2010) 1031–1044.

[94] J. Wang, J. Yang, K. Yu, F. Lv, T. Huang, Y. Gong, Locality-constrained linear coding for image classification, in: Proc. CVPR, 2010, pp. 3360–3367.

[95] P. Zhang, C. Wee, M. Nieghammer, D. Shen, P. Yap, Large deformation image classification using generalized locality-constrained linear coding, in: Proc. MICCAI, 2013, pp. 292–299.

[96] F. Xing, L. Yang, Robust selection-based sparse shape model for lung cancer image segmentation, in: Proc. MICCAI, 2013, pp. 404–412.

[97] Y. Wu, G. Liu, M. Huang, J. Guo, J. Jiang, W. Yang, W. Chen, Q. Feng, Prostate segmentation based on variant scale patch and local independent projection, IEEE Trans. Med. Imaging 33 (6) (2014) 1290–1303.

[98] X. Chen, X. Xu, J.Z. Huang, Y. Ye, TW-k-means: automated two-level variable weighting clustering algorithm for multiview data, IEEE Trans. Knowl. Data Eng. 25 (4) (2013) 932–944.

[99] J. Liu, S. Ji, J. Ye, SLEP: Sparse Learning with Efficient Projections, Arizona State University, 2009.

[100] J. Tropp, Greed is good: algorithmic results for sparse approximation, IEEE Trans. Inf. Theory 50 (2004) 2231–2242.

[101] J. Shi, J. Malik, Normalized cuts and image segmentation, IEEE Trans. Pattern Anal. Mach. Intell. 22 (8) (2000) 888–905.

[102] E. Elhamifar, R. Vidal, Sparse subspace clustering, in: Proc. CVPR, 2009, pp. 2790–2797.

CHAPTER 7

Rigid Motion Invariant Classification of 3D Textures and Its Application to Hepatic Tumor Detection

Sanat Upadhyay*, Saurabh Jain[†], Manos Papadakis[‡]
*University of Houston, Computational Biomedicine Lab, Houston, TX, United States
[†]THINK Surgical Inc., Fremont, CA, United States
[‡]University of Houston, Houston, TX, United States

Abstract

Soft tissue in 3D medical data sets is often associated with 3D textures. The granularity demonstrated by tissues in medical images due to natural tissue structure gives rise to 3D or 2D textures. In this chapter, we focus on 3D textures for the purpose of classifying tissue patches into healthy and pathological. In our case, both patch types come from the same organ but one is neoplastic. We view the problem of tissue segmentation/classification into pathological and normal as a problem of 3D texture segmentation/identification. Texture segmentation/identification requires two major components, which are often developed together: the representation of the original data and the segmentation/classification algorithm. We model textures as spatial stochastic processes in order to address the variability due to intratissue natural variation, and due to noise. To address this problem, we use isotropic multiresolution analysis to extract 3D rigid motion invariant texture patch signatures. Using a measure of statistical disparity combined with local average intensities, we create low dimensional features that can be used for distinguishing normal tissue patches from pathological ones. We describe how to eliminate the additional variability of 3D tissue texture models due to rigid motions. We also present experimental results on contrast enhanced CT scans of the liver to demonstrate our approach for tissue discrimination, using 3D rigid motion invariant texture classification.

Keywords

Isotropic MultiResolution Analysis (IMRA), Gaussian Markov Random Field (GMRF), Support vector machines, Liver cancer, 3D texture classification, Isotropic filter, Kullback–Leibler divergence, Feature space

7.1 INTRODUCTION

Over the course of the last two decades, a variety of deterministic or stochastic texture models and an even richer ensemble of texture discrimination methods have appeared in the literature, *e.g.*, [1–5]. Texture analysis has extensively been used to build assistive tools for diagnosis; some examples of which can be found in [6–10]. Additional work in this area can be found in Chapters 1 and 12 of this book. Our list is by no means exhaustive, but only indicative, restricted to some of the better known reviews of and

Biomedical Texture Analysis
DOI: 10.1016/B978-0-12-812133-7.00007-7

contributions in this modern line of research in pattern recognition. However, most of this work is exclusively devoted to 2D textures. These have been extensively used in medical image analysis and the development of diagnostic semantics in X-ray Computed Tomography (CT) and Magnetic Resonance Imaging (MRI) images. Although these image modalities produce 3D images of the human body, still radiologists and assistive analysis software work in 2D on a slice-by-slice basis and then the expert's brain tries to piece everything together in 3D, using prior knowledge of human anatomy and intuition. But could this process be carried out automatically? Tissues grow and live in 3D, and when they develop they do not necessarily pick orientations convenient to imaging analysis tools. This inconvenient truth motivates the use of genuine 3D texture analysis for developing tools for the semi or fully automated analysis of 3D X-ray CT or MRI images. Specializing in X-ray CT, hard tissue like bones are easily detectable and segmentable [11,12]. On the antipodal end, soft tissues are harder to distinguish due to the reduced variability of X-ray scattering properties of soft tissues. This practically eliminates boundaries between such tissues and organs. While organ boundaries can be identified using prior anatomy knowledge, distinguishing healthy from pathological tissue within the same organ is quite challenging in X-ray CT, due to the lack of variation of scattering properties between soft tissues. This is not the case with MRI, where small variations of water density result in observable variations of image intensity. Since both modalities are based on depicting intensities determined by tissue density, a common trick is to use contrast agents, which permeate in various soft tissues and enhance them by differentiating their density properties. While using them is not a complete solution to all of our CT-related image analysis problems, contrast agents are helpful (see Fig. 7.4), but do not solve all problems, since active lesions may not have distinguishable, hypodense, necrotic cores. Two such examples can be found in Fig. 7.4, top left and bottom right panels. This is because average local image intensities change in relation to neighboring tissues, and the permeation of the contrast agent, or lack thereof, generates patterns in 2D or in 3D enabling the segregation of pathologically suspicious from the healthy tissues. These patterns often exhibit large variability, and are harder for an expert's brain to pick than might be for an imaging algorithm to sense, particularly in 3D. A possible distinguishing pattern is the presence of a texture, which is the result of the spatial organization of the tissue and the contrast agent molecules in 2D or 3D patches. One of the challenges is that many of those textures are not structural, in the sense that they do not consist of periodically occurring patterns (see Section 1.2 of Chapter 1). Nevertheless, these patterns may exhibit statistical stability and be represented by stochastic textures. In particular, we model textures with stationary stochastic processes on a discrete lattice. Their structure is not deterministic, which our brains can immediately identify in 2D, but is governed by an unknown probability distribution that we typically do not even attempt to determine, *e.g.*, Gibbs random fields [13]. Natural local tissue variations modify 3D textures of the same tissue type; in addition, image acquisition "noise" adds another layer of variability to tissue textures.

The problem presented to us is to determine if two tissue samples (patches) are the same or not. To solve this problem, we do not need to know the underlying probability distributions of the stochastic textures associated with the different tissues. All we need is that pathological tissues are distinguishable from normal ones based on the statistical disparity of different probability models associated with each one of them. In other words, we are not looking for a generative model but a discriminative one. To this end, we need to develop quantities that can be computed from the imaging data, which if all are taken into account concurrently, can identify the disparity between the two probability models. This is pretty much what we do in hypothesis testing, where the quantity we use is called a statistic. In our work, we use a vector of random variables as *features*, which live in a multidimensional space, called the *feature space*. We hypothesize that a low dimensional feature space can be used to discriminate between textures. Following our belief, we pursue to determine these features and the metric of their statistical disparity for a given imaging application. We want to develop features suitable for discrimination of soft tissues, which may be positioned in any orientation. Therefore, feature extraction must be insensitive to all possible 3D orientations and the positioning of the tissue patch. This is precisely the motivation for the development of the theory and algorithms we present in this chapter. We also recommend Section 1.3.3 of Chapter 1 and Section 2.4 of Chapter 2 to the reader for a complementary treatise of this topic in 2D. Using a more formal language, we will present a number of theoretical methods and algorithmic tools for 3D rigid motion invariant soft-tissue discrimination. These methods work with the native 3D data instead of a slice-by-slice analysis of the 2D CT slices. We will discuss the challenges of this problem and the practical limitations of various solutions. The presentation herein, unifies our work in [14], [15], and [16] along with some more recent unpublished work.

Our work uses three main tools:

1. The Isotropic MultiResolution Analysis (IMRA) filters to generate projections of tomographic images at various scales.
2. A first-order Gaussian Markov Random Field (GMRF) model, which we fit to each of the images obtained via the multiscale decomposition algorithm induced by the IMRA.
3. Finally, using the fitted multiscale model we define a rotationally invariant statistical disparity distance measure which, due to the use of the IMRA decomposition algorithm, is 3D rigid motion invariant. In practice, we achieve this invariance only approximately using the proposed statistical disparity distance measure.

Combining the rotationally invariant distance between textures at different scales along with difference in mean intensities, we design low dimensional features for classifying different soft tissue types.

In summary, our contributions are the following:

1. A novel rotationally invariant distance between the high dimensional signatures of two arbitrary 3D textures **X** and **Y**; this distance is rigorously based on calculating the Haar integral (with respect to the measure of the 3D rotation group $SO(3)$) of the Kullback–Leibler divergences between pairs of GMRF models associated to each 3D rotation of textures **X** and **Y**.

2. IMRA-based implementation of 3D rotations of tissue patches in a computationally efficient manner, enabling discrimination between stochastic 3D textures having a broad variety of directional characteristics. We will also present the implementation of approximate rotations of image patches using filters with finite length.

3. Rules for the 3D rigid motion invariant texture discrimination and binary classification which take into account the proposed 3D rigid motion invariant distance between textures computed *at a range of scales*. These rules allow the use of simple GMRF models and avoid computationally costly parameter estimation of GMRFs with long range interactions. We experimentally establish that these rules enhance the sensitivity of 3D rigid motion invariant discrimination between 3D textures, and that they are applicable even to non-GMRF textures.

The early detection of liver cancer lesions can potentially improve the management of various forms of liver cancer. Typically, liver lesions are identified using contrast enhanced CT scans acquired at different phases of perfusion of the hepatic parenchyma by the infused contrast agent. The task of identifying the lesions is performed by a radiologist using a large number of images generated from this multiphase CT acquisition, and requires significant time and effort.

In this work, we present an algorithm and experimental results that demonstrate the feasibility of the development of a semiautomatic screening tool capable of detecting liver abnormalities in contrast enhanced X-ray CT-scans. Specifically, utilizing ideas proposed by Jain et al. [16], we develop 3D rigid motion invariant texture features. We experimentally establish that when these features are combined with mean attenuation intensity differences the new augmented features are capable of discriminating normal from abnormal liver tissue in arterial phase contrast enhanced X-ray CT-scans with high sensitivity and specificity.

When scans are acquired during different perfusion phases of the contrast agent, liver lesions result in hypodense or hyperdense Regions Of Interest (ROI) relative to normal hepatic parenchyma. Quite often, in the arterial phase hypodense ROIs are adjacent to denser ones. These denser regions may be hyperdense due to the increased vascularization of active cancerous lesions, or just normal parenchyma. More specifically, the driving assumption in our approach is that liver tissue ROIs can be represented in a contrast-enhanced X-ray CT scan by two components, *3D texture* and *local mean intensities*. The first of these two components captures the structure, while the second provides the average intensity of the ROI, which is a traditional feature for tissue discrimination.

Local attenuation intensity averages alone are not robust enough to discriminate normal from pathological tissue within the same organ [15].

Past literature has been mostly focused on 2D textures and has introduced a great variety of texture features ranging, for instance, from spatial frequencies based on Gabor filters [17–21] and wavelets [22–27] to autoregressive models such as Markov random fields [28–33]. For 2D texture characterization by local texture information linked to multipixel statistics, spatial interaction stochastic models such as Gibbs random fields have been widely studied [34,35,13]. In particular, Gaussian Markov Random Fields (GMRF) have been extensively applied for 2D texture classification [29,17,36,1], and more recently for 3D textures in [37]. Existing literature on 3D texture models and discrimination is still limited due to serious computational challenges native to 3D image analysis (see [38–42]).

Texture-based lesion segmentation has been successfully used in the past for the detection of cancerous hepatic lesions [43]. Texture features have been extracted in a slice-by-slice manner by combining first- and second-order moments [44]. Very similar approaches have also been used to segment the liver from neighboring organs [45–47]. Apart from the fact that our methods are natively designed to work in 3D, a fundamental difference between previous texture-based approaches and our work is that they use significantly more complex classifiers. Liver segmentation and detection of cancerous lesions has mostly been performed with nontexture based methods as in [48–53], where the difference in attenuation intensity between more and less contrast-perfused ROIs is used for traditional attenuation-based feature extraction, or as in [54,55], where deformable models are utilized to generate the boundaries between normal and abnormal tissues. Both of these approaches mostly limit the detection of cancer lesions to the hypo/hyperdense regions, because differences in average intensities are typically one of the dominant discriminative features. On the other hand, texture and local intensity average based features improve the discriminability between hypodense, hyperdense ROIs, and normal parenchyma, small or big lesions. If in addition, of 3D rigid motion invariance, or covariance is added to the generation of these features, then their robustness will be increased. Another interesting discussion on this topic is presented in Section 2.4.4 of Chapter 2, where directionally sensitive and insensitive features are compared. Before proceeding, we need to clarify that although our feature extraction utilizes isotropic filters, the features we extract have directionality. The isotropy of our filters helps to make our features, at least in their mathematical formulation, 3D rigid motion invariant.

Despite numerous notable successes in the field of hepatic tumor detection and segmentation, our work opens an unexplored direction. The proposed 3D rigid motion invariant features allow feature vectors from normal tissue samples to form clusters that are better defined than the clusters formed when the 3D texture features do not account for 3D rotations. This enhances the discriminatory power of the proposed features (Fig. 7.5). Our experimental results are not directly comparable with the results of others, because we only test the discriminative power of our features on sets of ROIs and we do not segment normal from abnormal liver tissue. However, it appears that the proposed features can be used for tissue discrimination with high sensitivity and specificity rates rendering them as a promising tool for developing segmentation algorithms. In Section 7.2, we present the fundamentals of Isotropic MultiResolution Analysis (IMRA) along with some of its theoretical highlights. IMRA is a multiresolution analysis ([56]) modified to be compatible with local rotations of an image patch. For the sake of the audience of this volume, a more detailed discussion on the how IMRA becomes compatible with local rotations is included in Section 7.2. In Section 7.4, we present the basic concepts of our approach and the ingredients of the algorithmic implementation of the proposed 3D rigid motion invariant texture distance. Section 7.5 discusses the use of this distance for multiscale 3D rigid motion invariant texture discrimination/classification, and includes a summary of the algorithmic implementation of our method. We briefly describe our experimental results in the Conclusions section.

7.2 ISOTROPIC MULTIRESOLUTION ANALYSIS

Isotropic MultiResolution Analysis (IMRA) [57] is an MRA for which the "core" subspace V_0 is invariant under rotations. An MRA is collection of closed subspaces of $L_2(\mathbb{R}^D)$ satisfying properties 1–4 of Definition 7.1. MRAs were introduced by Mallat and Meyer in the early 1980s as an attempt to formalize the connection between the analog and digital domains at various levels of resolution and provide a rigorous theoretical justification of the correspondence between mathematical operations applied on images in the physical domain and digital manipulations of their digital counterparts using filter banks. The introduction of MRA sparked an enormous amount of work both in the domain of pure mathematical analysis as well as in algorithmics and engineering applications.

The Fourier transform $\widehat{f} : \mathbb{R}^D \longrightarrow \mathbb{C}$ of $f \in L_1(\mathbb{R}^D)$ is defined by

$$\forall \boldsymbol{\omega} \in \mathbb{R}^D, \ \widehat{f}(\boldsymbol{\omega}) = \int_{\mathbb{R}^D} f(\boldsymbol{x}) e^{-2\pi i <\boldsymbol{x}, \boldsymbol{\omega}>} \, d\boldsymbol{x}.$$

Here, $L_p(\mathbb{R}^D)$, where $1 \leq p < \infty$, denotes the space of functions defined on the continuous domain $\mathbb{R}^D$ for which

$$\int_{\mathbb{R}^D} |f(\boldsymbol{x})|^p d\boldsymbol{x} < \infty \, ,$$

$<\boldsymbol{x}, \boldsymbol{\omega}>$ denotes the standard inner product on $\mathbb{R}^D \times \mathbb{R}^D$. The map $f \rightarrow \widehat{f}$ restricted to $L_1(\mathbb{R}^D) \cap L_2(\mathbb{R}^D)$ extends to a unitary map on $L_2(\mathbb{R}^D)$. Further down in this section, the reader will encounter $\mathbb{T}$ which denotes the interval $[-\frac{1}{2}, \frac{1}{2}]$, the fundamental domain of all 1-periodic functions. So, if $f \in L^p(\mathbb{T}^D)$, then f is a periodic function extended to the entire $\mathbb{R}^D$, but p-integrable inside $\mathbb{T}^D$ only.

A *frame* for a separable Hilbert space H is a sequence $\{f_i\}_{i \in \mathbb{I}} \subseteq H$, where $\mathbb{I}$ is a countable index set, for which there are constants $c_0, c_1 > 0$ such that

$$\forall f \in H, \ c_0 \|f\|^2 \leq \sum_{i \in \mathbb{I}} |\langle f, f_i \rangle|^2 \leq c_1 \|f\|^2.$$

c_0 and c_1 are called frame constants. The weaker version of the above inequality where $c_1 > 0$ such that the right-hand side of the above inequality is valid for all $f \in H$ is termed by referring to $\{f_i\}_{i \in \mathbb{I}}$ as a *Bessel* sequence. If $c_0 = c_1$, we say the frame is *tight*, and a tight frame with $c_0 = c_1 = 1$ is called a *Parseval* frame. A frame is a Riesz basis for H if it is also a minimal family generating H. We use the terms frame (Riesz) sequence when we refer to a countable frame or a Riesz basis of a subspace of a Hilbert space. For $x \in \mathbb{R}^D$ and $r > 0$, $B(x, r)$ will denote the ball centered at x with radius r. By supp$f$ we mean the set of points x such that $f(x) \neq 0$ and $e_\mathbf{n}(\boldsymbol{\omega}) := e^{-2\pi i <\mathbf{n}, \boldsymbol{\omega}>}$, with $\boldsymbol{\omega} \in \mathbb{R}^D$. We define the unitary rotation operator by $\mathcal{R}_\mathbf{A}f(x) := f(\mathbf{A}^T x)$, where $f \in L_2(\mathbb{R}^D)$, $x \in \mathbb{R}^D$, $\mathbf{A} \in SO(D)$ and $\mathbf{A}^T$ denotes the transpose of the matrix $\mathbf{A}$. We also define a translation by $y \in \mathbb{R}^D$ via $T_y f(x) = f(x - y)$ for all x.

Given $\Omega \subset \mathbb{R}^D$ with positive measure, PW_Ω is the closed subspace of $L_2(\mathbb{R}^D)$ defined as

$$PW_\Omega = \{f \in L_2(\mathbb{R}^D) : \ \mathrm{supp}\widehat{f} \subseteq \Omega\}.$$

The subspace PW_Ω is called a *Paley–Wiener* subspace of $L_2(\mathbb{R}^D)$ associated to Ω. In the special case when Ω is a ball of radius r centered at the origin $\mathbf{0}$, we denote the associated *Paley–Wiener* subspace of $L_2(\mathbb{R}^D)$ by PW_r.

One of the major achievements of the MRA theory is the extension of Shannon's sampling theorem for band-limited functions, which bridges mathematically the digital and analog domains. The interested reader, who wishes to dig a bit more may find that in Property 4 of Definition 7.1 we do not mention anything about the generators of the core subspace V_0. This is not a mistake. It can be proved that there is always a set of functions φ_i whose integer translates $T_\mathbf{n}$ generate V_0. The most used cases of MRA are those who are generated by a single function φ, often called scaling function or father wavelet. The MRA theory allows the easy construction of wavelets, which became quite popular in the 1980s and 1990s. One of major discussion items in MRA theory is its invariance to 2D–3D rigid motions. This problem turns out to be subtle. As we see from Property 4 of Definition 7.1, MRA-core subspaces are invariant only by integer translates. The only core subspaces which can be invariant to all translates are the Paley–Wiener ones, but those suffer from the fact that the induced filter banks are implemented by filters with infinite length and significant oscillations. Such filters are not nice for applications, especially for texture analysis. One way to fix this problem is to smooth their Fourier transforms (Fig. 7.2B shows the Fourier transform of such a filter). The length of the filter will still be infinite but the oscillations will almost go away (Fig. 7.2A). Then the filter can be truncated in order to be used. Or one can choose to implement the filter with fast convolution via the Fast Fourier Transform (FFT), which is what we do here. Then we retain the property of translation invariance of V_0. This implies that if $f \in V_0$ then $T_y f \in V_0$. This allows to recalculate the coefficients of the expansion of $T_y f$ with respect to the $T_n \varphi$'s, using only the coefficients of the expansion of f with respect to the $T_n \varphi$'s. This is precisely what we refer to as translation covariance of V_0. This property is inherited to every V_j. General MRAs have only translation covariance, to translate only by dyadic rational increments. Using the dilation operator, we can move a dilation of f in V_j in which we can calculate the coefficients of the dyadic rational translation of f: Take $f \in V_0$ and let $f = \sum_n c_n T_n \varphi$. Then

$$T_{2^{-j} \mathbf{n}_0} f = \sum_{\mathbf{n}} c_{\mathbf{n}} T_{2^{-j} \mathbf{n}_0} T_{\mathbf{n}} \varphi. \tag{7.1}$$

Now, the refinability of φ implies

$$\varphi = \sum_{\mathbf{k}} d_{\mathbf{k}} T_{2^{-j} \mathbf{k}} (\varphi(2^j \cdot)) \tag{7.2}$$

which, together with (7.1), imply

$$T_{2^{-j} \mathbf{n}_0} f = \sum_{\mathbf{k}} \left(\sum_{\mathbf{n}} c_{\mathbf{n}} d_{\mathbf{k} - \mathbf{n}_0} \right) T_{2^{-j} \mathbf{k}} (\varphi(2^j \cdot)). \tag{7.3}$$

The key for deriving Eq. (7.3) is that a translate of φ retains the shape of φ and $T_{2^{-j} \mathbf{n}_0} (\varphi(2^j \cdot))$ is one of the generators of V_j, thus allowing to derive an easy, intuitive formula for expressing $T_{2^{-j} \mathbf{n}_0} (\varphi(2^j \cdot))$ in terms of $T_{2^{-j} \mathbf{k}} (\varphi(2^j \cdot))$. The best thing about the previous calculation is that it is exact. In conclusion, there are several ways to facilitate translation covariance in 2D or 3D and any D, with the simplest one being to stick with dyadic rational increments.

On the other hand, the situation is subtle when dealing even with simple rotations, other than those which are multiples of 90°. One of the reasons is that, even if rotations are multiples of 90°, a rotation of φ cannot be expressed in terms of the translates $T_{2^{-j} \mathbf{k}} (\varphi(2^j \cdot))$, unless φ remains invariant under rotations. This is one of the reasons that led us to adopt the IMRA approach and the isotropic refinable functions. The other reason is that a rotation of the sampling grid leaves it invariant only if the rotation is a multiple of 90° (see Fig. 7.1). The latter issue does not pose a challenge, as well as a rotation of φ remains inside V_0 (inspect Figs. 7.2B and 7.2A to get a feeling why this is

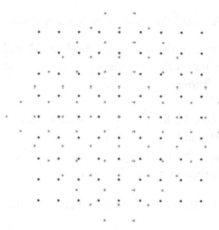

Figure 7.1 Blue dots are the points of the sampling lattice $\mathbb{Z}^2$ and the red dots are the points of this lattice rotated by 45°. Notice that the red dots almost never come close to the blue ones. (For interpretation of the references to color in this figure legend, the reader is referred to the web version of this chapter.)

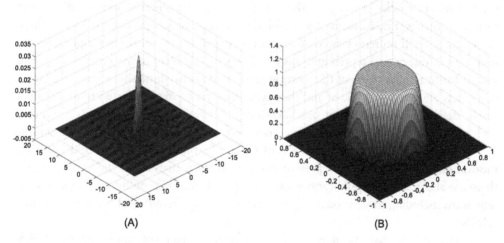

(A)   **(B)**

Figure 7.2 A representation of low-pass IMRA filter in (A) space domain (h_0) and (B) frequency domain ($\widehat{h}_0$).

true), because then it can be expressed by its translates, pretty much in the same way as we did in Eq. (7.2).

Let Ω be a measurable subset of $\mathbb{R}^D$.

1. We say that $\Omega \subset \mathbb{R}^D$ is *radial* if for all $\mathbf{A} \in SO(D)$, $\mathbf{A}(\Omega) = \Omega$.
2. A $D \times D$ matrix is *expansive* if it has integer entries and if all of its eigenvalues have absolute value greater than 1.

3. An expansive matrix, $\mathbf{A}$, is *radially expansive* if $\mathbf{A}$ is a multiple of a rotation.

Definition 7.1. An Isotropic MultiResolution Analysis (IMRA) of $L_2(\mathbb{R}^D)$ with respect to a radially expansive matrix $\mathbf{A}$ is a sequence $\{V_j\}_{j \in \mathbb{Z}}$ of closed subspaces of $L_2(\mathbb{R}^D)$ satisfying the following conditions:

1. $\forall j \in \mathbb{Z}$, $V_j \subset V_{j+1}$,
2. $f \in V_j$, if and only if $f(2^{-j} \cdot) \in V_0$ for every $j \in \mathbb{Z}$ (dyadic dilations),
3. $\cup_{j \in \mathbb{Z}} V_j$ is dense in $L_2(\mathbb{R}^D)$ and $\cap_{j \in \mathbb{Z}} V_j = \{0\}$,
4. V_0 is invariant under the action of the group $\{T_\mathbf{n} : \mathbf{n} \in \mathbb{Z}^D\}$,
5. If P_0 is the orthogonal projection onto V_0, then

$$\mathcal{R}_\mathbf{A} P_0 = P_0 \mathcal{R}_\mathbf{A} \qquad \text{for all } \mathbf{A} \in SO(D). \qquad (7.4)$$

First, this structure is not a multiresolution analysis in the classical sense, because the existence of an orthonormal or Riesz basis for its "core" subspace V_0 is not required. The problem that concerns wavelet practitioners is the covariance of wavelet representations of images, and signals in general, under shifts or other transformations of the coordinate system. Specializing this problem to images, it is crucial for a significant number of tasks in image analysis to have covariant representations with respect to rigid motions, *i.e.*, rotations and translations of an image especially in two or three dimensions. Such representations are also known as *steerable* and there are previous attempts with directional representations to address this problem in the context of digital images for finite subgroups of the rotation group and in two dimensions [58–70]. Those constructs are not MRA-based and they typically refer to steerability without attempting to define it. To the best of our knowledge, 3D rigid motion covariant multiscale representations are not as widely utilized (see Section 1.3.3 of Chapter 1). At this point, we wish to mention that Section 2.4.3 of Chapter 2 offers an excellent and inspiring motivation and references to representations for texture analysis. In the same chapter, there are also numerous publications on the topic of steerable multidimensional representations including more modern approaches such as Convolutional Neural Networks (CNN).

Specializing on MRAs, it is natural to impose weaker conditions on the MRA subspaces and in particular on the core subspace V_0 for MRAs defined with respect to dilations given by radially expansive matrices before even attempting to construct MRA-wavelet decompositions that are covariant with respect to rigid motions. In our first design, we used radial ideal filters in order to achieve a high degree of symmetry in wavelet filter design [71], generalizing thus the Shannon MRA (for a definition see [72]) in multidimensions. In order to achieve radiality, we relaxed the classical Riesz basis of the integral translates of the scaling function assumption to allow for this set to be a nonminimal frame. All these ideas led us to the addition of the last property in Definition 7.1. The proofs of all results in this section can be found in [57].

Theorem 7.1. *([57]) Let V be an invariant subspace of $L_2(\mathbb{R}^D)$ under the action of the translation group induced by $\mathbb{Z}^D$. Then V remains invariant under all rotations if and only if $V = PW_\Omega$ for some radial measurable subset Ω of $\mathbb{R}^D$.*

As an immediate consequence, we obtain a characterization of IMRAs generated by a single function, in the sense that there exists a function φ such that V_0 is the closed linear span of $\{T^{\mathbf{n}}\varphi : \mathbf{n} \in \mathbb{Z}^D\}$.

Proposition 7.1. *([57]) Let $\mathbf{A}$ be a radially expansive matrix and $\mathbf{C} := \mathbf{A}^*$, the transpose of $\mathbf{A}$. A sequence $\{V_j\}_{j\in\mathbb{Z}}$ is an IMRA with respect to $\mathbf{A}$ if and only if $V_j = PW_{\mathbf{C}^j(\Omega)}$, where Ω is radial and satisfies:*

(i) $\Omega \subset \mathbf{C}(\Omega)$.

(ii) *The set-theoretic complement of $\cup_{j=1}^{\infty} \mathbf{C}^j(\Omega)$ is null.*

(iii) $\lim_{j\to\infty} |\mathbf{C}^{-j}(\Omega)| = 0$.

Moreover, the only singly generated IMRAs are precisely $V_j = PW_{\mathbf{C}^j(\Omega)}$, where (Ω) is a radial subset of $\mathbb{T}^D$ satisfying (i), (ii), and (iii).

The previous result has more theoretical than practical value. MRAs play a significant role in the theory because they model the transition from analog to digital and backward. However, for practical purposes, it will be desirable to work with a single generator function of an MRA. The nesting of the subspaces V_j forces us to choose such a function to be refinable: A function φ in $L_2(\mathbb{R}^D)$ is called *refinable*, with respect to dilations induced by an expansive matrix $\mathbf{A}$, if there exists a measurable, essentially bounded, $\mathbb{Z}^D$-periodic function $\widehat{h}_0$ such that $\widehat{\varphi}(\mathbf{A}^*\boldsymbol{\omega}) = \widehat{h}_0(\boldsymbol{\omega})\widehat{\varphi}(\boldsymbol{\omega})$, a.e. $\boldsymbol{\omega} \in \mathbb{R}^D$. The function $\widehat{h}_0$ is called the low-pass filter or mask corresponding to φ, which also plays the role of the sampling kernel. Scaling functions are refinable. The latter property assumption can be formalized by requiring $\{T^{\mathbf{n}}\varphi : \mathbf{n} \in \mathbb{Z}^D\}$ to be a Bessel family. When φ satisfies this property and it is refinable, then it is a scaling function.

A naturally arising question is which are the refinable functions φ that give rise to IMRAs. An obvious choice is a function whose Fourier transform is the characteristic function of set Ω described in the previous proposition, whose discontinuities destroy its spatial localization. The characterization of isotropic refinable functions is nontrivial and it is thoroughly studied in [57]. Here, we include only two relevant results indicating that such functions are not the most desirable for applications.

Theorem 7.2. *([57]) Let $\mathbf{A}$ be a radially expansive matrix and $\varphi \in L_2(\mathbb{R}^D)$ be a refinable function with mask $\widehat{h}_0 \in L_\infty(\mathbb{T}^D)$. If φ is isotropic (Fig. 7.2B), then it is not compactly supported in the spatial domain. Moreover, an IMRAs cannot be defined by φ, if $\widehat{\varphi}$ does not vanish outside the ball of radius $1/2$.*

The previous theorem dictates, in order to have full 3D rigid motion steerability in a MRA setting we must work with functions which are band-limited within the ball

$B(0, 1/2)$ in the frequency domain if we want to work with dyadic dilations. Translation invariance can be achieved by avoiding downsampling at every decomposition level. However, the lack of natively finite length filters for IMRA implementations imposes the need to truncate isotropic low and high pass filters and in some applications this may be a problem. In Section 7.3, we will suggest a method on how to partially overcome this problem by settling for a concept of rotational steerability limited to some finite orientations. There is another aspect of the IMRA that we need to elaborate on before wrapping up this section. Isotropy may be viewed as potentially losing sensitivity to directional characteristics of an image. This is actually not true. The autocovariance function of the probability model we will use for texture feature extraction is primarily a model of three fundamental directions of correlation. The IMRA does not smooth out those directionalities or wave front orientations. On the contrary, it preserves them. Its subbands show the same sensitivity regardless of the orientation of a singularity or of the directions in which the correlations appear to be stronger or weaker. What the IMRA misses is the ability to identify these directions, because its subbands depend only on scale and they are rotationally invariant.

7.3 IMPLEMENTING ROTATIONS WITH COMPACTLY SUPPORTED REFINABLE FUNCTIONS

The conclusion, Theorem 7.2, of the previous section naturally leads to consider MRA structures where the core subspace invariance under rotations. In fact, one of the big problems is that the rotation of a refinable function is no longer refinable. This is because rotations do not preserve the integer sampling grid (Fig. 7.1), so $\hat{h}_0(\mathbf{A}\cdot)$, the low pass filter corresponding to $\varphi(\mathbf{A}\cdot)$, the rotation of a refinable φ by $\mathbf{A}$, is not necessarily a periodic function. But why should we care about refinable functions? The reason is that those functions allow the use of filter banks for implementing fast multiscale decompositions of images. Our attempt to circumvent this particular problem is to abandon classical rotations and instead use orthogonal matrices with integer entries. As we later on argue, this trick will help to use compactly supported refinable functions and implement "approximate rotations" of them. But why should we care about this class of refinable functions? The answer is they lead to MRAs implementable with convolutional finite length filter. This prevents ringing, approximation numerical errors, and distant correlations which infinite length filters, even truncated, suffer from. For clarity purposes, our discussion is restricted to two dimensions, but all arguments can be easily extended to 3D. To simplify the notation, we also use only dyadic dilations.

Consider $\mathbf{A}_i = \begin{pmatrix} a_i & -b_i \\ b_i & a_i \end{pmatrix}$ with $a_i, b_i \in \mathbb{Z}^+$ and $i = 1, 2, \dots, n_0$. The columns of $\mathbf{A}_i$ are orthogonal with norm $\sqrt{a_i^2 + b_i^2}$. Essentially each

$$\mathbf{A}_i = \sqrt{a_i^2 + b_i^2} \begin{pmatrix} \dfrac{a_i}{\sqrt{a_i^2 + b_i^2}} & \dfrac{-b_i}{\sqrt{a_i^2 + b_i^2}} \\ \dfrac{b_i}{\sqrt{a_i^2 + b_i^2}} & \dfrac{a_i}{\sqrt{a_i^2 + b_i^2}} \end{pmatrix}$$

can be thought of as the multiple of a rotation matrix by an angle $\theta = \arctan(b_i/a_i)$.

Suppose $f \in L^2(\mathbb{R}^2)$ is an image such that $\widehat{f}(\boldsymbol{\omega}) = \widehat{\mu}(\boldsymbol{\omega})\widehat{\varphi}(\boldsymbol{\omega})$, where $\widehat{\mu} \in L^2(\mathbb{T}^2)$, and φ is a compactly supported refinable function. Images have finite extent, so $\widehat{\mu}$ is a trigonometric polynomial. In other words, the image f belongs to the zero-scale resolution space defined by the integer translates of φ. We called this space V_0 in the previous section. If we now rotate the image f by $\theta = \arctan(-b_i/a_i)$ and normalize accordingly inside the argument of f, i.e., to take $f(\det(\mathbf{A}_i)^{-1}\mathbf{A}_i^T \cdot)$, we have

$$\widehat{f}(\mathbf{A}_i\boldsymbol{\omega}) = \widehat{\mu}(\mathbf{A}_i\boldsymbol{\omega})\widehat{\varphi}(\mathbf{A}_i\boldsymbol{\omega}) .$$

If $\mathbf{A}_0 = \begin{pmatrix} 2^J & 0 \\ 0 & 2^J \end{pmatrix}$, and we set $\widehat{\varphi_0}(\boldsymbol{\omega}) = \widehat{\varphi}(\mathbf{A}_0\boldsymbol{\omega})$, with $\boldsymbol{\omega} \in \mathbb{R}^2$, then we obtain

$$\widehat{f}(\mathbf{A}_i\boldsymbol{\omega}) = \widehat{\mu}(\mathbf{A}_i\boldsymbol{\omega})\widehat{\varphi}(\mathbf{A}_i\boldsymbol{\omega}) = \widehat{\mu}(\mathbf{A}_i\boldsymbol{\omega})\widehat{\varphi}(\mathbf{A}_0(\mathbf{A}_0^{-1}\mathbf{A}_i\boldsymbol{\omega})) = \widehat{\mu}(\mathbf{A}_i^T\boldsymbol{\omega})\widehat{\varphi_0}(\mathbf{A}_0^{-1}\mathbf{A}_i\boldsymbol{\omega}) \qquad (7.5)$$

where

$$\widehat{\varphi_0}(\mathbf{A}_0^{-1}\mathbf{A}_i\boldsymbol{\omega}) = \widehat{\varphi}(\mathbf{A}_0(\mathbf{A}_0^{-1}\mathbf{A}_i\boldsymbol{\omega})) = \widehat{\varphi}(\mathbf{A}_i\boldsymbol{\omega}).$$

On the other hand, we have $\widehat{f_0}(\boldsymbol{\omega}) = \widehat{f}(\mathbf{A}_0\boldsymbol{\omega}) = \widehat{\mu}(\mathbf{A}_0\boldsymbol{\omega})\widehat{\varphi_0}(\boldsymbol{\omega})$. Then applying the transformation $\boldsymbol{\omega} \longmapsto \mathbf{A}_0^{-1}\mathbf{A}_i\boldsymbol{\omega}$, yields $\widehat{f}(\mathbf{A}_i\boldsymbol{\omega}) = \widehat{f_0}(\mathbf{A}_0^{-1}\mathbf{A}_i\boldsymbol{\omega})$. So,

$$\widehat{f_0}(\mathbf{A}_0^{-1}\mathbf{A}_i\boldsymbol{\omega}) = \widehat{\mu}(\mathbf{A}_i\boldsymbol{\omega})\widehat{\varphi_0}(\mathbf{A}_0^{-1}\mathbf{A}_i\boldsymbol{\omega}), \quad a.e. \qquad (7.6)$$

But

$$\mathbf{A}_0^{-1}\mathbf{A}_i = \frac{1}{2^J}\begin{pmatrix} a_i & -b_i \\ b_i & a_i \end{pmatrix} = \frac{\sqrt{a_i^2 + b_i^2}}{2^J}\begin{pmatrix} \dfrac{a_i}{\sqrt{a_i^2 + b_i^2}} & \dfrac{-b_i}{\sqrt{a_i^2 + b_i^2}} \\ \dfrac{b_i}{\sqrt{a_i^2 + b_i^2}} & \dfrac{a_i}{\sqrt{a_i^2 + b_i^2}} \end{pmatrix}. \qquad (7.7)$$

If we take $(a_i, b_i) \in \mathbb{Z}^+ \times \mathbb{Z}^+$, and J to be such that $\frac{\sqrt{a_i^2+b_i^2}}{2^J}$ is close to 1, then we have that $\mathbf{A}_0^{-1}\mathbf{A}_i$ is an approximate rotation by $\arctan\left(-\frac{b_i}{a_i}\right)$. Thus Eqs. (7.6) and (7.7) imply that in order to apply the approximate rotation $\mathbf{A}_0\mathbf{A}_i^{-1}$ to f_0, we have to switch to an appropriate scale. In the space domain, this operation amounts to oversampling. As we know, the accurate approximation of 2D curves requires high resolution as anyone who holds this volume with the intent to review it knows. The rotation/scaling $x \to \det(\mathbf{A}_i)^{-1}\mathbf{A}_i^T x$ amounts to changing the scale from the nominal to a coarse one, approximately by 2^{-J} and then implementing an approximate rotation by a certain angle, which will leave the integer sampling grid invariant. Since we do not leave the integer grid, all computations are accurate to the extent that the approximate rotation is satisfactory enough. Moreover, all implementations are with finite length filters (no need for filter truncations).

This method allows us to implement with high accuracy a finite number of rotations of images in 2D and using Euler angles even in 3D. What we have shown so far is that the approximate rotation (given that this statement $\frac{\sqrt{a_i^2+b_i^2}}{2^J}$ is close to 1 is satisfactory for the time being), of f_0 by $\arctan\left(-\frac{b_i}{a_i}\right)$ can be computed with precise operations in the discrete domain. Therefore to implement approximately rotations by $\arctan\left(-\frac{b_i}{a_i}\right)$, $i = 1, 2, ..., n_0$, we need to select (a_i, b_i) in such a way that $\frac{\sqrt{a_i^2+b_i^2}}{2^J}$, where J is fixed suitable scale, is close to 1. Finally, we must mention that, in practice, we work with the samples of f interlaced by J zeros per dimension. Of course, this causes aliasing and in order to remove this aliasing we need to reverse to extreme oversampling by low passing and by downsampling J times. The opposite of this procedure will be used in the next section when rotations will be implemented approximately using the nonfinite IMRA filters.

7.4 CONNECTING IMRA WITH STOCHASTIC TEXTURE MODELS USING GAUSSIAN MARKOV RANDOM FIELDS (GMRF)

Stochastic textures are often modeled by random fields. A *random field* $\mathbf{X} = \{X_s : s \in S\}$ is a set of real-valued random variables defined on the same probability space. Such a field is called *Gaussian* if any finite linear combination of the $\{X_s\}$ has a Gaussian distribution. A first step in texture discrimination or classification is to define a *texture signature*, specified as a vector of computable texture features, belonging to a fixed finite or infinite dimensional Feature Vector Space (FVS). High dimensionality of FVS formally facilitates texture discrimination by their signatures at the expense of computational cost and higher sensitivity to noise.

We work with 3D textures, which we model as random fields defined on the continuous domain $\mathbb{R}^3$. After image acquisition, realizations of these textures are digital 3D images, where we assume, that image intensities at gray levels are given only for points in a discrete sampling lattice. Nonetheless, our approach easily extends to multicomponent images. Specifically, given an arbitrary, stationary 3D texture $\mathbf{X}$ with zero mean defined on a fixed 3D rectangular lattice $\Lambda \subset \mathbb{R}^3$ similar to $\mathbb{Z}^3$, we assume that $\mathbf{X}$ is the restriction to Λ of a "continuous parent" 3D texture $\{X_s, \ s \in \mathbb{R}^3\}$ is generated by a stationary Gaussian random field which we will also denote by $\mathbf{X}$ for brevity. The autocovariance function ζ of $\mathbf{X}$ is defined by $\zeta(\boldsymbol{u}) = \mathrm{Cov}(X_s, X_{s+u})$ for all $\boldsymbol{u}$ and $\boldsymbol{s}$ in $\mathbb{R}^3$ and we assume that it satisfies the following properties:

1. The Fourier transform of ζ is C^∞.
2. The support of $\widehat{\zeta}$, is contained in a ball centered at the origin.

The second property above is used to place ζ in a IMRA subspace. The first property imposes a natural property for ζ, the rapid decay in space, which prevents correlations between distant voxels.

The **A**-rotation of a continuous 3D texture $\mathbf{X}$, denoted by $\mathbf{X^A}$ is the continuous texture such that the random variable $(X^A)_s$ corresponds to the random variable $X_{A^T s}$. Obviously, the "continuous" parent texture is not affected by 3D rigid motions. While 3D translates induced by the action of $\mathbb{Z}^3$ do not affect the discrete texture $\mathbf{X}$, due to the spatial homogeneity assumption, the discrete texture $\mathbf{X}$ is not invariant under random 3D rotations, in general. Hence, for each 3D rotation $\mathbf{A} \in SO(3)$, $\mathbf{X^A}$ is defined by first applying the rotation $\mathbf{A}$ to the continuous "parent" texture $\mathbf{X}$, and then by restricting this rotated continuous texture to Λ.

Let $b_0 > 0$ be the radius of the ball in Item 2 above and $0 < b < 1/2$ satisfies $2^{j-1}b < b_0 < 2^j b$ for some scale $j \in \mathbb{Z}^+$. We use an IMRA $\{V_j\}_j$ generated by a radial refinable function φ with C^∞ Fourier transform $\widehat{\varphi}$ such that $\widehat{\varphi}(\boldsymbol{\omega}) = 1$ for all $|\boldsymbol{\omega}| \le 2^{-j}b_0$ and $\widehat{\varphi}(\boldsymbol{\omega}) = 0$ for $|\boldsymbol{\omega}| \ge b$. Then V_0 contains all L_2-functions whose Fourier transform is supported in the ball of radius b, therefore $\zeta \in V_j$. With no loss of generality, we may and will always assume $j = 0$, so that $b/2 < b_0 < b$. Hence $\widehat{\varphi} = 1$ on the support of $\widehat{\zeta}$, which implies that for all $\boldsymbol{k} \in \mathbb{Z}^3$, $\langle \zeta, T_k \varphi \rangle = \zeta(\boldsymbol{k}) = Cov(X_k, X_0)$.

The properties of ζ imply that the autocovariance $\zeta_\mathbf{A}$ of the continuous texture $\mathbf{X^A}$ is given by $\zeta_\mathbf{A} = \mathcal{R}_\mathbf{A}\zeta$. Our goal is to calculate the autocovariance of the discrete texture $\mathbf{X^A}$ which is the restriction of its continuous counterpart on the lattice $\mathbb{Z}^3$. Equivalently, we want to estimate the sequence $\{\zeta_\mathbf{A}(\boldsymbol{k})\}_{k \in \mathbb{Z}^3}$ for all rotations $\mathbf{A}$ from the known input $\{\zeta(\boldsymbol{k})\}_{k \in \mathbb{Z}^3}$. Since $\zeta \in V_0$ we conclude that $\zeta_\mathbf{A}$ belongs to V_0 as well, hence $\zeta_\mathbf{A}$ is accurately represented by its values $\{\zeta_\mathbf{A}(\boldsymbol{k})\}_{k \in \mathbb{Z}^3}$ on the same lattice regardless of the rotation $\mathbf{A}$. Therefore the $\boldsymbol{k}$th value of $\zeta_\mathbf{A}$ is given by

$$\zeta_\mathbf{A}(\boldsymbol{k}) = \langle \mathcal{R}_\mathbf{A}\zeta, T_k\varphi \rangle = \langle \zeta, T_{\mathbf{A}k}\mathcal{R}_{\mathbf{A}^T}\varphi \rangle = \langle \zeta, T_{\mathbf{A}k}\varphi \rangle, \qquad \boldsymbol{k} \in \mathbb{Z}^3. \qquad (7.8)$$

Eq. (7.8) is, in fact, a simple steerability rule:

If $\mathbf{A}_1, \mathbf{A}_2$ are two rotations and $(\zeta_{\mathbf{A}_2})_{\mathbf{A}_1}$ is the autocovariance of the rotated texture $\mathbf{X}^{\mathbf{A}_2}$ by $\mathbf{A}_1$, then Eq. (7.8) implies

$$(\zeta_{\mathbf{A}_2})_{\mathbf{A}_1}(\boldsymbol{k}) = \langle \zeta_{\mathbf{A}_2}, T_{\mathbf{A}_1 \boldsymbol{k}}\varphi \rangle = \langle \zeta, T_{\mathbf{A}_2 \mathbf{A}_1 \boldsymbol{k}}\varphi \rangle = \zeta_{\mathbf{A}_2 \mathbf{A}_1}(\boldsymbol{k}), \qquad (7.9)$$

which shows why Eq. (7.8) is a steerability rule. The computational implementation of this rule is carried out by approximating $\zeta_{\mathbf{A}}(\boldsymbol{k}) = \langle \zeta, T_{\mathbf{A}\boldsymbol{k}}\varphi \rangle$ by $\langle \zeta, T_{2^{-j_0}\boldsymbol{k}'}\varphi \rangle$, where $\boldsymbol{k}' \in \mathbb{Z}^3$, by taking $j_0 > 0$ to be high enough so that points $2^{-j_0}\boldsymbol{k}'$ and $\mathbf{A}\boldsymbol{k}$ are sufficiently close. This computation is performed by iteratively applying j_0-steps of the reconstruction algorithm of the fast IMRA-transform (see Section 7.2).

Discrimination between two generic 3D textures $\mathbf{X}$ and $\mathbf{Y}$ relies on the fitting of multiscale GMRF models after applying arbitrary rotations to $\mathbf{X}$ and $\mathbf{Y}$. Fitting does not require $\mathbf{X}$ or $\mathbf{Y}$ to be Gaussian or stationary. When we fit a statistical model to a texture, we focus on the discriminability of the model, and we ignore that the model we use may not be the actual model of the texture. Hence parameter estimation of GMRFs is used only as a tool to extract local statistically discriminative characteristics of textures. This estimation is part of our feature extraction process. Its discriminability depends on the class of 3D textures that we deal with, and remains to be established with experimental validation. It is quite conceivable that for other types of textures more complex GMRF models may be necessary. The complexity of these parameter estimations increases quickly when one increases the size of the basic neighborhoods defining these Gibbs random fields [73]. Thus to minimize computational costs we deliberately restricted the class of GMRFs considered here to those with order-one neighborhoods. But since such models essentially capture only near correlations, we first implement a multiscale decomposition of each texture $\mathbf{X}$ by a fast IMRA decomposition algorithm and we then fit GMRFs with order-one neighborhoods to each one of these monoscale texture outputs. This multiscale parametrization does capture medium and long distance correlations between voxels. We will skip all the details about GMRF models, because they will only tire the reader of this volume. The reader may refer to [16] for all of these details.

The next step is to define a 3D rigid motion covariant signature for a discrete texture. A natural assumption for textures is that their statistical properties of interest are invariant under translations within a stationary region, which in our case should be interpreted as tissue homogeneity. Therefore the problem of 3D rigid motion invariant texture discrimination reduces to that of 3D rotationally invariant texture discrimination. Rotationally invariant discrimination between textures has been studied mostly in 2D. For instance, in [74] two circularly symmetric autoregressive models are fitted to the 2D texture data, one for the four nearest neighbors, and one for the four diagonal nearest neighbors. Circularly symmetric models with higher order neighborhoods were used in [17], where the L_1-norm of the outputs of circularly symmetric 2D Gabor filters was a key feature. A basic shortcoming of these models is that rotationally invariant discrimination between textures is achieved by handling nonisotropic textures as if they were isotropic. A mathematically rigorous treatment of isotropic textures can be found in [75]. However, generic textures often exhibit directional characteristics, which isotropic models cannot capture. The setback of these approaches is recognized in [1], where 2D textures are modeled by continuous stationary GMRFs and 2D image texture rotations are generated via a continuous version of the discrete power spectral density defined by the digitized image data. The best rotation angle matching the given unknown texture to each texture from a predefined gallery of textures is estimated by maximizing a likelihood function derived from the continuous power spectral densities. Texture classification is then carried out by comparing model parameters. This approach worked well in 2D, but a 3D version would be computationally extremely expensive, since the likelihood function in [1], is a product of terms evaluated at each node in a 3D lattice. The approach of [76], based on local binary patterns, seems to be the first attempt to build a 3D texture classification scheme robust to 3D rotations.

Our key idea is to define a computable rotationally invariant distance between any pair of adequately defined 3D texture signatures.

If $\mathbf{X}$ is a discrete texture, we fit a zero mean GMRF $G_{\mathbf{X}}(\mathbf{A}, j)$ from the class introduced in the previous section, to the rotated textures $\mathbf{X}^{\alpha}$ at multiple scales j. There is no harm to assume that mean intensities are zero. The texture is created by intensity variations, occurring in a statistically stable manner to form a texture. Later, in our statistical disparity metric between two tissues we take into account local average differences of density values.

Given two discrete 3D textures $\mathbf{X}$ and $\mathbf{Y}$ with zero intensity means our first choice for a texture "distance" at scale j would be the minimum taken over $\mathbf{C} \in SO(3)$ of the Kullback–Leibler (KL)-divergence of the GMRF models of $G_{\mathbf{X}}(\mathbf{I_3}, j)$ from $G_{\mathbf{Y}}(\mathbf{C}, j)$, where $\mathbf{I_3}$ is the 3×3 identity matrix. It seems reasonable to think that this model should in principle be able to give the best rotation matching angle between the two textures. If it does so, then $G_{\mathbf{X}}(\mathbf{A}, j)$ and $G_{\mathbf{Y}}(\mathbf{CA}, j)$ should appear statistically indistinguishable with $G_{\mathbf{X}}(\mathbf{I_3}, j)$ from $G_{\mathbf{Y}}(\mathbf{C}, j)$, respectively, as $\mathbf{A}$ runs throughout $SO(3)$. At least the KL-divergence between the first and second pair should be identical. This may not be true because the GMRF model is fitted and the $\mathbf{X}$ and $\mathbf{Y}$ are discrete textures rotated by rotations, which do not leave the sampling grid $\mathbb{Z}^3$ invariant.

In particular, the heart of the problem is not to classify textures as "same" or "not the same" by finding their best matching angle but to develop a decision function uniquely designating the value "same" or "different" to all possible configurations of their 3D rigid motions. That is to group 3D textures via an equivalence relation: Two textures $\mathbf{X}$ and $\mathbf{Y}$ are considered to be the same if any 3D rigid motion of $\mathbf{X}$ is declared to be the "same" with a 3D rigid motion of $\mathbf{Y}$. These considerations lead us to choose the orbit $\{G_{\mathbf{X}}(\mathbf{A}, j) : \mathbf{A} \in SO(3)\}$ of the fitted GMRF model of $\mathbf{X}$. Those orbits are covariant to 3D rotations and invariant to 3D translates (recall that downsampling is suspended in order to maintain invariance with respect to integer translations). Thus the problem that the first choice of texture distance appears to have, is solved if we instead find a "distance" between the orbits $\{G_X(A, j) : A \in SO(3)\}$ and $\{G_Y(CA, j) : A \in SO(3)\}$. The average over $SO(3)$ of the KL-divergences between the orbits for the fitted Gaussian fields $G_X(A, j)$ and $G_Y(A, j)$ is an obvious choice of such a "distance." In fact, we define as the 3D rigid motion invariant distance between X and Y at the scale j we the minimum taken over $C \in SO(3)$ of the average of the KL-divergences of the orbits of the GMRF models $G_X(A, j)$ and $G_Y(CA, j)$ as A traverses $SO(3)$.

The strictly stationary GMRF X is assumed to have an autocovariance function ζ tending to 0 at infinity, hence X is ergodic. Thus (see [77, Theorems III.4.2 and 4.4]) for $|\Lambda|$ large enough, each $\zeta(k)$ can be approximated with sufficient accuracy by the empirical autocovariance $\zeta_0(k)$. To compare pairs of stationary GMRFs, we use the KL-divergence between stochastic processes. The symmetrized KL-divergence of two N-dimensional Gaussian probability distributions G_1 and G_2, with zero mean and invertible covariance matrices Σ_1 and Σ_2, is given by

$$\text{KLdist}(\Sigma_1, \Sigma_2) = \frac{1}{2}\text{Trace}(\Sigma_2^{-1}\Sigma_1 + \Sigma_1^{-1}\Sigma_2 - 2I_{N\times N}). \qquad (7.10)$$

In light of (7.8), we define the *monoscale 3D rigid motion covariant texture signature of the observed texture* X *at the scale corresponding to the density of the lattice* Λ *by* $\Gamma_X(A) = (\widehat{\Sigma})_A$, $A \in SO(3)$, where $(\widehat{\Sigma})_A$ is the autocovariance matrix of the fitted GMRF model extracted from $\{(\zeta_0)_A(k)\}_{k\in\Lambda}$ by computing it from the rotated texture patch of X^A, using the IMRA filters using k as the center of the rotation. A A-rotation of the discrete texture X is the restriction of the A-rotation of its continuous parent on Λ. Rotating the discrete texture X amounts to rotating the autocovariance ζ of its continuous parent. This leads us to propose 3D texture distance:

$$Rdist(X, Y) := \min_{C \in SO(3)} \int_{SO(3)} \text{KLdist}(\Gamma_X(A), \Gamma_Y(CA)) \, dA. \qquad (7.11)$$

Since empirical autocovariances are only approximations of the true autocovariances and a very simple GMRF model fitted to the rotations of the texture, we assert that the 3D rotational covariance of Γ_X is true only approximately. This implies that the average of the symmetrized KL-divergence between two texture signatures Γ_X and Γ_Y expressed by $Rdist$ is only approximately 3D rotationally invariant [16]. The 3D shift invariance of $Rdist$ is implicit in the definition of the texture signature, so it is inherited by $Rdist$.

By taking into account the assumed symmetries of the GMRF-model [16], we select a finite 3D grid of points $(\alpha, \beta, \gamma) \in EB = [0, \pi] \times [0, \frac{\pi}{2}] \times [0, \pi]$ by defining a partition of EB into "cubes" having the same Haar measure. This is achieved by defining a grid

Π_N of points $\mathbf{A}_{(i,l,k)} = (\alpha_i, \beta_l, \gamma_k) \in EB$ where

$$\alpha_i = \left\{\frac{i\pi}{N}\right\} \quad i = 0, 1, \ldots, N-1,$$

$$\beta_l = \left\{\arccos\left(1 - \frac{l+0.5}{N}\right)\right\} \quad l = 0, 1, \ldots, N-1,$$

$$\gamma_k = \left\{\frac{k\pi}{N}\right\} \quad k = 0, 1, \ldots, N-1.$$

The discrete values of β start with 0.5 to avoid the unwieldy *gimbal lock* point $\beta_0 = 0$. Using the discrete set of rotations just defined, a computationally implementable version of the distance *Rdist* defined in Eq. (7.11) is given by

$$\min_{\mathbf{C} \in SO(3)} \frac{1}{N^3} \sum_{i=0}^{N-1} \sum_{l=0}^{N-1} \sum_{k=0}^{N-1} \mathrm{KLdist}\left(\Gamma_{\mathbf{X}}(\mathbf{A}_{(i,l,k)}), \Gamma_{\mathbf{Y}}(\mathbf{C}\mathbf{A}_{(i,l,k)})\right). \tag{7.12}$$

Once again the symmetry properties of the fitted GMRF model imply that the numerical minimization in (7.12) can be restricted to $\mathbf{C} \in EB$ [16]. In our experiments, we chose $N = 5$ or $N = 4$ for synthetic data and $N = 5$ or $N = 6$ for the experiments with the X-ray CT-data.

7.5 FEATURE SPACE FOR ROTATIONALLY INVARIANT TEXTURE DISCRIMINATION

To implement the multiscale *Rdist*-based comparison of two textures, we begin with two realizations $\mathbf{x}^{(0)}$ and $\mathbf{y}^{(0)}$ of the arbitrary 3D textures $\mathbf{X}$ and $\mathbf{Y}$ as above. Both $\mathbf{x}^{(0)}$ and $\mathbf{y}^{(0)}$ are given on an initial finite lattice $\Lambda^{(0)}$. We apply the IMRA-decomposition algorithm [75,78] to $\mathbf{x}^{(0)}$ and $\mathbf{y}^{(0)}$ in order to generate two corresponding multiscale realizations $\mathbf{x}^{(j)}$ and $\mathbf{y}^{(j)}$ defined for each of the coarser scales $j = -1, -2 - 3, -J$. Convolutions with h_0 (the inverse Fourier transform of $\widehat{h_0}$) can again be performed as fast convolution in the frequency domain. In order to maintain $\mathbb{Z}^3$-translation invariance when we apply the IMRA-decomposition algorithm, we skip the downsampling by 2. In this case, for a single application of the decomposition algorithm, the order-1 neighbors at this scale are the order-3 neighbors, *i.e.*, two voxels apart on each of the x, y, z-directions. This alternative allows to collect more data with the convention that $\Lambda^{(j)}$ is $\Lambda^{(0)}$ itself reduced by an appropriate boundary so that order-one neighborhoods at each level j are contained in $\Lambda^{(0)}$.

Subsequently, for $j = 0, -1, -2 - 3, -J$, we define $Rdist_j(\mathbf{X}, \mathbf{Y}) = Rdist(\mathbf{x}^{(j)}, \mathbf{y}^{(j)})$. The vector of these $J+1$ rotationally invariant distances $Rdist_j(\mathbf{x}, \mathbf{y})$, where $j = 0, -1, \ldots -J$, will be our multiscale 3D texture discrimination tool.

7.5.1 Self-distance for 3D textures

Theoretically the self-distance $Rdist_j(\mathbf{X}, \mathbf{X}^{\mathbf{A}})$ of the 3D texture $\mathbf{X}$ from every rotation of itself at scale j must be zero. However, when we compute the distance of a 3D texture from a rotation of itself, we only achieve an approximate minimization in formula (7.12). Furthermore, note that we implement texture rotations via a rotation of their empirical autocovariance function which adds another level of approximation to $Rdist_j(\mathbf{X}, \mathbf{X}^{\mathbf{A}})$. Hence, in practice, "self-distances" are nonzero. Therefore, discrimination between 3D textures must take these nonzero "self-distances" into account as detailed further down.

To estimate the self-distance of a 3D texture in the extensive numerical tests presented below, we generate rotations of $\mathbf{X}^{(j)}$ by 20 randomly selected $\mathbf{A}_k \in SO(3)$ ($k = 1, 2, \ldots, 20$) and define $diam_{\mathbf{X}}(j)$ to be the 80th percentile of the 20 distances $Rdist(\mathbf{X}^{(j)}, (\mathbf{X}^{(j)})^{\mathbf{A}_k})$. The choice of 20 random rotations and of the 80th percentile were reached after numerical testing which showed that these choices provided a reasonably accurate estimate of the self-distance of the texture whose realization is $\mathbf{X}$. In a nutshell, $diam_{\mathbf{X}}(j)$ is a Monte Carlo estimate of the baseline computational errors that will affect the computation of $Rdist_j(\mathbf{X}, \mathbf{Y})$, the computable distance of $\mathbf{X}$ from any other texture whose realization is $\mathbf{Y}$. This interpretation of $diam_{\mathbf{X}}(j)$ is the key to defining discrimination between 3D textures.

7.6 3D TEXTURE-BASED FEATURES

Our tissue classification scheme consists of two levels of classifiers. The design of both levels is traditional. The first level consists of an ensemble of Support Vector Machines (SVM) classifiers, which at the second level decide by majority voting whether a tissue ROI is normal or abnormal. The SVM classifiers use low-dimensional feature vectors whose components express the statistical disparity at one or more scales of the 3D texture corresponding to a given liver tissue ROI for the 3D texture of a normal reference ROI and the difference of average intensities between the two ROIs. To develop these classifiers the human operator selects a small number of reference normal ROIs from an X-ray CT-scan that is examined. The proposed feature design takes into account that a liver consists of soft tissues with varying 3D orientations thus requiring features to be invariant under 3D rotations and translations. In particular, cancer will tend to develop along blood vessels, which themselves appear with a varying degree of 3D orientations. Moreover, malignancies form their vasculature with an even richer orientation variation in 3D. Attenuation intensity local average-based features would automatically be invariant to 3D rigid motions and, therefore, insensitive to the variety of 3D orientations of the patterns formed by tissues of interest, but 3D texture features must be specifically designed to be 3D rigid motion invariant.

We stress that the 3D texture corresponding to a tissue ROI is almost never an order-one GMRF. Yet, we carry out our computations as if it were such a GMRF. We use this computa-

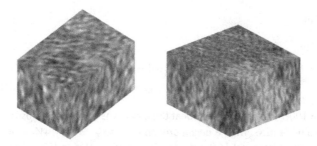

Figure 7.3 Typical 3D view of the texture of a normal liver ROI (left) and of an abnormal (neoplastic) liver ROI (right).

tionally simple and numerically efficient stochastic model as a probe for tissue discrimination and not as a texture model for soft tissues. The 3D texture features for the promised liver tissue SVM classifiers are constructed via (7.11).

Let $\{\mathbf{s}_k\}_{k=1,2,\ldots,N}$ be 3D volumes from normal or abnormal liver tissues and $\mathbf{\Lambda}_k :=$ $\mathbf{s}_k \cap \mathbf{\Lambda}$. In Fig. 7.3, we see two examples of such tissue samples. We stress that those samples do not need to have rectangular shape. Now, fix a sample $\mathbf{n}$ which is known to be normal to be the *reference normal*. For each one of the $\{\mathbf{s}_k\}_{k=1,2,\ldots,N}$, we derive the feature vector $f(\mathbf{s}_k; \mathbf{n})$ relative to $\mathbf{n}$ according to the following algorithm.

The first component of the feature vector $f(\mathbf{s}_k; \mathbf{n})$ is

$$D_M(\mathbf{s}_k, \mathbf{n}) := \frac{|\overline{\mathbf{s}_k} - \overline{\mathbf{n}}|}{\sqrt{var(\mathbf{s}_k) + var(\mathbf{n})}}.$$

D_M standardizes the statistical disparity due to the difference in the average attenuation intensity between $\mathbf{s}_k$ and $\mathbf{n}$. To form the remaining components of the feature vector $f(\mathbf{s}_k; \mathbf{n})$, we use the 3D rigid motion invariant statistical disparity $RD_j(\mathbf{s}_k, \mathbf{n})$ at scales $j = 0, -1$, between the 3D textures corresponding to $\mathbf{s}_k$ and $\mathbf{n}$ where

$$RD_j(\mathbf{s}_k, \mathbf{n}) = max \left\{ 0, \frac{Rdist_j(\mathbf{s}_k, \mathbf{n}) - diam_{\mathbf{n}}(j)}{diam_{\mathbf{n}}(j)} \right\}. \qquad (7.13)$$

[16]. If for a given sample $\mathbf{s}_k$ and at some scale j, $RD_j(\mathbf{s}_k, \mathbf{n})$ is large, then we can conclude that the sample $\mathbf{s}_k$ has a different 3D texture than the reference normal $\mathbf{n}$ $\mathbf{s}_k$, thus more the tissue from which $\mathbf{s}_k$ originated is likely to be abnormal.

BOX 7.1 Algorithm for Computation of $RD_j(\mathbf{s}_k, \mathbf{n})$

(i) Adjust intensity values in both $\mathbf{s}_k$ and $\mathbf{n}$ to have zero mean.

(ii) Upsample each $\mathbf{s}_k$ to a twice the denser grid as in [14], to approximate each $\mathbf{s}_k$ at a finer scale. The upsampled 3D texture sample is convolved with the isotropic low-pass synthesis filter $\widehat{h}_0$:

$$\widehat{h}_0(\omega) = \begin{cases} 1 & |\omega| \leq \frac{1-b}{2t}, \\ \frac{1}{2}\left[1 + \cos\left(\frac{\pi t}{b}\left(|\omega| - \frac{1-b}{2t}\right)\right)\right] & \frac{1-b}{2t} < |\omega| \leq \frac{1+b}{2t}, \\ 0 & \text{otherwise}, \end{cases}$$

where $t = 100/84$ and $b = 1/7$. Repeat the process as needed (we upscale twice).

(iii) Let τ be a node in $\mathbf{s}_k$, then a *neighborhood* of τ is $\eta_\tau := (\tau + W) \subset \Lambda_k$ where W is a symmetric neighborhood of the origin. We set $W = W^+ \cup W^-$ where $W^- = -W^+$ and $W^+ = \{(2^{-j+1}, 0, 0), (0, 2^{-j+1}, 0), (0, 0, 2^{-j+1})\}$. The order-one GMRF model limits interactions within W. Also define $\Lambda'_k \subset \Lambda_k$ to satisfy $(\eta_\tau + W) \subset \Lambda_k$ for every $\tau \in \Lambda'_k$. We extract our statistics from $\mathbf{s}_k|_{\Lambda'_k}$.

(iv) Compute the empirical autocovariance matrix ζ_0 of $\mathbf{s}_k|_{\Lambda'_k}$ via:

$$\zeta_0(\tau) = \frac{1}{|\Lambda'_k|} \sum_{r \in \Lambda'_k} s_{k_r} s_{k_{r+\tau}}, \quad \text{for all } \tau \text{ such that } \|\tau\|_\infty \leq 2^{-j+2},$$

where $|\Lambda'_k|$ denotes the number of voxels in Λ'_k; ζ_0 is of size $(2^{-j+3} + 1)^3$.

(v) Any of the 3D textures need not satisfy $\zeta_0(\tau) = \zeta_0(-\tau)$. So, we artificially symmetrize ζ'_0 by setting $\zeta'_0(\tau) := \frac{1}{2}\left[\zeta_0(\tau) + \zeta_0(-\tau)\right]$ for all $\tau \in \Lambda'_k$ such that $\|\tau\|_\infty \leq 2^{-j+2}$. To simplify the notation from now on, we use $\zeta'_0 = \zeta_0$.

(vi) Let $\mathbf{y}_r = [s_{k_l} + s_{k_{-l}}], l \in (r + W^+)$. Define $\mathbf{Y} = [\mathbf{y}_r], r \in \Lambda'_k$. The least squares estimates $\widehat{\sigma}$ and $\widehat{\boldsymbol{\beta}}$ of the order-one GMRF model that fits the data are given by the statistics: $\widehat{\boldsymbol{\beta}}(\mathbf{s}_k) = (\mathbf{Y}^T\mathbf{Y})^{-1}\mathbf{Y}^T\mathbf{s}_k$ and $\widehat{\sigma^2}(\mathbf{s}_k) = \frac{1}{|\Lambda'_k|}\left(\mathbf{s}_k^T\mathbf{s}_k - \widehat{\boldsymbol{\beta}}^T\mathbf{Y}^T\mathbf{s}_k\right)$.

(vii) $\mathbf{Y}^T\mathbf{s}_k$ and $\mathbf{Y}^T\mathbf{Y}$ are given by $(\mathbf{Y}^T\mathbf{s}_k)_r = |\Lambda'_k|(\zeta_0(r) + \zeta_0(-r)) \quad \forall r \in W^+$

$$(\mathbf{Y}^T\mathbf{Y})_{(\tau,r)} = |\Lambda'_k|[\zeta_0(r - \tau) + \zeta_0(r + \tau) + \zeta_0(-r - \tau) + \zeta_0(-r + \tau)] \; (\tau, r) \in W^+ \times W^+.$$

(viii) The computation of $(\mathbf{Y}^T\mathbf{Y})^{-1}$ is performed as in [14], using a least square solver.

(ix) Take a finite set of 3D rotations $\mathbf{A}_{(i,l,k)}$, over a grid $EB = [0, \pi] \times [0, \frac{\pi}{2}] \times [0, \pi]$, that is uniform with respect to Haar measure. Find rotations of ζ_0 on this grid to obtain $\zeta_0^{\mathbf{A}}$ for each i, l, k, using bilinear interpolation. It is important to avoid *gimbal lock* by choosing the grid points as prescribed in Section 7.4.

(x) $\Gamma_{\mathbf{X}}(\mathbf{A}) = (\widehat{\Sigma})_{\mathbf{A}}, \mathbf{A} \in SO(3)$. Hence order-one GMRF model parameters fitted on the rotated textures $\mathbf{s}_k^{\mathbf{A}}$ and $\mathbf{n}^{\mathbf{A}}$, is estimated by the correspondingly autocovariance matrix $\zeta_0^{\mathbf{A}}$ using the corresponding rotations by $\mathbf{A}$ of both $\mathbf{s}_k$ and $\mathbf{n}$.

(xi) Compute $Rdist_j(\mathbf{s}_k, \mathbf{n})$ by minimizing $F(v)$ over $v \in EB$, using a pattern search algorithm [79], where $F(v)$ is given by

$$F(v) = \frac{1}{N^3} \sum_{i=0}^{N-1} \sum_{l=0}^{N-1} \sum_{k=0}^{N-1} \text{KLdist}_j\left(\Gamma_{\mathbf{s}_k}(\mathbf{A}_{(i,l,k)}), \Gamma_{\mathbf{n}}(v\mathbf{A}_{(i,l,k)})\right).$$

A computationally efficient way to compute KLdist is described in [14].

(xii) Finally, $RD_j(\mathbf{s}_k, \mathbf{n})$ is computed for $j = 0, -1$ using the isotropic low-pass filter [16] with $t = \frac{2^{-j} \cdot 100}{84}$. We keep the low-pass output at the original resolution. To extract the order-one GMRF statistics, we use the interactions of pixels 2^{-j+1} apart and we repeat all of the previous steps with the exception of step (ii).

As noted in [16], the fitting of the same order-one GMRF model on decomposition of the same discrete texture at more than one scale, allows to incorporate texture information in the discriminative features from voxels that are not order-one neighbors. For classification we used SVMs, but it would be interesting to use a neural network with hidden layers.

7.7 CONCLUSION

In this chapter, we aimed to summarize various methods for the analysis of 3D textures with applications in biomedical imaging. We highlighted one particular method, which extracts second-order Gaussian features at multiple scales and average local intensities to generate an ensemble of discriminative features suitable for soft-tissue discrimination. The proposed features can be used with any binary classifier or even neural networks. The results and the details of the process of the experimental verification of the discriminability of the proposed features can be found in [16,80,15].

Our experiments only establish the capability of the proposed features for soft-tissue discrimination. Samples of tissue ROIs used for our experiments can be seen in Fig. 7.4. We do not claim that our method for hepatic tumor detection is a method that has been evaluated beyond the proof-of-concept stage. Experiments with synthetic textures, artificially rotated, show that combining the disparities between textures at different scales lead to rigid motion invariant classification of textures with high accuracy [16]. Experiments conducted with ROIs taken from gall bladder and muscle regions indicate that the textural disparity function has the capability of identifying tissue types that are significantly different, for instance, two organs or diseased from healthy tissue in contrast-enhanced X-ray CT scans [15]. We noticed that the rotational covariance of our features improves the clustering of normal tissues (Fig. 7.5) and that discrimination improves with the use of the local average disparity component D_M. On the other hand, for X-ray CT, D_M alone, is not sufficient for discrimination, while the vector of the textural disparities at various scales works better than D_M alone [15]. Furthermore, we wish to emphasize that stochastic textures often have statistical directional characteristics (*e.g.*, the eigendirections of the autocovariance matrix), which may not always be discernible by (convolutional) directional filters. The rotational covariance the IMRA filters improves the consistency of the extraction of these characteristics independently of the texture patch orientation (see evidence of this in Fig. 7.5 and in the results of

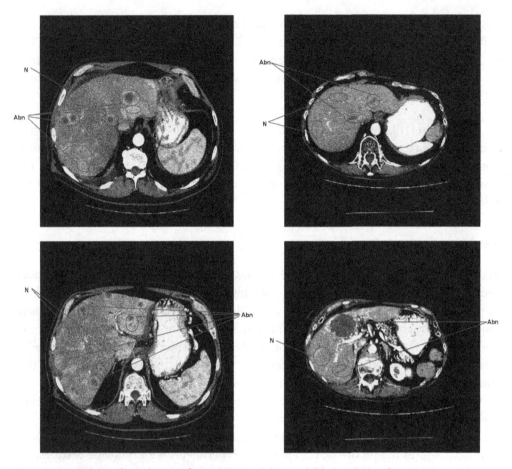

Figure 7.4 Examples of our choices of normal (N) and diseased (Abn) regions, whose cross-sections are shown in the above 2D slices. We selected many different kind of abnormalities for our experiments, which includes tumors of different sizes and from different stages

simulated texture experiments in [16]). To the best of our knowledge, there is no systematic study on how directional representations, *e.g.*, those listed in Section 2.4.2 of Chapter 2, affect the extraction of statistical directional characteristics of a stochastic texture when its orientation varies.

We have not examined how the method works with MRI or other modalities. However, the method is sufficiently broad to be customizable for other modalities.

Since the pathways for propagation can be absolutely random, we cannot expect the textures corresponding to cancerous ROIs to show any similarity. However, the normal regions of the same soft tissue type must be relatively similar to one another (see Fig. 7.4). So, the method we proposed can be utilized as an one–class classification method, where the proposed feature vectors from normal tissue will cluster, thus their

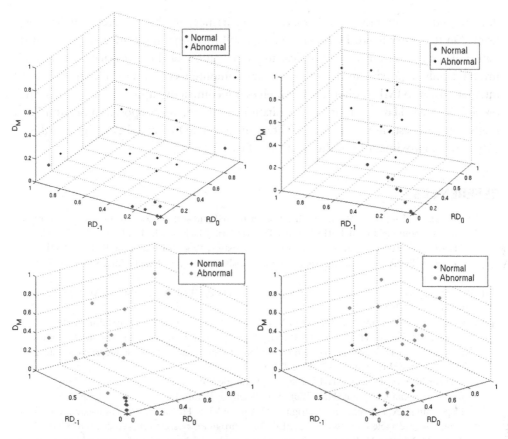

Figure 7.5 Scatterplots show that the use of 3D rotationally invariant texture disparities improves normal tissue ROI feature vector clustering. Left: RD_j's given by texture distance $Rdist_j$ (Eq. (7.11)). Most feature vectors from normal tissue ROIs cluster around the origin and clearly away from their abnormal ROI counterparts. Right: texture disparities are computed using $KLdist$ only, without averaging over 3D rotations. The normal tissue ROI cluster is less pronounced and is more proximal to abnormal tissue ROI feature vectors.

disparities will be smaller than those of abnormal regions. Clearly, what we propose is a system for screening ROIs, as if one is looking with a torchlight, and not a diagnostic system. Once an ROI is flagged, an expert will have to inspect it and render a diagnosis. Our method has the ability to combine the classical approach based on the different attenuation values resulting from tissue variation, which is what all radiologists have been trained to look for, but also 3D features which exploit the full power of the 3D information modern scanners provide.

Since we proposed this method a number of other approaches aiming to incorporate rotational invariance in the classification have been developed, exploiting the power of directional filters. However, directional filters are not compactly supported in the space domain and do not provide full rotational invariance, in their discrete implementations. On the other hand, IMRA filters offer full rotational invariance to the resulting statistical disparities, but the numerical implementation of these disparities makes this invariant only approximate. KL-distances are computed via numerically implemented integrations. Also IMRA filters are themselves not compactly supported in space as Theorem 7.2 shows. Having all of these limitations in mind, we proposed in Section 7.3 a method to, at least, use finite length filters in 2D or 3D to avoid ringing in order to extract the approximations of the rotations of each s_k.

REFERENCES

[1] F.S. Cohen, Z. Fan, M.A. Patel, Classification of rotated and scaled textured images using Gaussian Markov Random Field models, IEEE Trans. Pattern Anal. Mach. Intell. 13 (2) (1991) 192–202.

[2] R. Haralick, Statistical and structural approaches to texture, Proc. IEEE 67 (5) (1979) 610–621.

[3] S.Y. Lu, K.S. Fu, A syntactic approach to texture analysis, Comput. Graph. Image Process. 7 (1978) 303–330.

[4] R. Chin, C. Harlow, Automated visual inspection: a survey, IEEE Trans. Pattern Anal. Mach. Intell. PAMI-4 (6) (1982) 557–573.

[5] F. Tomita, S. Tsuji, Computer Analysis of Visual Textures, Kluwer Academic Publishers, 1990.

[6] P. Cirujeda, Y.D. Cid, H. Müller, D. Rubin, T.A. Aguilera, B.W. Loo, M. Diehn, X. Binefa, A. Depeursinge, A 3-D Riesz-covariance texture model for prediction of nodule recurrence in lung CT, IEEE Trans. Med. Imaging 35 (12) (Dec 2016) 2620–2630.

[7] A. Depeursinge, A. Foncubierta-Rodriguez, D. Van de Ville, H. Müller, Multiscale Lung Texture Signature Learning Using the Riesz Transform, Springer, Berlin, Heidelberg, 2012, pp. 517–524.

[8] A. Depeursinge, A. Foncubierta-Rodriguez, D. Van de Ville, H. Müller, Rotation-covariant texture learning using steerable Riesz wavelets, IEEE Trans. Image Process. 23 (2) (Feb 2014) 898–908.

[9] F. Orlhac, C. Nioche, M. Soussan, I. Buvat, Understanding changes in tumor textural indices in pet: a comparison between visual assessment and index values in simulated and patient data, J. Nucl. Med. (2016).

[10] A. Kassner, R.E. Thornhill, Texture analysis: a review of neurologic MR imaging applications, Am. J. Neuroradiol. 31 (5) (2010) 809–816.

[11] A. Depeursinge, A. Foncubierta-Rodríguez, A. Vargas, D. Van de Villey, A. Platon, P.-A. Poletti, H. Müller, Rotation-covariant texture analysis of 4D dual-energy CT as an indicator of local pulmonary perfusion, in: 2013 IEEE 10th International Symposium on Biomedical Imaging (ISBI), IEEE, 2013, pp. 145–148.

[12] A. Depeursinge, A. Foncubierta-Rodriguez, D. Van De Ville, H. Müller, Three-dimensional solid texture analysis in biomedical imaging: review and opportunities, Med. Image Anal. 18 (January 2014) 176–196.

[13] I.M. Elfadel, R.W. Picard, Gibbs random fields, cooccurrences, and texture modeling, IEEE Trans. Pattern Anal. Mach. Intell. 16 (1) (1994) 24–37.

[14] S. Upadhyay, S. Jain, M. Papadakis, R. Azencott, 3D-rigid motion invariant discrimination and classification of 3D-textures, Proc. SPIE 8138 (2011) 813821.

[15] S. Upadhyay, M. Papadakis, S. Jain, G.W. Gladish, I.A. Kakadiaris, R. Azencott, Semi-automatic discrimination of normal tissue and liver cancer lesions in contrast enhanced X-ray CT-scans, in: H. Yoshida, D. Hawkes, M. Vannier (Eds.), Proc. Computational and Clinical Applications in Abdominal Imaging, Nice, France, vol. 7601, Springer, October 1–5, 2012, pp. 158–167.

[16] S. Jain, M. Papadakis, S. Upadhyay, R. Azencott, Rigid motion invariant classification of 3D-textures, IEEE Trans. Image Process. 21 (5) (May 2012) 2449–2463.

[17] R. Porter, N. Canagarajah, Robust rotation-invariant texture classification: wavelet, Gabor filter and GMRF based schemes, IEE Proc., Vis. Image Signal Process. 144 (3) (1997) 180–188.

[18] A. Teuner, O. Pichler, B.J. Hosticka, Unsupervised texture segmentation of images using tuned Gabor filters, IEEE Trans. Image Process. 4 (6) (1995) 863–870.

[19] R. Azencott, J.P. Wang, L. Younes, Texture classification using Gabor filters, IEEE Trans. Pattern Anal. Mach. Intell. 19 (2) (1997) 148–152.

[20] D. Dunn, W.E. Higgins, Optimal Gabor filters for texture segmentation, IEEE Trans. Pattern Anal. Mach. Intell. 16 (1994) 947–964.

[21] T.S. Lee, Image representation using 2D Gabor wavelets, IEEE Trans. Pattern Anal. Mach. Intell. 18 (1996) 959–971.

[22] K. Etemad, R. Chellappa, Separability based tree-structured local basis selection for texture classification, in: ICIP 1994, vol. 3, Austin, TX, 1994.

[23] T. Chang, C. Kuo, Texture analysis and classification with tree-structured wavelet transform, IEEE Trans. Image Process. 2 (4) (1995) 429–441.

[24] M. Unser, Texture classification and segmentation using wavelet frames, IEEE Trans. Image Process. 2 (4) (1995) 429–441.

[25] M. Do, M. Vetterli, Rotation invariant texture characterizaton and retrieval using steerable wavelet domain hidden Markov models, IEEE Trans. Multimed. 4 (4) (2002) 517–527.

[26] H. Choi, R. Baraniuk, Multiscale image segmentation using wavelet domain hidden Markov models, in: Conf. Mathem. Model., Bayes Est., Inv. Prob., Proc. SPIE 3816 (1999) 306–320.

[27] S.K. Alexander, R. Azencott, M. Papadakis, Isotropic multiresolution analysis for 3D-textures and applications in cardiovascular imaging, in: D. Van Der Ville, V. Goyal, M. Papadakis (Eds.), Wavelets XII, Proc. SPIE 6701 (2007) 67011S.

[28] R.L. Kashyap, Analysis and synthesis of image patterns by spatial interaction models, in: Progress in Pattern Recognition, 1981, pp. 149–186.

[29] R. Chellappa, Two-dimensional discrete Gaussian Markov random field models for image processing, in: Progress in Pattern Recognition, vol. 2, 1985, pp. 79–112.

[30] J. Mao, A.K. Jain, Texture classification and segmentation using multiresolution simultaneous autoregressive models, Pattern Recognit. 25 (2) (1992) 173–188.

[31] G.R. Cross, A.K. Jain, Markov random field texture models, IEEE Trans. Pattern Anal. Mach. Intell. PAMI (5) (1983) 25–39.

[32] C. Bouman, B. Liu, Multiple resolution segmentation of textured images, IEEE Trans. Pattern Anal. Mach. Intell. PAMI-13 (2) (1991) 99–113.

[33] R. Azencott, C. Graffigne, C. Labourdette, Edge detection and segmentation of textured images using Markov fields, in: Lecture Notes in Statistics, vol. 74, Springer-Verlag, 1992, pp. 75–88.

[34] S. Geman, D. Geman, Stochastic relaxation, Gibbs distributions, and the Bayesian restoration of images, IEEE Trans. Pattern Anal. Mach. Intell. 6 (1984) 721–741.

[35] H. Derin, H. Elliott, Modeling and segmentation of noisy and textured images using Gibbs random fields, IEEE Trans. Pattern Anal. Mach. Intell. PAMI-9 (1) (1987) 39–55.

[36] R.L. Kashyap, R. Chellappa, Estimation and choice of neighbors in spatial-interaction models of images, IEEE Trans. Inf. Theory 1 (1983) 60–72.

[37] E. Boykova Ranguelova, Segmentation of Textured Images on Three-Dimensional Lattices, PhD Thesis, University of Dublin, Trinity College, 2002.

[38] V.A. Kovalev, F. Kruggel, H.J. Gertz, D.Y. von Cramon, Three-dimensional texture analysis of MRI brain datasets, IEEE Trans. Med. Imaging 20 (5) (2001) 424–433.

[39] A.S. Kurani, D.H. Xu, J.D. Furst, D.S. Raicu, Co-occurrence matrices for volumetric data, in: 7th IASTED International Conference on Computer Graphics and Imaging, 2004.

[40] D.H. Xu, A.S. Kurani, J.D. Furst, D.S. Raicu, Run-length encoding for volumetric texture, in: 4th IASTED International Conference on Visualization, Imaging and Image Processing, 2004.

[41] A. Madabhushi, M. Feldman, D. Metaxas, D. Chute, J. Tomaszewski, A novel stochastic combination of 3D texture features for automated segmentation of prostatic adenocarcinoma from high resolution MRI, in: Medical Image Computing and Computer-Assisted Intervention, vol. 2878, 2003, pp. 581–591.

[42] C.C. Reyes-Aldasoro, A. Bhalerao, Volumetric feature selection for MRI, Inf. Process. Med. Imag. 2732 (2003) 282–293.

[43] D. Pescia, N. Paragios, S. Chemouny, Automatic detection of liver tumors, in: 5th IEEE International Symposium on Biomedical Imaging: From Nano to Macro, May 2008, pp. 672–675.

[44] R.M. Haralick, K. Shanmugam, I. Dinstein, Textural features for image classification, IEEE Trans. Syst. Man Cybern. 3 (1973) 610–621.

[45] J. Liu, Q. Hu, Z. Chen, P. Heng, Adaptive liver segmentation from multi-slice CT scans, in: Y. Peng, X. Weng (Eds.), 7th Asian-Pacific Conference on Medical and Biological Engineering, in: Proc. IFMBE Proceedings, vol. 19, Springer, 2008, pp. 381–384.

[46] R. Susomboon, D. Raicu, J. Furst, T.B. Johnson, A co-occurrence texture semi-invariance to direction, distance and patient size, in: J.M. Reinhardt, J.P.W. Pluim (Eds.), Proc. Medical Imaging 2008: Image Processing, Proc. SPIE 6914 (2008).

[47] M. Pham, R. Susomboon, T. Disney, D. Raicu, J. Furst, A comparison of texture models for automatic liver segmentation, in: J.P.W. Pluim, J.M. Reinhardt (Eds.), Proc. Medical Imaging 2007: Image Processing, Proc. SPIE 6512 (2007).

[48] N.H. Abdel-massieh, M.M. Hadhoud, K.M. Amin, Automatic liver tumor segmentation from CT scans with knowledge-based constraints, in: Proc. Biomedical Engineering Conference (CIBEC), 2010 5th Cairo International, Cairo International Biomedical Engineering Conference, IEEE, December 2010, pp. 215–218.

[49] A. Militzer, T. Hager, F. Jager, C. Tietjen, J. Hornegger, Automatic detection and segmentation of focal liver lesions in contrast enhanced CT images, in: 2010 20th International Conference on Proc. Pattern Recognition (ICPR), IEEE, August 2010, pp. 2524–2527.

[50] L. Massoptier, S. Casciaro, A new fully automatic and robust algorithm for fast segmentation of liver tissue and tumors from CT scans, Eur. Radiol. 18 (8) (2008) 1658–1665, http://dx.doi.org/10.1007/s00330-008-0924-y.

[51] Z.H. Zhou, X.Y. Liu, On multi-class cost-sensitive learning, Comput. Intell. 26 (3) (August 2010) 232–257.

[52] K.S. Seo, Automatic hepatic tumor segmentation using composite hypothesis, in: Second International Conference on Image Analysis and Recognition (ICIAR '05), in: Proc. Lecture Notes in Computer Science, vol. 3656, Springer, 2005.

[53] S.J. Park, K.S. Seo, J.A. Park, Automatic hepatic tumor segmentation using statistical optimal threshold, in: V. Sunderam, G. van Albada, P. Sloot, J. Dongarra (Eds.), 5th International Conference On Computational Science (ICCS '05), in: Proc. Lecture Notes in Computer Science, vol. 3514, 2005.

[54] M.P. Jolly, L. Grady, 3D general lesion segmentation in CT, in: Proc. Biomedical Imaging: From Nano to Macro. 5th IEEE International Symposium on Biomedical Imaging, IEEE, 2008.

[55] T. Chen, D. Metaxas, A hybrid framework for 3D medical image segmentation, Med. Image Anal. 9 (6) (2005) 547–565.

[56] I. Daubechies, Ten Lectures on Wavelets, CBMS, vol. 61, SIAM, 1992.

[57] J.R. Romero, S.K. Alexander, S. Baid, S. Jain, M. Papadakis, The geometry and the analytic properties of isotropic multiresolution analysis, Adv. Comput. Math. 31 (1) (2009) 283–328.

[58] E.H. Adelson, E. Simoncelli, R. Hingoranp, Orthogonal pyramid transforms for image coding, in: Visual Communications and Image Processing II, Proc. SPIE 845 (1987) 50–58.

[59] E.P. Simoncelli, W.T. Freeman, E.H. Adelson, D.J. Heeger, Shiftable multi-scale transforms, IEEE Trans. Inf. Theory 38 (2) (1992) 587–607.

[60] A. Karasaridis, E. Simoncelli, A filter design technique for steerable pyramid image transforms, in: Acoustics Speech and Signal Processing, ICASP, Atlanta, GA, May 7–10 1996.

[61] W.T. Freeman, E.H. Adelson, The design and use of steerable filters, IEEE Trans. Pattern Anal. Mach. Intell. 13 (9) (1991) 891–906.

[62] J. Portilla, V. Strela, M.J. Wainwright, E.P. Simoncelli, Image denoising using Gaussian scale mixtures in the wavelet domain, IEEE Trans. Image Process. 12 (2003) 1338–1351, Computer Science techn. rep. nr. TR2002-831, Courant Institute of Mathematical Sciences.

[63] J.M. Coggins, A.K. Jain, A spatial filtering approach to texture analysis, Pattern Recognit. Lett. 3 (3) (1985) 195–203.

[64] A.K. Jain, F. Farrokhnia, Unsupervised texture segmentation using Gabor filters, IEEE Trans. Pattern Anal. Mach. Intell. 18 (2) (1991) 1167–1186.

[65] P. Heinlein, J. Drexl, W. Schneider, Integrated wavelets for enhancement of microcalcifications in digital mamography, IEEE Trans. Med. Imaging 22 (3) (2003) 402–413.

[66] G. Tzgakarakis, B. Beferull-Lozano, P. Tsakalides, Rotation-invariant texture retrieval with Gaussianized steerable pyramids, IEEE Trans. Image Process. 15 (9) (2006) 2702–2718.

[67] Y. Wan, R.D. Nowak, Quasi-circular rotation invariance in image denoising, in: ICIP, vol. 1, 1999, pp. 605–609.

[68] M. Unser, N. Chenouard, A unifying parametric framework for 2D steerable wavelet transforms, SIAM J. Imaging Sci. 6 (1) (2013) 102–135.

[69] N. Chenouard, M. Unser, 3D steerable wavelets in practice, IEEE Trans. Image Process. 21 (11) (2012) 4522–4533.

[70] M. Unser, N. Chenouard, D. Van De Ville, Steerable pyramids and tight wavelet frames in $l_2(\mathbb{R}^d)$, IEEE Trans. Image Process. 20 (10) (October 2011) 2705–2721.

[71] M. Papadakis, G. Gogoshin, I.A. Kakadiaris, D.J. Kouri, D.K. Hoffman, Non-separable radial frame multiresolution analysis in multidimensions, Numer. Funct. Anal. Optim. 24 (2003) 907–928.

[72] E. Hernández, G. Weiss, A First Course on Wavelets, CRC Press, Boca Raton, FL, 1996.

[73] S. Lakshmanan, H. Derin, Valid parameter space for 2-D Gaussian Markov Random Fields, IEEE Trans. Inf. Theory 39 (2) (1993) 703–709.

[74] R.L. Kashyap, A. Khotanazad, A model-based method for rotation invariant texture classification, IEEE Trans. Pattern Anal. Mach. Intell. 8 (4) (1986) 472–481.

[75] M. Papadakis, B.G. Bodmann, S.K. Alexander, D. Vela, S. Baid, A.A. Gittens, D.J. Kouri, S.D. Gertz, S. Jain, J.R. Romero, X. Li, P. Cherukuri, D.D. Cody, G.W. Gladish, I. Aboshady, J.L. Conyers, S.W. Casscells, Texture-based tissue characterization for high-resolution CT-scans of coronary arteries, Commun. Numer. Methods Eng. 25 (2009) 597–613.

[76] J. Fehr, H. Burkhardt, 3D rotation invariant local binary patterns, in: ICPR 2008, Tampa, FL, 2008.

[77] R. Azencott, D. Dacunha-Castelle, Series of Irregular Observations, Springer-Verlag, 1986.

[78] B.G. Bodmann, M. Papadakis, et al., Frame isotropic multiresolution analysis for micro CT scans of coronary arteries, in: M. Papadakis, A. Laine, M. Unser (Eds.), Wavelets XI, Proc. SPIE 5914 (2005) 59141O.

[79] C. Audet, J.E. Dennis Jr., Analysis of generalized pattern searches, SIAM J. Optim. 13 (3) (2002) 889–903.

[80] S. Upadhyay, S. Jain, M. Papadakis, R. Azencott, 3D-rigid motion invariant discrimination and classification of 3D-textures, in: M. Papadakis, D. Van De Ville, V. Goyal (Eds.), Wavelets and Sparsity XIV, Proc. SPIE 8138 (2011), http://dx.doi.org/10.1117/12.891721.

CHAPTER 8

An Introduction to Radiomics: An Evolving Cornerstone of Precision Medicine

Sara Ranjbar*, J. Ross Mitchell†
*Arizona State University, Tempe, AZ, USA
†Mayo Clinic Arizona, Phoenix, AZ, USA

Abstract

Over the past decade a compelling body of literature has emerged suggesting a more pivotal role for imaging in the diagnosis, prognosis, and monitoring of diseases. These advances have facilitated the rise of an emerging practice known as Radiomics: the extraction and analysis of large numbers of quantitative features from medical images to improve disease characterization and prediction of outcome. Our aim is to clarify the potential clinical utility of radiomics specifically in solid tumor cancer care. First, we discuss the clinical processes of cancer management (*i.e.*, diagnosis, monitoring, and treatment). Second, the limitations of the current process are discussed followed by a discourse of the benefits of radiomics. Third, we provide a high-level description of the workflow of radiomics and explain how it relates to biomedical texture analysis. We conclude with a review of the literature in this field and the remaining challenges ahead.

Keywords

Imaging, Radiomics, Solid tumors, Radiogenomics, Quantitative features

8.1 INTRODUCTION

The rise of high-throughput tools capable of massive processing of genomics data has led to exciting advances in the field of computational biology. The ability to translocate gene expression profiles of various tissues has provided us with an unprecedented snapshot of disease signatures on a molecular and cellular level [1]. These exciting advances are slowly influencing how we view other data sources in health care. Imaging has traditionally been viewed as a largely qualitative diagnostic tool; a means to provide important anatomical and morphological information about the clinical question at hand. The vast amount of available high quality imaging data is an excellent candidate for the analytical high-throughput approach.

Numerous studies have considered imaging as a source of quantitative mineable data to advance diagnosis, prognosis, and the monitoring of various diseases, leading to the birth of a new field termed "radiomics" [1–4]. Radiomics is an area of quantitative imaging research designed to maximize the extraction of mineable high-dimensional features from diagnostic imaging studies. The body of research advocating radiomics has a common promise: that the extracted information from images might prove to be complementary or interchangeable with other sources, *i.e.*, demographics, pathology, blood biomarkers, and genomics [3]. By analyzing the content of imaging data, we can potentially capture the underlying intratumoral and molecular structure of the lesions and their environment [5]. Hence radiomics could potentially enhance the clinical decision making process of cancer care [2].

One of the emerging branches of radiomics research, known as "radiogenomics," focuses on mapping the quantitative imaging input to the findings of computational biology. A major motivation behind radiogenomics studies is that imaging is already part of routine diagnostics, an advantage that gene expression profiling is only beginning to develop [1,3]. Also, in contrast to genomics profiling, virtually all patients go through imaging multiple times during their treatment [2] and, given the noninvasive nature of the procedure, imaging can be repeated as often as required. Establishing association maps between radiomics patterns and previously determined treatment response or gene-expression findings can lead to noninvasive imaging signatures for the genomics pathways and a shift toward individualized medicine in routine care.

The aim of this chapter is to provide an improved understanding of the clinical utility of radiomics – and by extension radiogenomics – in solid tumor cancer care. Section 8.2 provides a review of the overall clinical decision making process of solid tumor cancers, along with its limitations. Next, in Section 8.3, we discuss the potential contributions of radiomics to the improvement of cancer care at various stages of the decision-making process. This section is followed by a brief overview of the radiomics workflow in Section 8.4, a glance at some recent advancement in Section 8.5, and the challenges ahead in Section 8.6.

8.2 BACKGROUND ON CANCER CARE

Cancer care of solid tumors involves the detection, diagnosis, characterization, treatment, and monitoring of the disease. Detection begins when, upon an initial indication, a radiology exam is requested for a patient. The radiologist then prompts an exam by determining the content of the order based on the clinical context for the initial request, including the appropriate imaging modality (Computed Tomography or CT, Magnetic Resonance Imaging or MRI, etc.), body part (breast, lung, brain, etc.), and image acquisition protocol (*e.g.*, with/without injection of contrast agent) [6]. Upon receiving the exam request, a radiology session is scheduled and the images are acquired. The radiologist reviews the images, and prepares a report containing the image findings. Image findings typically contain detailed information regarding the key attributes of lesion(s) including anatomic location, qualitative, and quantitative description of the lesion. The

RECIST criteria

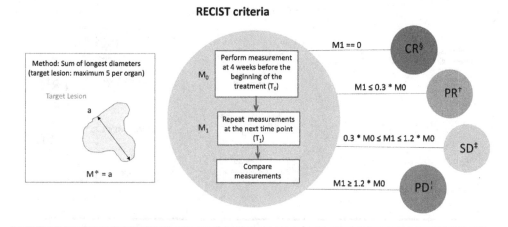

*Measurement, § complete response, †partial response, ‡stable disease, ⁱprogressive disease

Figure 8.1 Solid tumor response assessment in RECIST.

reported qualitative attributes of the lesion(s) include the radiologist's subjective visual description of the lesion such as lesion shape, margin, and border or density.

Quantitative attributes of lesions are measures of tumor burden estimation and are rarely reported in standard of care practice at initial detection [6], but are rather reported in the following imaging studies during monitoring and the treatment response assessment process. The contents of a radiology report at this stage includes the above-mentioned attributes, as well as an estimation of comparison of lesion attributes (qualitative and quantitative) over time. The criteria for measuring the lesion quantitative attributes are defined by the Response Evaluation Criteria In Solid Tumors (RECIST), and/or Wold Health Organization (WHO). As shown in Figs. 8.1 and 8.2, estimation of tumor burden in clinical practice relies on human measurement of one- or two-dimensional descriptors of tumor size [7,8].

Following the initial diagnosis, the next step is characterization of the detected lesion. The gold standard for characterization of lesions is the histopathologic analysis of tissue samples acquired through surgical biopsy or resection [9]. Initial diagnosis (benign or malignant, as well as grading) and genetic status of solid tumors are determined from tissue samples obtained via biopsy.

The next step is establishing the treatment plan which may involve a combination of three options: tumor resection through surgery, radiation therapy, and chemotherapy. Radiotherapy is a major part of the treatment process for roughly half of cancer patients [10]. It is particularly useful for cancers of the head and neck, lung, bladder, and prostate [10]. It is also used as a presurgery treatment to shrink the tumor and control the spread of the disease (*i.e.*, in breast cancer) [10]. Radiotherapy is sometimes the only

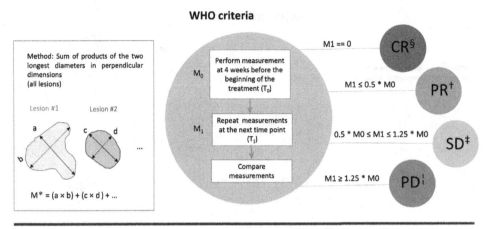

WHO criteria

Method: Sum of products of the two longest diameters in perpendicular dimensions (all lesions)

Lesion #1   Lesion #2

$M^* = (a \times b) + (c \times d) + ...$

M_0 — Perform measurement at 4 weeks before the beginning of the treatment (T_0)

M_1 — Repeat measurements at the next time point (T_1)

Compare measurements

M1 == 0 → CR§

M1 ≤ 0.5 * M0 → PR†

0.5 * M0 ≤ M1 ≤ 1.25 * M0 → SD‡

M1 ≥ 1.25 * M0 → PDⁱ

*Measurement, § complete response, † partial response, ‡ stable disease, ⁱ progressive disease

Figure 8.2 Solid tumor response assessment in WHO.

treatment option for locally advanced cancers where the location of the lesion renders surgery an unsuitable option (*i.e.*, cancer of the larynx) [10]. The prescribed radiation dose depends of the type of cancer, the tolerance of the surrounding normal tissue, and the amount of radiotherapy toxicity up to five years past treatment [10].

Treatment of cancer via chemotherapy heavily depends on the information provided via the acquired tissue samples. For decades, malignant lesion chemotherapy treatment has involved targeting rapidly dividing cancer cells using intravenous cytotoxic chemotherapy. Due to the blind nature of this type of treatment, many rapidly dividing normal cells (*e.g.*, hair, gastrointestinal epithelium, bone marrow) are also affected leading to classic toxicities for the patients (*i.e.*, of alopecia, gastrointestinal symptoms, and myelosuppression) [11]. Cancer treatment has experienced a dramatic change due to the new advances toward targeted therapies. In the age of targeted therapies, specific monoclonal antibodies and small molecule inhibitors are targeted to disrupt distinct mechanisms of tumor cells. Elements such as the lesion cellular environment, including presence/absence of enzyme and protein expression, cell type, presence/absence of gene alterations, level of oxygenation within the cell, and growth factors, all affect the course of treatment [12].

8.2.1 Biomarkers and cancer care

The advancements toward targeted therapies go hand-in-hand with the advancements of biomarker research. The high prevalence of cancer has prompted significant promotion in biomarker research, even though the utility of biomarkers in the clinical domain does not start or end with cancer.

Table 8.1 Biomarker terminology reference (Hodgson et al. [14])

Term	Definition
Biological marker (biomarker)	A characteristic that is objectively measured and evaluated as an indicator of normal biologic processes, pathogenic processes, or pharmacological responses to a therapeutic intervention.
Clinical endpoint	A characteristic or variable that reflects how a patient feels or functions, or how long a patient survives.
Surrogate endpoint	A biomarker intended to substitute for a clinical endpoint. A clinical investigator uses epidemiologic, therapeutic, pathophysiologic, or other scientific evidence to select a surrogate endpoint that is expected to predict clinical benefit, harm, or lack of benefit.
Useful biomarker	(1) Informs risk/benefit ratio when decisions are to be made. (2) Does so in a better/safer/faster/earlier/cheaper way than existing approaches. (3) Generally applicable: sample and technology must be available/accessible. (4) Has a known identity.

As presented in Table 8.1, a *biomarker* is a test that can be objectively measured and evaluated as an indicator of normal (or abnormal) biologic processes [13]. Within the context of solid tumors, biospecimen samples such as biopsies serve as the main source of biomarkers used in the treatment and monitoring of cancer.

Biomarkers play major roles in oncology. They help identify cancer stage, monitor the disease progress, forecast the effect of treatment, and predict outcome (*e.g.*, time to relapse, survival) [15]. Due to their direct correlation with clinical endpoints, biomarkers can help determine optimal dosing for treatment [9] and can lead to smaller sample sizes in clinical trials [15]. Biomarker development offers many benefits including faster, cheaper, and better diagnosis and treatment, as well as efficiency in the decision making process [15,16].

An imaging biomarker has been defined as "an imaged characteristic that is objectively measured and evaluated as an indicator of normal biological processes, pathogenic processes, or a response to a therapeutic intervention" [17]. The lesion size remains the only recognized imaging biomarker used in monitoring the response to treatment.

8.2.2 Limitations of response assessment process

Tumor response criteria solely rely on the measurements of lesion size and disregard the complexity of the lesion morphology and changes in tumor behavior [4]. However, lesion size change is not always a sufficient measure of response to treatment as the changes in tumor dynamics are considered to start much earlier than the subsequent change in lesion size. Moreover, change in tumor size often does not predict patient prognosis such as overall or progression-free survival [18,6].

Other downsides of this process are the interreader and intrareader measurement variability [19], along with inconsistency between readers in the selection of target and nontarget lesions for inclusion in the calculation [6]. These issues are so widely known that the pharmaceutical companies often employ "independent review panels" to standardize tumor response measurements in clinical trials [8]. Based on a report by Thiesse et al. [20] on a large multicenter trial in oncology, the independent review panels and the case radiologists were found to have major disagreement in 40% of the cases and minor disagreements in a further 10.5% of cases. Qualitative descriptions of key characteristics such as margin, shape, border, and intensity patterns (enhancement pattern, vascularity of breast lesions) also suffer from subjectivity and measurement variability across readers.

8.2.3 Limitations of characterization process

Biopsy samples are obtained through invasive procedures that extract tissue from a single location within the lesion. These invasive procedures cannot be performed with too much frequency; in fact, oncologists normally rely on the information provided via two or three biopsies for their decisions during the whole course of treatment. Thus biopsies are narrow in scope and do not reflect the variability of molecular structure of tumors [3]. In fact, they can only serve as a snapshot of the tumor's molecular structure at the time and location of biopsy sampling. This sets a limitation on the utility of biopsy samples given that the heterogeneity of solid tumors has been established at the phenotypic, physiologic, and genomics levels [21–23].

Another drawback of biopsies is the narrow temporal scope of the information that they can provide. Given that cancer is a phenotypically and molecularly progressive disease, the molecular context of the disease changes over time [14]. Hence a single biopsy at a single time point may not be able to convey the changes in the molecular context of the lesion.

Biopsy procedures are susceptible to sampling error. Closed biopsies (with a needle, endoscopy, the preferred biopsy procedure in complicated cases) have been shown to have a lower rate of accuracy compared to open biopsies (with an incision), especially for soft-tissue tumors [24]. The histopathology reports may contain errors as well. In one study [25], prostate pathology reports were found to have an error rate of 44% in the Gleason scores. The authors found that in 10% of the pathology reports the errors were significant enough to change the course of treatment. Another study [26] found that 7.8% of the breast cancer pathology reports included errors significant enough to affect the choice between a lumpectomy and mastectomy for the patients.

8.2.4 Limitations of current biomarkers

Targeted therapies are by no means one size fits all; some treatments may only be effective in certain groups within the affected population whose cancers have a specific molecular target, but not be effective in the absence of such a target [11]. Further advancement of targeted therapies depends on multiple factors: targeting molecularly defined subsets of several tumor types, identifying patients with those tumor types, and developing efficient strategies to evaluate targeted agents in patients [9]. All the above, in one way or another, depend on advancement in biomarker assay development.

Most new potential treatments fail to obtain regulatory approval partly due to clinical trial biomarkers that are poor predictors of treatment effectiveness. An ideal cancer biomarker should follow the changes of the molecular context in which they find themselves. However, biospecimen biomarkers inherit the limitations of their source: biopsy samples. They offer limited *longitudinal granularity*: a quality that is critical for assessing the impact of therapeutic interventions on the molecular context of the disease [27].

8.3 THE POTENTIAL AREAS OF RADIOMICS UTILITY

Quantitative imaging has a well-established footprint in cancer research. Imaging traits have been linked to a variety of diagnostic applications including characterization of healthy and pathologic tissues, differentiation of tumor regions, as well as in tumor grading. Some examples in this regard are the imaging studies in brain [28–31], breast [32–37], lung [38,39], and prostate [40] cancers. Imaging qualities have also been used to predict prognosis markers such as time to progression and survival rate of brain tumors [41], nonsmall cell lung cancer [42], colorectal cancer [43–45], and renal cell cancer [46].

The radiomics approach is comparable to the methodology of the above studies in the exploration of imaging data treatment monitoring, outcome prediction, or as potential biomarkers. However, it differs from them in the high throughput approach to quantification of tumor phenotypes. By facilitating maximal extraction of information from diagnostic imaging, radiomics not only leads to comprehensive assessment of the clinical utility of the imaging features, but also serves as a bench mark to standardize feature extraction and analysis processes. Hence this approach has been used to assess robustness and reproducibility of features [47], to determine the significance of inter-scanner variability [48] and the optimal image intensity quantization method [49], and to compare across feature selection and classification methods [50].

Radiomics data has the promise to reflect the complexity of tumor morphology, composition, and behavior. By promptly detecting small and early focal responses, this data can lead to enhanced tumor characterization and response assessment. Advocates of radiomics are motivated by the fact that imaging is nearly noninvasive (completely when no contrast agent is injected) and can shed light on the lesion state in their entirety even in complicated cases where biopsy procedures are risky. Through enabling earlier termination of wrong treatment plans, the radiomics approach can induce faster decisions, lower the cost of treatment, and enhance the patient quality of life during treatment.

Growth of radiomics research can potentially lead to the progress of imaging biomarker research. Alizadeh et al. [51] identify blood-borne biomarkers as the main noninvasive physical source for gaining insight into the molecular composition of tumors. Cell-free DNA and circulating tumor cells can be accessed through blood samples to provide insight into the molecular composition of tumors. However, blood samples have a fundamental limitation as they discuss: "It is unclear whether primary tumors or metastases contribute more to the pool of circulating cancer material. It seems clear, however, that even if circulating material is found to faithfully reflect the tumor itself, there is still a need for more efficient ways of isolating the cells and nucleic acids from the blood and for data analysis tools that can more faithfully reconstruct the parent tumor." Upon validation, radiomics imaging traits can breed noninvasive biomarkers that can be identified as surrogates for important clinical outcomes [52]. These biomarkers can be utilized in various stages of cancer care such as detection, diagnosis, assessment of prognosis, prediction of response to treatment, and monitoring [2].

Another potential benefit for this research is advancing targeted therapies toward a more genetically informed paradigm through radiogenomics [53]. The term "radiogenomics" has been used in the literature to describe two quite different concepts: (1) the practice of analyzing whole genome to enhance our understanding of molecular pathogenesis and risk factors of variation in radiotherapy toxicity [54,10,55], (2) establishing association maps between radiomics data and genomics (or other -omics) data. To clarify, we mainly refer to the latter in this review. The term "imagenomics" has also been used in the literature to describe this field [56]. Radiogenomics has elicited great interest among imaging and oncology researchers. Given that genomics profiling is not available for all the patients that go through cancer treatment, radiogenomics can offer two potential uses. First, radiogenomics offers the potential to provide association maps of genomics markers across the whole lesion. Considering that genomics markers are local and temporal, radiogenomics can suggest where certain genes are expressed or mutated in other parts of the lesion. This can lead to the introduction of *radiophenotypes*, specific image phenotypes that associate with presence/absence of genes [52]. These characteristics can serve as surrogates for gene expression signatures [1]. If such association maps are established and validated, radiogenomics can prompt biopsy procedures to target more specific areas of tumors [53]. Second, radiogenomics can serve as an independent source of additional information to inform on how a biological process is reflected in images [52]. Such information could be merged with genomics data leading to enhanced diagnostic and prognostic power [2].

8.4 WORKFLOW OF RADIOMICS

Fig. 8.3 shows the simplified workflow of the training and application of the radiomics approach. Step 1 shows the training phase of the workflow in a conventional learning scheme which is bound by preparation of feature sets. In this learning scheme, a high number of features are extracted from diagnostic images via a radiomics engine (part a). These features are then analyzed in the prediction engine (part b) using machine learning techniques. The output of the prediction engine divides the high dimensional feature space of radiomics to a much simpler classification space (part c) to assess the clinical hypothesis of the dataset. Step 2 shows how this pipeline can be applied for prediction of the result on a previously unseen case.

As we move toward deep learning schemes, yesterday's types of radiomics workflows are deemed obsolete. Deep learning schemes do not require hand crafted feature extraction or analysis and as such are able to arrive at the classification space directly. We mainly focus on the conventional learning schemes in this chapter. For more information regarding deep learning the interested reader is referred to Chapters 4 and 9.

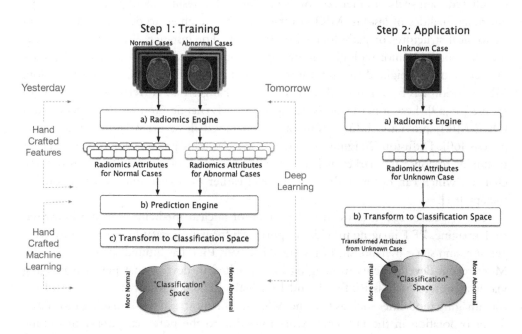

Figure 8.3 A simplified radiomics workflow. See the text for additional information.

Implementation of the radiomics approach (in the traditional learning scheme) involves the following discrete processes: (a) image acquisition, (b) segmentation, (c) feature extraction, and (d) feature analysis and model generation. In this section, we briefly describe each of these steps.

8.4.1 Image acquisition

Just as cancer therapy research has advanced, so too has cancer imaging, resulting in images with higher temporal resolution (number of frames per unit time), spatial resolution (number of image pixels per spatial unit), and contrast resolution (number of bits per pixel). Depending upon the cancerous organ, the location of metastasis, and patient tolerance to the procedure, various imaging modalities are ordered. Some well-known image modalities defined earlier are Computed Tomography (CT), Magnetic Resonance Image (MRI), or Positron Emission Tomography (PET).

CT scans can provide high resolution structural information about the location and shape of lesions [57]. When combined with contrast injection, CT scans exhibit strong contrast reflecting the presence of underlying disease or injury in the organs, blood vessels, and/or tissue types [18]. CT perfusion shows which areas are perfused adequately with blood and provides detailed functional information on delivery of blood or blood flow to specific organs.

Tumor functional and molecular characteristics are assessed using MRI. The use of different spin-echo sequences for capturing MRI results in exhibition of different functionalities of lesions. MRI contrast may be weighted to demonstrate different anatomical structures or pathologies (see Fig. 8.4). For instance, T1-weighed MRI is used for obtaining morphological information while MR diffusion weighted, Flair sequences and T2-weighted are useful as measures of cellularity, edema, and inflammation. MR diffusion has been reported to detect early changes that correlate with tumor response [58,59]. Lesion perfusion is captured using Dynamic Contrast Enhanced-MRI (DCE-MRI) providing information on lesion blood flow, permeability, and angiogenesis [60]. Diffusion-Weighted MRI (DW-MRI) holds promise for use as a cancer treatment response biomarker as it is sensitive to macromolecular and microstructural changes, which can occur at the cellular level earlier than anatomical changes during therapy [61].

Tumor metabolism is assessed using a type of nuclear medicine imaging known as PET imaging. PET imaging provides important information regarding blood flow, oxygen use, and sugar metabolism of lesions. However, PET has limited spatial resolution. More recent developments in imaging have resulted in devices that perform imaging via dual sources such as PET/CT and PET/MR [62,63]. However, the standard-of-care imaging modalities such as CT and MRI are the most explored imaging modalities for incorporation in the radiomics workflows due to the potential impact in routine care [64–66].

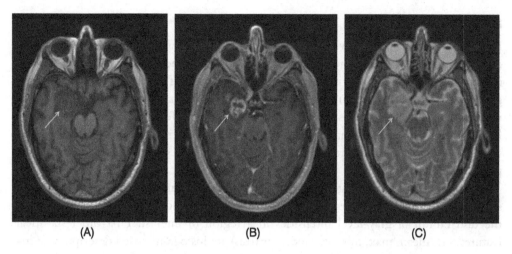

Figure 8.4 Conventional MRI images viewed in axial plane. The arrows show the location of glioblastoma multiforme (GBM) on (A) standard precontrast T1-weighted, (B) post-contrast T1-Weighted, and (C) precontrast T2-Weighted sequences.

8.4.2 Segmentation

Next, lesion segmentation is performed to precisely define the spatial extent of the lesion on the images. The validity of the segmentation result is of utmost importance because of its direct impact on the radiomics features and the utility of their content. Prior to segmentation, several image preprocessing steps are requisite, including spatial registration of different image sequences and modalities, noise reduction, and intensity calibration to account for the differences in the anatomical and intensity variations of the images.

Manual outlining by an expert is considered the *ground truth* in tumor segmentation. However, the result of manual segmentation can be unreliable due to inter and intrareader segmentation variability [67]. Moreover, outlining a 3D volumetric object manually is a tedious and time–consuming task rendering it impractical. In comparison, semiautomatic methods involve a minimal amount of input from an expert user (*e.g.*, a seed point to initialize the segmentation or manual editing of the results). Region growing methods, level set methods, graph cut methods, active contours algorithms and semiautomatic segmentations such as livewires are among the most widely used segmentation approaches [4]. Automatic and semiautomatic segmentation methods are the preferred method of segmentation in radiomics workflows due to their robustness and significantly higher levels of reproducibility [68].

8.4.3 Feature extraction

Radiomics features cover a wide range of quantitative characteristics. Lesion shape, location, and vascularity are among the features with established prognostic value, and are already a part of the radiologists' lexicons for description of lesions [2]. Studies investigating this set of features aim to reduce the effects of human error and inter-observer variability in the clinical decision making process [35,69]. Texture features on the other hand, aim to capture and quantify the heterogeneity patterns inside a lesion. The research for this group of features is still in its infancy, but has shown promising predictive power in focused studies.

Texture descriptors are generally divided into three major categories: statistical, structural, and spectral. Statistical texture features are nondeterministic descriptors of the distribution of gray-level intensities in a region of interest. First-order statistical features (*i.e.*, mean, max, min, variance, kurtosis) are histogram-based descriptors of intensity distribution and can only provide global information regarding the target region intensity pattern. Second-order statistical features describe interrelationships of gray-level values within a region and are the most widely used feature extraction method in medical pattern recognition tasks. Gray Level Cooccurrence Matrices (GLCM) [70], and Gray Level Run Length Matrices (GLRLM) [71], are two instances of this group (see Section 3.3 of Chapter 3). In GLCM, a variety of features (see Fig. 8.5) are used to describe a matrix generated from the observed frequency of gray-level combinations within neighboring pixels, while in GLRLM the focus is on describing coarseness of texture in predefined directions. The utility and efficacy of GLRLM in comparison with GLCM or spectral methods were debated in the early stages [72,73], however, it has been increasingly investigated for clinical applications in the last decade [74–76]. Gray-scale and rotationally invariant Local Binary Patterns (LBP) [77] is another statistical descriptor of texture focusing on the patterns of intensity transition within the sub regions of an area of interest (see Section 3.4 of Chapter 3). The radius of the subregions determines the scale of the texture described by the LBP features.

Structural descriptors view texture in terms of texture primitives such as micro and macrotextures (described in Chapter 1). Model-based features describe the region of

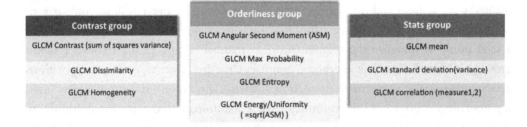

Figure 8.5 Grouping of GLCM features (courtesy of GLCM tutorial [80]).

interest through mathematical models such as fractals or stochastic models [78]. There are fewer reported radiomics applications for structural and model-based methods (an example: [79]), as they are viewed more suitable for synthesis than analysis. Detailed discussion on this group of features can be found in Chapter 5.

Spectral descriptors rely on the frequency or scale domain representation of the region of interest for calculation of texture features. By adjusting the window/filter expansion in space or frequency, these transform-based methods allow for multiresolution, multiscale representation of texture (see Chapter 7). The Wavelet Transform (WT) [81], the Riesz transform [82], the Stockwell Transform (ST) [83], the Discrete Orthonormal Stockwell Transform (DOST) [41] and the Polar Stockwell Transform (PST) [84], Gabor Filter Banks (GFB) [85], and Laplacian of Gaussian Histograms (LoGHist) [86] are only some examples of this group (see Section 3.2 of Chapter 3). Chapter 3 provides a detailed discussion on the biomedical texture operators. Chapter 12 discusses an online tool for 3D radiomics feature extraction in PET/CT.

8.4.4 Analysis and validation

Feature selection is an invaluable part of the radiomics workflow. The high throughput nature of radiomics results in an expected level of redundancy among features. By removing correlated and nondiscriminative features, feature selection avoids fitting to noise. Features can be selected prior to classification training based on a criterion or as part of training in a wrapper-based approach. It should be noted that there is no inherent superiority among different feature selection methods and the choice should be made as a consequence of sample distributions of the data at hand and the classification method [87].

Depending on the hypothesis of the study, other data types (*i.e.*, demographics, pathology, blood biomarkers, genomics data) can serve as the *class label*, or can be merged with radiomics features to build models that predict clinical endpoints (survival, progress, tumor stage, tumor diagnosis). Performance metrics such as accuracy, sensitivity, specificity, positive and negative predictive values are reported as measures of model performance. Successful models are maintained in a database for future validation studies.

A certain classification approach might be better suited for a type of data based on the characteristics of the data and the manner that the classification approach is utilized. Designing the best classification approach that can lead to stable and reproducible models requires a substantial amount of knowledge in machine-learning and statistical inference. The interested reader is referred to Chapters 6 and 11, and [88–90].

8.5 EXAMPLES OF RADIOMICS LITERATURE

Aerts et al. [18], showed that it is possible to identify imaging traits that are consistent across multiple datasets and associate with tumor prognosis. They conducted a study in which 440 features were extracted to quantify tumor intensity, shape, and texture, on the CT images of patients with lung or head-and-neck cancer. Features were found to have significant association with tumor histology, and to have prognostic power in seven independent datasets. The authors also found associations of radiomics features with the gene-expression profiles of lung cancer patients in one dataset. Along these lines, Parmar et al. [91] investigated cancer-specific radiomic feature clusters in four independent lung and head-and-neck cancer cohorts. Several feature clusters, extracted from the pretreatment CT images, exhibited stability across the datasets. The authors found strong associations of these features with lung prognosis and staging of head and neck lesions. Coroller et al. [92] extracted 635 radiomic features from pre-treatment CT-scans of lung adenocarcinoma patients. 35 features were predictive of Distant Metastasis (DM), and 12 features showed correlation with survival. The authors concluded that given the strong power of their radiomics signature in predicting DM, the result could be used as a prognostic biomarker for factors such as distant metastasis.

Hu et al. [93] highlighted how radiomics can serve as a noninvasive means to potentially predict histopathology. They extracted 100+ texture features from MRI images of GBM lesions at the biopsy locations. The authors generated a classification model that predicted the percentage of tumor content from the texture features. Fig. 8.6 shows the

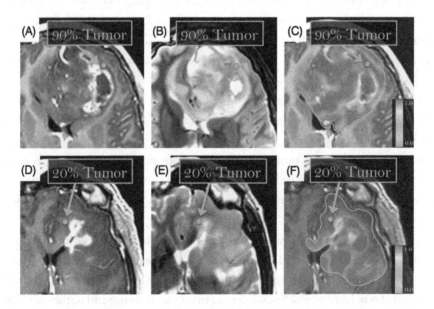

Figure 8.6 Prediction of tumor content in GBM using radiomics data [93].

colormap overlay of their predictions across the entire lesions in two validation cases. A similar approach was used by Ramkumar et al. [94] to distinguish sinonasal Inverted Papilloma (IP) and Squamous Cell Carcinoma (SCC). Hu et al. [53] also predicted local gene expression data through multivariate analysis of Copy Number Variant (CNV) status, multiparametric MRI, and texture analysis. This study proposed that a radiomics model with an acceptable level of success in the prediction of the presence of a certain gene can provide invaluable information to guide further biopsy procedures. They reported high accuracies in prediction of PDGFRA, EGFR, CDKN2A, and RB1 driver genes. Fig. 8.7 shows the proposed tree model of imaging features for prediction of PDGFRA amplification (Part A), along with color map overlay of prediction (Part C) on 2 stereotactic biopsies (Part B) for a validation case.

Zinn and others [5] identified and corroborated genomic correlates of edema/cellular invasion in GBM with MRI phenotypes. Gevaert and others [95] extracted quantitative image features from the enhancing necrotic portions of GBM tumor and peritumoral edema on MR images. Their radiogenomics map linked 56% of the 50+ imaging features with biologic processes, while 3 and 7 imaging features were significantly correlated with survival and molecular subgroups, respectively. Focusing on GBMs, Diehn et al. [96] showed initial proof that a radiogenomics approach could potentially be used in selecting patients for targeted therapies. They identified an imag-

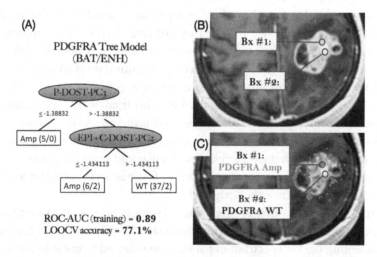

Figure 8.7 Radiogenomics model for GBM (Figure courtesy of [53]). (A) (P = isotropic diffusion; EPI + C = T2*W signal loss; DOST = discrete orthonormal Stockwell transform; PC3, PC2 derived from principal component analysis (PCA), AUC, area under curve). (B) the locations of 2 stereotactic biopsies (Bx#1, Bx#2) on CE-MRI. DNA CNVs demonstrated amplification (Amp, Bx#1) and diploid/wild-type status (WT, Bx#2). (C) regions (ROIs) of predicted PDGFRA amplification (red voxels) using tree model classification. (For interpretation of the references to color in this figure legend, the reader is referred to the web version of this chapter.)

ing biomarker of EGFR expression from MR images; an image trait termed "contrast necrosis" highly associated with EGFR gene expression. They effectively utilized this image trait to select GBM patients with overexpressed EGFR protein.

In a clear cell renal cell carcinoma application, Karlo and others [97] found associations between known genetic mutations and imaging traits describing the tumor margin, size, vascularity, and enhancement on pretreatment CT images. Segal et al. [98] associated distinctive imaging traits in CT scans with global gene expression profiles of liver cancer related to angiogenesis (*i.e.* vascular endothelial growth factor VEGF). Their results showed association maps between 28 features and 78% of the global gene expression profiles. In a preliminary breast cancer radiogenomics study [99], the authors provided an association map linking MR image phenotypes to underlying global gene expression patterns. They found 21 (out of 26) imaging traits on breast MRI as globally correlated with 71% of the total genes measured in patients with breast cancer.

8.6 CHALLENGES OF RADIOMICS

Radiomics is still in infancy. Further development of this field is expected to have major standardization, characterization, and technical obstacles including introduction of harmonized image acquisition protocols, inclusion of robust segmentation methods with minimum user input, extraction of robust and optimally informative features with minimal level of redundancy, and finally, development of suitable approaches to feature analysis and classification [4]. We believe that utilizing deep learning approaches, discussed in Chapters 4 and 9, will help address many of these issues. Also, developing a consistent radiomics framework requires effective communication among the stakeholders of all the building blocks of the process (*i.e.*, image acquisition, segmentation, feature extraction, model generation). Upon establishing this framework, the actual potential of the radiomics features can be critically assessed using an evidence-based approach. The critical steps in this regard involve evaluating features for exhibiting predictable and systematic relationships with the underlying molecular and genomic diversity of tumors, and evaluating the relationship between imaging features and molecular and genomic pathways for consistency across the images [100].

Patient risk factor data (such as subject age, biology) also need to be incorporated into radiomics as these factors can influence the image features [101]. Large standardized datasets representing the full spectrum of patients are required to result in unbiased inferences. The Cancer Imaging Archive (TCIA),[1] Quantitative Imaging Data Warehouse (QIDW),[2] Alzheimer's Disease Neuroimaging Initiative (ADNI),[3] and Osteoarthritis

[1] http://www.cancerimagingarchive.net.
[2] http://qidw.rsna.org.
[3] http://www.adni-info.org.

Initiative (OAI)[4] are examples of targeted data collections that aim to generate a consensus among the stakeholders on how to optimize, harmonize, and standardize data collection and analysis processes.

Development of clinical markers has been described in a roadmap that involves *qualification assessment* and *validation* of potential biomarkers [102]. Biomarker validation studies – including imaging biomarker studies – should be conducted to assess two key factors: (a) that the presence of the biomarker is closely linked to the presence of the target disease or condition, and (b) the process of detection and measurement of the biomarker is accurate, reproducible, and feasible over time [103,104,102]. The former factor calls for conducting studies that assess imaging-pathology, imaging-molecular, or imaging-genomics correlations. The latter relies on standardization of the feature extraction process and consensus on the interpretation of the features. Initiatives such as Quantitative Imaging Biomarkers Alliance (QIBA), sponsored by Radiological Society of North America (RSNA) and the National Institute for Biomedical Imaging and Bioengineering [105], are among promising steps that encourage collaboration among the stakeholders and advance the field through provision of a consensus on imaging protocols, and the measurement accuracy of imaging biomarkers.

8.7 CONCLUSIONS

Despite the major advances in computational biology, gaps still exist in our understanding of cancer dynamics. Radiomics – and by extension radiogenomics – offer the potential for a noninvasive diagnostic tool to complement existing methods of assessing disease characteristics and behavior. This potentially valuable tool can enhance diagnosis, monitoring, and prediction of response to treatment for various cancers. This field is still young and requires rigorous technical and clinical validation studies. As the field grows, we can expect radiomics to bridge the gaps of knowledge and help shift medicine closer to a fundamental genetic-based and truly personalized practice.

REFERENCES

[1] A.M. Rutman, M.D. Kuo, Radiogenomics: creating a link between molecular diagnostics and diagnostic imaging, Eur. J. Radiol. 70 (2) (2009) 232–241.
[2] R.J. Gillies, P.E. Kinahan, H. Hricak, Radiomics: images are more than pictures, they are data, Radiology 278 (2) (2015) 563–577.
[3] P. Lambin, E. Rios-Velazquez, R. Leijenaar, S. Carvalho, R.G.P.M. van Stiphout, P. Granton, C.M.L. Zegers, R. Gillies, R. Boellard, A. Dekker, et al., Radiomics: extracting more information from medical images using advanced feature analysis, Eur. J. Cancer 48 (4) (2012) 441–446.

[4] http://oai.epi-ucsf.org/datarelease.

[4] V. Kumar, Y. Gu, S. Basu, A. Berglund, S.A. Eschrich, M.B. Schabath, K. Forster, H.J.W.L. Aerts, A. Dekker, D. Fenstermacher, et al., Radiomics: the process and the challenges, Magn. Reson. Imaging 30 (9) (2012) 1234–1248.

[5] P.O. Zinn, B. Majadan, P. Sathyan, S.K. Singh, S. Majumder, F.A. Jolesz, R.R. Colen, Radiogenomic mapping of edema/cellular invasion MRI-phenotypes in glioblastoma multiforme, PLoS ONE 6 (10) (2011) e25451.

[6] M.A. Levy, D.L. Rubin, Current and future trends in imaging informatics for oncology, Cancer J. 17 (4) (2011) 203.

[7] A.J. Buckler, J.L. Mulshine, R. Gottlieb, B. Zhao, P.D. Mozley, L. Schwartz, The use of volumetric CT as an imaging biomarker in lung cancer, Acad. Radiol. 17 (1) (2010) 100–106.

[8] A.R. Padhani, L. Ollivier, The RECIST criteria: implications for diagnostic radiologists, Br. J. Radiol. 74 (887) (2001) 983–986.

[9] C.L. Sawyers, The cancer biomarker problem, Nature 452 (7187) (2008) 548–552.

[10] C.M. West, G.C. Barnett, Genetics and genomics of radiotherapy toxicity: towards prediction, Gen. Med. 3 (8) (2011) 52.

[11] D.E. Gerber, Targeted therapies: a new generation of cancer treatments, Am. Fam. Phys. 77 (3) (2008) 311–319.

[12] M.E. Gaguski, P. Begyn, Advances in cancer therapy: targeted agents, Nurs. Manage. 39 (5) (2008) 45–46.

[13] V.G. De Gruttola, P. Clax, D.L. DeMets, G.J. Downing, S.S. Ellenberg, L. Friedman, M.H. Gail, R. Prentice, J. Wittes, S.L. Zeger, Considerations in the evaluation of surrogate endpoints in clinical trials: summary of a national institutes of health workshop, Control. Clin. Trials 22 (5) (2001) 485–502.

[14] R.D. Hodgson, R.D. Whittaker, A. Herath, D. Amakye, G. Clack, Biomarkers in oncology drug development, Mol. Oncol. 3 (1) (2009) 24–32.

[15] J.K. Aronson, Biomarkers and surrogate endpoints, Br. J. Clin. Pharmacol. 59 (5) (2005) 491–494.

[16] J.A. Wagner, Biomarkers: principles, policies, and practice, Clin. Pharmacol. Ther. 86 (1) (2009) 3–7.

[17] J.W. Prescott, Quantitative imaging biomarkers: the application of advanced image processing and analysis to clinical and preclinical decision making, J. Digit. Imaging 26 (1) (2013) 97–108.

[18] H.J.W.L. Aerts, E. Rios Velazquez, R.T.H. Leijenaar, C. Parmar, P. Grossmann, S. Carvalho, J. Bussink, R. Monshouwer, B. Haibe-Kains, D. Rietveld, et al., Decoding tumour phenotype by noninvasive imaging using a quantitative radiomics approach, Nat. Commun. 5 (2014).

[19] R. Ford, L. Schwartz, J. Dancey, L.E. Dodd, E.A. Eisenhauer, S. Gwyther, L. Rubinstein, D. Sargent, L. Shankar, P. Therasse, et al., Lessons learned from independent central review, Eur. J. Cancer 45 (2) (2009) 268–274.

[20] P. Thiesse, L. Ollivier, D. Di Stefano-Louineau, S. Négrier, J. Savary, K. Pignard, C. Lasset, B. Escudier, Response rate accuracy in oncology trials: reasons for interobserver variability. Groupe Français d'immunothérapie of the Fédération Nationale des Centres de Lutte Contre le Cancer, J. Clin. Oncol. 15 (12) (1997) 3507–3514.

[21] S. Yachida, S. Jones, I. Bozic, T. Antal, R. Leary, B. Fu, M. Kamiyama, R.H. Hruban, J.R. Eshleman, M.A. Nowak, et al., Distant metastasis occurs late during the genetic evolution of pancreatic cancer, Nature 467 (7319) (2010) 1114–1117.

[22] M. Gerlinger, A.J. Rowan, S. Horswell, J. Larkin, D. Endesfelder, E. Gronroos, P. Martinez, N. Matthews, A. Stewart, P. Tarpey, et al., Intratumor heterogeneity and branched evolution revealed by multiregion sequencing, N. Engl. J. Med. 2012 (366) (2012) 883–892.

[23] A. Sottoriva, I. Spiteri, S.G.M. Piccirillo, A. Touloumis, V.P. Collins, J.C. Marioni, C. Curtis, C. Watts, S. Tavaré, Intratumor heterogeneity in human glioblastoma reflects cancer evolutionary dynamics, Proc. Natl. Acad. Sci. 110 (10) (2013) 4009–4014.

[24] M.C. Skrzynski, J.S. Biermann, A. Montag, M.A. Simon, Diagnostic accuracy and charge-savings of outpatient core needle biopsy compared with open biopsy of musculoskeletal tumors, J. Bone Jt. Surg. Am. 78 (5) (1996) 644–649.

[25] P.L. Nguyen, D. Schultz, A.A. Renshaw, R.T. Vollmer, W.R. Welch, K. Cote, A.V. D'Amico, The impact of pathology review on treatment recommendations for patients with adenocarcinoma of the prostate, Urol. Oncol. 22 (2004) 295–299.

[26] V.L. Staradub, K.A. Messenger, N. Hao, M. Morrow, Changes in breast cancer therapy because of pathology second opinions, Ann. Surg. Oncol. 9 (10) (2002) 982–987.

[27] M. Dowsett, Preoperative models to evaluate endocrine strategies for breast cancer, Clin. Cancer Res. 9 (1) (2003) 502s–510s.

[28] S. Herlidou-Meme, J.M. Constans, B. Carsin, D. Olivie, P.A. Eliat, L. Nadal-Desbarats, C. Gondry, E. Le Rumeur, I. Idy-Peretti, J.D. De Certaines, MRI texture analysis on texture test objects, normal brain and intracranial tumors, Magn. Reson. Imaging 21 (9) (2003) 989–993.

[29] D. Mahmoud-Ghoneim, G. Toussaint, J.-M. Constans, D. Jacques, Three dimensional texture analysis in MRI: a preliminary evaluation in gliomas, Magn. Reson. Imaging 21 (9) (2003) 983–987.

[30] R.A. Lerski, K. Straughan, L.R. Schad, D. Boyce, S. Blüml, I. Zuna, VIII. MR image texture analysis – an approach to tissue characterization, Magn. Reson. Imaging 11 (6) (1993) 873–887.

[31] E.I. Zacharaki, S. Wang, S. Chawla, D.S. Yoo, R. Wolf, E.R. Melhem, C. Davatzikos, Classification of brain tumor type and grade using MRI texture and shape in a machine learning scheme, Magn. Reson. Med. 62 (6) (2009) 1609–1618.

[32] N.R. Mudigonda, R. Rangayyan, J.E.L. Desautels, Gradient and texture analysis for the classification of mammographic masses, IEEE Trans. Med. Imaging 19 (10) (2000) 1032–1043.

[33] P. Gibbs, L.W. Turnbull, Textural analysis of contrast-enhanced MR images of the breast, Magn. Reson. Med. 50 (1) (2003) 92–98.

[34] Y. Zheng, S. Englander, S. Baloch, E.I. Zacharaki, Y. Fan, M.D. Schnall, D. Shen, STEP: spatiotemporal enhancement pattern for MR-based breast tumor diagnosis, Med. Phys. 36 (7) (2009) 3192–3204.

[35] D. Newell, K. Nie, J.-H. Chen, C.-C. Hsu, J.Y. Hon, O. Nalcioglu, M.-Y. Su, Selection of diagnostic features on breast MRI to differentiate between malignant and benign lesions using computer-aided diagnosis: differences in lesions presenting as mass and non-mass-like enhancement, Eur. Radiol. 20 (4) (2010) 771–781.

[36] S.C. Agner, S. Soman, E. Libfeld, M. McDonald, K. Thomas, S. Englander, M.A. Rosen, D. Chin, J. Nosher, A. Madabhushi, Textural kinetics: a novel dynamic contrast-enhanced (DCE)-MRI feature for breast lesion classification, J. Digit. Imaging 24 (3) (2011) 446–463.

[37] S.A. Waugh, C.A. Purdie, L.B. Jordan, S. Vinnicombe, R.A. Lerski, P. Martin, A.M. Thompson, Magnetic resonance imaging texture analysis classification of primary breast cancer, Eur. Radiol. 26 (2) (2016) 322–330.

[38] A. Depeursinge, D. Van De Ville, A. Platon, A. Geissbuhler, P.-A. Poletti, H. Müller, Near-affine-invariant texture learning for lung tissue analysis using isotropic wavelet frames, IEEE Trans. Inf. Technol. Biomed. 16 (4) (jul 2012) 665–675.

[39] O.S. Al-Kadi, D. Watson, et al., Texture analysis of aggressive and nonaggressive lung tumor CE CT images, IEEE Trans. Biomed. Eng. 55 (7) (2008) 1822–1830.

[40] A. Wibmer, H. Hricak, T. Gondo, K. Matsumoto, H. Veeraraghavan, D. Fehr, J. Zheng, D. Goldman, C. Moskowitz, S.W. Fine, et al., Haralick texture analysis of prostate MRI: utility for differentiating non-cancerous prostate from prostate cancer and differentiating prostate cancers with different Gleason scores, Eur. Radiol. 25 (10) (2015) 2840–2850.

[41] S. Drabycz, R.G. Stockwell, J.R. Mitchell, Image texture characterization using the discrete orthonormal S-transform, J. Digit. Imaging 22 (6) (2009) 696.

[42] B. Ganeshan, S. Abaleke, R.C. Young, C.R. Chatwin, K.A. Miles, Texture analysis of non-small cell lung cancer on unenhanced computed tomography: initial evidence for a relationship with tumour glucose metabolism and stage, Cancer Imaging 10 (1) (2010) 137–143.

[43] F. Ng, B. Ganeshan, R. Kozarski, K.A. Miles, V. Goh, Assessment of primary colorectal cancer heterogeneity by using whole-tumor texture analysis: contrast-enhanced CT texture as a biomarker of 5-year survival, Radiology 266 (1) (2013) 177–184.

[44] S.-X. Rao, D.M.J. Lambregts, R.S. Schnerr, W. van Ommen, T.J.A. van Nijnatten, M.H. Martens, L.A. Heijnen, W.H. Backes, C. Verhoef, M.-S. Zeng, et al., Whole-liver CT texture analysis in colorectal cancer: does the presence of liver metastases affect the texture of the remaining liver?, United Eur. Gastroenterol. J. 2 (6) (2014) 530–538.

[45] K.A. Miles, B. Ganeshan, M.R. Griffiths, R.C.D. Young, C.R. Chatwin, Colorectal cancer: texture analysis of portal phase hepatic CT images as a potential marker of survival 1, Radiology 250 (2) (2009) 444–452.

[46] V. Goh, B. Ganeshan, P. Nathan, J.K. Juttla, A. Vinayan, K.A. Miles, Assessment of response to tyrosine kinase inhibitors in metastatic renal cell cancer: CT texture as a predictive biomarker, Radiology 261 (1) (2011) 165–171.

[47] R.T.H. Leijenaar, S. Carvalho, E. Rios Velazquez, W.J.C. Van Elmpt, C. Parmar, O.S. Hoekstra, C.J. Hoekstra, R. Boellaard, A.L.A.J. Dekker, R.J. Gillies, et al., Stability of FDG-PET Radiomics features: an integrated analysis of test-retest and inter-observer variability, Acta Oncologica 52 (7) (2013) 1391–1397.

[48] D. Mackin, X. Fave, L. Zhang, D. Fried, J. Yang, B. Taylor, E. Rodriguez-Rivera, C. Dodge, A.K. Jones, et al., Measuring computed tomography scanner variability of radiomics features, Invest. Radiol. 50 (11) (2015) 757–765.

[49] R.T.H. Leijenaar, G. Nalbantov, S. Carvalho, W.J.C. van Elmpt, E.G.C. Troost, R. Boellaard, H.J.W.L. Aerts, R.J. Gillies, P. Lambin, The effect of SUV discretization in quantitative FDG-PET radiomics: the need for standardized methodology in tumor texture analysis, Sci. Rep. 5 (2015) 11075.

[50] C. Parmar, P. Grossmann, J. Bussink, P. Lambin, H.J.W.L. Aerts, Machine learning methods for quantitative radiomic biomarkers, Sci. Rep. 5 (2015) 13087.

[51] A.A. Alizadeh, V. Aranda, A. Bardelli, C. Blanpain, C. Bock, C. Borowski, C. Caldas, A. Califano, M. Doherty, M. Elsner, et al., Toward understanding and exploiting tumor heterogeneity, Nat. Med. 21 (8) (2015) 846–853.

[52] M.D. Kuo, N. Jamshidi, Behind the numbers: decoding molecular phenotypes with radiogenomics – guiding principles and technical considerations, Radiology 270 (2) (2014) 320–325.

[53] L.S. Hu, S. Ning, J.M. Eschbacher, L.C. Baxter, N. Gaw, S. Ranjbar, J. Plasencia, A.C. Dueck, S. Peng, K.A. Smith, et al., Radiogenomics to characterize regional genetic heterogeneity in glioblastoma, Neuro-Oncol. 19 (1) (2017) 128–137.

[54] Z. Guo, Y. Shu, H. Wang, H. Zhou, W. Zhang, Radiogenomics helps to achieve personalized therapy by evaluating patient responses to radiation treatment, Carcinogenesis (2015) bgv007.

[55] P.M.S.N. Carol, Radiogenomics: the promise of personalized treatment in radiation oncology?, Clin. J. Oncol. Nurs. 18 (2) (2014) 185.

[56] H. Yang, E. Kundakcioglu, J. Li, T. Wu, J.R. Mitchell, A.K. Hara, W. Pavlicek, L.S. Hu, A.C. Silva, C.M. Zwart, et al., Healthcare intelligence: turning data into knowledge, IEEE Intell. Syst. 29 (3) (2014) 54–68.

[57] R.J. van Klaveren, M. Oudkerk, M. Prokop, E.T. Scholten, K. Nackaerts, R. Vernhout, C.A. van Iersel, K.A.M. van den Bergh, S. van't Westeinde, C. van der Aalst, et al., Management of lung nodules detected by volume CT scanning, N. Engl. J. Med. 361 (23) (2009) 2221–2229.

[58] Y. Wang, Z.E. Chen, P. Nikolaidis, R.J. McCarthy, L. Merrick, L.A. Sternick, J.M. Horowitz, V. Yaghmai, F.H. Miller, Diffusion-weighted magnetic resonance imaging of pancreatic adenocarcinomas: association with histopathology and tumor grade, J. Magn. Reson. Imaging 33 (1) (2011) 136–142.

[59] J.-L. Van Laethem, C. Verslype, J.L. Iovanna, P. Michl, T. Conroy, C. Louvet, P. Hammel, E. Mitry, M. Ducreux, T. Maraculla, et al., New strategies and designs in pancreatic cancer research: consensus guidelines report from a European expert panel, Ann. Oncol. (2011) mdr351.

[60] O. Wu, L. Østergaard, R.M. Weisskoff, T. Benner, B.R. Rosen, A.G. Sorensen, Tracer arrival timing-insensitive technique for estimating flow in MR perfusion-weighted imaging using singular value decomposition with a block-circulant deconvolution matrix, Magn. Reson. Med. 50 (1) (2003) 164–174.

[61] H.C. Thoeny, B.D. Ross, Predicting and monitoring cancer treatment response with diffusion-weighted MRI, J. Magn. Reson. Imaging 32 (1) (2010) 2–16.

[62] T. Beyer, D.W. Townsend, T. Brun, P.E. Kinahan, et al., A combined PET/CT scanner for clinical oncology, J. Nucl. Med. 41 (8) (2000) 1369.

[63] W.-D. Heiss, The potential of PET/MR for brain imaging, Eur. J. Nucl. Med. Mol. Imaging 36 (1) (2009) 105–112.

[64] A. Larroza, D. Moratal, A. Paredes-Sánchez, E. Soria-Olivas, M.L. Chust, L.A. Arribas, E. Arana, Support vector machine classification of brain metastasis and radiation necrosis based on texture analysis in MRI, J. Magn. Reson. Imaging 42 (5) (2015) 1362–1368.

[65] A.L. Stockham, A.L. Tievsky, S.A. Koyfman, C.A. Reddy, J.H. Suh, M.A. Vogelbaum, G.H. Barnett, S.T. Chao, Conventional MRI does not reliably distinguish radiation necrosis from tumor recurrence after stereotactic radiosurgery, J. Neuro-Oncol. 109 (1) (2012) 149–158.

[66] S.T. Chao, M.S. Ahluwalia, G.H. Barnett, G.H.J. Stevens, E.S. Murphy, A.L. Stockham, K. Shiue, J.H. Suh, Challenges with the diagnosis and treatment of cerebral radiation necrosis, Int. J. Radiat. Oncol. Biol. Phys. 87 (3) (2013) 449–457.

[67] C. Tanner, M. Khazen, P. Kessar, M.O. Leach, D.J. Hawkes, Classification improvement by segmentation refinement: application to contrast-enhanced MR-mammography, in: International Conference on Medical Image Computing and Computer-Assisted Intervention, Springer, 2004, pp. 184–191.

[68] C. Parmar, E. Rios Velazquez, R. Leijenaar, M. Jermoumi, S. Carvalho, R.H. Mak, S. Mitra, B.U. Shankar, R. Kikinis, B. Haibe-Kains, et al., Robust radiomics feature quantification using semiautomatic volumetric segmentation, PLoS ONE 9 (7) (2014) e102107.

[69] W.K. Moon, C.-M. Lo, J.M. Chang, C.-S. Huang, J.-H. Chen, R.-F. Chang, Quantitative ultrasound analysis for classification of BI-RADS category 3 breast masses, J. Digit. Imaging 26 (6) (2013) 1091–1098.

[70] R.M. Haralick, Statistical and structural approaches to texture, Proc. IEEE 67 (5) (1979) 786–804.

[71] M.M. Galloway, Texture analysis using gray level run lengths, Comput. Graph. Image Process. 4 (2) (1975) 172–179.

[72] J.S. Weszka, C.R. Dyer, A. Rosenfeld, A comparative study of texture measures for terrain classification, IEEE Trans. Syst. Man Cybern. (4) (1976) 269–285.

[73] R.W. Conners, C.A. Harlow, A theoretical comparison of texture algorithms, IEEE Trans. Pattern Anal. Mach. Intell. (3) (1980) 204–222.

[74] A.K. Mohanty, S. Beberta, S.K. Lenka, Classifying benign and malignant mass using GLCM and GLRLM based texture features from mammogram, Int. J. Eng. Res. Appl. 1 (3) (2011) 687–693.

[75] K.R. Krishnan, R. Sudhakar, Automatic classification of liver diseases from ultrasound images using GLRLM texture features, in: Soft Computing Applications, Springer, 2013, pp. 611–624.

[76] H. Wibawanto, A. Susanto, T.S. Widodo, S.M. Tjokronegoro, Discriminating cystic and non cystic mass using GLCM and GLRLM-based texture features, Int. J. Electron. Eng. Res. 2 (2010) 569–580.

[77] T. Ojala, M. Pietikäinen, T. Mäenpää, Gray scale and rotation invariant texture classification with local binary patterns, in: European Conference on Computer Vision, Springer, 2000, pp. 404–420.

[78] B.B. Chaudhuri, N. Sarkar, Texture segmentation using fractal dimension, IEEE Trans. Pattern Anal. Mach. Intell. 17 (1) (1995) 72–77.

[79] M. Seppä, M. Hämäläinen, Visualizing human brain surface from T1-weighted MR images using texture-mapped triangle meshes, NeuroImage 26 (1) (2005) 1–12.

[80] M. Hall-Beyer, GLCM texture: a tutorial, in: National Council on Geographic Information and Analysis Remote Sensing Core Curriculum, 2000.

[81] S.G. Mallat, A theory for multiresolution signal decomposition: the wavelet representation, IEEE Trans. Pattern Anal. Mach. Intell. 11 (7) (1989) 674–693.

[82] A. Depeursinge, A. Foncubierta-Rodriguez, D. Ville, H. Müller, Lung texture classification using locally-oriented Riesz components, in: Medical Image Computing and Computer-Assisted Intervention – MICCAI 2011, January 2011.

[83] R.G. Stockwell, L. Mansinha, R.P. Lowe, Localization of the complex spectrum: the S transform, IEEE Trans. Signal Process. 44 (4) (1996) 998–1001.

[84] H. Zhu, Y. Zhang, X. Wei, L.M. Metz, A.G. Law, R. Mitchell, MR multi-spectral texture analysis using space-frequency information, in: Proceedings of the International Conference on Mathematics and Engineering Techniques in medicine and Biological Sciences, METMBS'04, 2004.

[85] A.K. Jain, F. Farrokhnia, Unsupervised texture segmentation using Gabor filters, Pattern Recognit. 24 (12) (1991) 1167–1186.

[86] F. Davnall, C.S.P. Yip, G. Ljungqvist, M. Selmi, F. Ng, B. Sanghera, B. Ganeshan, K.A. Miles, G.J. Cook, V. Goh, Assessment of tumor heterogeneity: an emerging imaging tool for clinical practice?, Insights Imaging 3 (6) (2012) 573–589.

[87] P.M. Szczypiński, M. Strzelecki, A. Materka, A. Klepaczko, MaZda – a software package for image texture analysis, Comput. Methods Programs Biomed. 94 (1) (2009) 66–76.

[88] R.A. Johnson, D.W. Wichern, et al., Applied Multivariate Statistical Analysis, vol. 5, Prentice Hall, Upper Saddle River, NJ, 2002.

[89] P.-N. Tan, Introduction to Data Mining, Pearson Education India, 2006.

[90] I.H. Witten, E. Frank, M.A. Hall, C.J. Pal, Data Mining: Practical Machine Learning Tools and Techniques, Morgan Kaufmann, 2016.

[91] C. Parmar, R.T.H. Leijenaar, P. Grossmann, E. Rios Velazquez, J. Bussink, D. Rietveld, M.M. Rietbergen, B. Haibe-Kains, P. Lambin, H.J.W.L. Aerts, Radiomic feature clusters and prognostic signatures specific for lung and head & neck cancer, Sci. Rep. 5 (2015) 11044.

[92] T.P. Coroller, P. Grossmann, Y. Hou, E. Rios Velazquez, R.T.H. Leijenaar, G. Hermann, P. Lambin, B. Haibe-Kains, R.H. Mak, H.J.W.L. Aerts, CT-based radiomic signature predicts distant metastasis in lung adenocarcinoma, Radiother. Oncol. 114 (3) (2015) 345–350.

[93] L.S. Hu, S. Ning, J.M. Eschbacher, N. Gaw, A.C. Dueck, K.A. Smith, P. Nakaji, J. Plasencia, S. Ranjbar, S.J. Price, et al., Multi-parametric MRI and texture analysis to visualize spatial histologic heterogeneity and tumor extent in glioblastoma, PLoS ONE 10 (11) (2015) e0141506.

[94] S. Ramkumar, S. Ranjbar, S. Ning, D. Lal, C.P. Wood, C.M. Zwart, T. Wu, J.R. Mitchell, J. Li, J.M. Hoxworth, Multiparametric magnetic resonance imaging and texture analysis to distinguish Sinonasal Inverted Papilloma from Squamous Cell Carcinoma, J. Neurol. Surg. Part B: Skull Base 77 (S 01) (2016) A041.

[95] O. Gevaert, L.A. Mitchell, A.S. Achrol, J. Xu, S. Echegaray, G.K. Steinberg, S.H. Cheshier, S. Napel, G. Zaharchuk, S.K. Plevritis, Glioblastoma multiforme: exploratory radiogenomic analysis by using quantitative image features, Radiology 273 (1) (2014) 168–174.

[96] M. Diehn, C. Nardini, D.S. Wang, S. McGovern, M. Jayaraman, Y. Liang, K. Aldape, S. Cha, M.D. Kuo, Identification of noninvasive imaging surrogates for brain tumor gene-expression modules, Proc. Natl. Acad. Sci. 105 (13) (2008) 5213–5218.

[97] C.A. Karlo, P. Luigi Di Paolo, J. Chaim, A.A. Hakimi, I. Ostrovnaya, P. Russo, H. Hricak, R. Motzer, J.J. Hsieh, O. Akin, Radiogenomics of clear cell renal cell carcinoma: associations between CT imaging features and mutations, Radiology 270 (2) (2014) 464–471.

[98] E. Segal, C.B. Sirlin, C. Ooi, A.S. Adler, J. Gollub, X. Chen, B.K. Chan, G.R. Matcuk, C.T. Barry, H.Y. Chang, et al., Decoding global gene expression programs in liver cancer by noninvasive imaging, Nat. Biotechnol. 25 (6) (2007) 675–680.

[99] S. Yamamoto, D.D. Maki, R.L. Korn, M.D. Kuo, Radiogenomic analysis of breast cancer using MRI: a preliminary study to define the landscape, Am. J. Roentgenol. 199 (3) (2012) 654–663.

[100] M.D. Kuo, S. Yamamoto, Next generation radiologic–pathologic correlation in oncology: Rad-Path 2.0, Am. J. Roentgenol. 197 (4) (2011) 990–997.

[101] A. Depeursinge, D. Racoceanu, J. Iavindrasana, G. Cohen, A. Platon, P.-A. Poletti, H. Müller, Fusing visual and clinical information for lung tissue classification in high-resolution computed tomography, Artif. Intell. Med. 50 (1) (September 2010) 13–21.

[102] J.C. Waterton, L. Pylkkanen, Qualification of imaging biomarkers for oncology drug development, Eur. J. Cancer 48 (4) (2012) 409–415.

[103] J.J. Smith, A.G. Sorensen, J.H. Thrall, Biomarkers in imaging: realizing radiology's future 1, Radiology 227 (3) (2003) 633–638.

[104] J.A. Wagner, S.A. Williams, C.J. Webster, Biomarkers and surrogate end points for fit-for-purpose development and regulatory evaluation of new drugs, Clin. Pharmacol. Ther. 81 (1) (2007) 104–107.

[105] A.J. Buckler, L. Bresolin, N.R. Dunnick, D.C. Sullivan, A collaborative enterprise for multi-stakeholder participation in the advancement of quantitative imaging, Radiology 258 (3) (2011) 906–914.

CHAPTER 9

Deep Learning Techniques on Texture Analysis of Chest and Breast Images

Jie-Zhi Cheng*, Chung-Ming Chen[†], Dinggang Shen[‡]
*Shenzhen University, School of Biomedical Engineering, Shenzhen, Guangdong, PR China
[†]National Taiwan University, Institute of Biomedical Engineering, Taipei, Taiwan
[‡]University of North Carolina at Chapel Hill, Biomedical Research Imaging Center, Chapel Hill, NC, United States

Abstract

The deep learning techniques can automatically learn texture and image context features from training data without the need of explicit feature engineering. With proper and sufficient training data, useful features can be learned to support various medical image analysis applications. In this chapter, we discuss recent deep learning studies on the analysis of chest and breast images. The specific elaborated computerized techniques are computer-aided detection, computer-aided diagnosis, and automatic semantic mapping. We show that the feature learned with the deep learning techniques can be helpful to boost the performances of these computerized applications for the analysis of medical images.

Keywords

Computer-aided detection, Computer-aided diagnosis, Semantic mapping, Breast imaging, Lung imaging, Deep learning

9.1 INTRODUCTION

Recent advance in deep learning techniques [1–5] has reshaped the design paradigm of the computerized analysis for medical images. Deep learning techniques are able to automatically learn image features from training data and are free of explicit feature engineering. The automatically extracted features by deep learning techniques may have more descriptive power for characterizing the object of interest than the conventional feature extraction methods if sufficient training images are given. Thus the image analysis procedure can be significantly simplified with deep learning techniques, compared to the procedure used in the standard pattern recognition framework. Under a deep learning context, the steps of feature extraction and classification/regression can be seamlessly realized in the same architecture without the need of ad hoc image processing steps. Meanwhile, since the intermediate results, for example, learned in the middle layers of neurons, of a deep learning architecture can be mostly visualized, it will be easier to monitor the learning status. Therefore the main research stream of image analysis field like computer vision, image processing, medical image analysis, and so on is now mainly occupied with deep learning studies. In this chapter, we introduce the recent advance

Biomedical Texture Analysis
DOI: 10.1016/B978-0-12-812133-7.00009-0

of deep learning techniques on the texture analysis of biomedical images. We focus on chest and breast diseases and also discuss three major computerized applications of computer-aided diagnosis, computer-aided detection, and computer-aided attributing.

9.2 COMPUTER-AIDED DETECTION

Computer-aided detection scheme is an important and highly demanded computerized application to boost the performance of clinical image reading. With advance and popularization of modern medical imaging techniques, the medical image examination and checkup are now becoming common practice and routine tools to provide the visualization of internal organs for the support of diagnostic workup. However, this may also lead to significant increase of medical images and impose more workload on the staff in the radiology department. Therefore the manpower shortage can be possibly expected. The purpose of the computer-aided detection scheme is to automatically retrieve the locations of possible lesions, nodules, and so on, in the 3D medical images, such as computed tomography (CT) scans, to assist the image reading process. For example, since a lung CT scan may be composed nearly a hundred of slices, the reading of all CT images can be a tiring process. Specifically, the reading of a CT may involve the adjustment of the intensity window and back-and-forth traverse of CT slices for differentiation of normal tissues/organs and potential lesions or nodules. When the number of medical images goes formidably large, it may be possible to miss some subtle lesions or nodules in the images, even for a senior radiologist. Accordingly, to improve the efficiency of clinical medical image reading, many commercial or laboratory computer-aided detection tools have been introduced and shown to be helpful in increasing the performance of nodule/lesion detection [6–8].

The conventional computer-aided detection scheme is mainly composed of several intermediate steps such as image preprocessing of denoising and segmentation, candidate proposal, feature extraction, feature selection, and classification. The goodness of an intermediate result at each step in the workflow pipeline highly depends on the results of previous steps. Therefore the algorithmic design for every step must be carefully elaborated. Particularly, the extraction of morphological and textural features is very crucial for false-positive reduction for the results of candidate proposal. Also, the feature extraction should account for the high appearance variation of lesion/nodule while still preserving the discrimination power for the differentiation of true lesion/nodule and nonlesion/nodule. Therefore the feature extraction can be one of the important issues for the algorithmic development of a computer-aided detection scheme. On the other hand, the recent progress of deep learning techniques may change the design paradigm of the computer-aided detection scheme. The deep learning techniques have been shown successfully in many detection contexts [9–11] with promising results. In

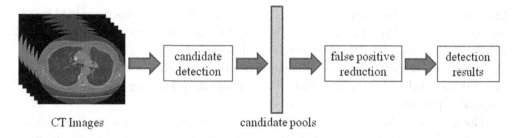

Figure 9.1 Flowchart for a conventional computer-aided detection scheme.

this section, we discuss several recent deep learning techniques developed for the detection problems.

9.2.1 Lung nodule detection in CT scans

Lung nodule detection for the CT scans is a very important problem due to huge demand of health checkups. It is because lung cancer remains one of the leading causes of cancer mortality in the world [12]. To achieve the goal of early detection [13], the thoracic CT is one of major imaging tools for screening of pulmonary nodules. However, a CT scan contains numerous axial images that need to be interpreted slice-by-slice by the radiologists. There are two general problems in the clinical manual reading. One is that some small and subtle nodules are likely to be missed because of their small size, low contrast, and location and of potential human errors. The other is that reading outcomes may vary with radiologists. For example, in an earlier evaluation study, a radiologist only found 518 nodules out of 628 nodules from 150 patients in low-dose CT scans [14]. Another comparative study reported a high variation of reading results in which three radiologists only identified 68%, 78%, and 82% of the consensus readings on 116 nodules [8]. In a National Lung Screening Trail (NLST) study, the average percentage of positive and negative agreements for all reader pairs among 16 radiologists was only 82% [15]. To alleviate the potential human error and improve the efficiency of image reading, in the past decades, many endeavors have been made for the computer-aided detection approaches, both in the commercial and laboratory products [16,17]. The challenges of lung nodule detection lie in that pulmonary nodules can develop isolatedly in the lung parenchyma or be attached to other pulmonary organs like vessels, fissure, chest wall, and airway. Therefore, *not only* the high appearance variation of nodule, *but also* the complicated anatomical context will be well considered for the algorithm design. The conventional flowchart for a computer-aided detection of pulmonary nodules is shown in Fig. 9.1 for illustration. The major issues are on candidate proposal and false-positive reduction, as detailed further.

The step of candidate proposal aims to retrieve nodule candidate objects as many as possible with high sensitivity, but without the care of false-positive findings. In the literature, the candidate proposal step can be realized with the features computed based on shape index [18,19], Hessian matrix [18,20], gradient divergence [20], intensity [20, 21], local binary pattern [21], Haar wavelet [21], morphology [19–21], and so on. Most candidate proposals need image segmentation to obtain the aforementioned features. To our best knowledge, there is less study that employs deep learning techniques for the proposal of potential nodule candidates.

The goal of a false-positive reduction step is eliminating the false-positive candidates from the object pool collected by the candidate proposal step. The elimination process should keep the sensitivity reasonably high. Therefore the false-positive reduction step should be equipped with good discrimination power for the differentiation of nodule and nonnodule objects. As discussed earlier, the high intraclass variation and low interclass difference can render the false-positive reduction difficult. The deep learning techniques have been recently shown to make significant progress for the false-positive reduction in the context of computer-aided detection of pulmonary nodules [11,22]. The most methods focus on the usage of Deep Convolutional Neural Network (DCNN) for the false-positive reduction.

The DCNN architecture is a very popular deep learning technique and has been successfully applied to address many medical image analysis problems [23–25,10,9]. It is a biologically inspired neural network architecture, which can exploit the spatial patterns from the training image data in a layer-by-layer fashion with the aid of convolutional kernels. Provided with sufficiently large training data and suitable parameter setting, the DCNN can address many complicated pattern recognition problems. A typical DCNN model can be constituted of pairs of convolutional (C) and max-pooling (M) layers and commonly ended with fully connected layers on the top of the network architecture. Several feature maps can be obtained by convolving the learnable linear filters at the convolutional layer with the previous layer. Specifically, given that the $h_j^l(x)$ is the jth feature map in the lth layer and $h_n^{l-1}(x)$ ($n = 1, ..., N$) is the feature map in the $(j-1)$th layer, the feature map h_j^l can be computed by

$$h_j^l(\boldsymbol{x}) = \sigma \left(\sum_{n=1}^{N} (W_{jn}^l * h_n^{l-1})(\boldsymbol{x}) + b^l \right), \tag{9.1}$$

where $W_{jn}^l(\boldsymbol{x})$ is the convolutional kernel (linear filter) connected to the nth feature map of the previous layer, b^l is the bias term for the lth layer, and $\sigma(\cdot)$ is the nonlinear neural activation function of the Rectified Linear Unit (ReLU) and can be defined as $\sigma(x) = max(x, 0)$. The ReLU was shown to be capable of creating sparse representation with hard linearity and may help to attain better classification performance than the conventional sigmoid function [26].

The M layers perform suppression of nonmaximal values to offload the computation for the latter layers and are robust to translation effect. The fully connected layer, where neurons are fully connected, is the conventional neural network model of multilayer perceptron. The fully connected layer aims to conduct high-level reasoning and nonlinear feature combinations. For the classification purpose, a DCNN model can be ended with soft-max functions at the final output layer. In the training process, the internal parameters can be adjusted with the minimization of the loss function of all training samples x_i $(i = 1, ..., N)$. The loss function can be defined as

$$\mathcal{L}(x_i, y_i; \theta) = - \sum_{i}^{N} \sum_{k=1}^{K} (1\{y_i = k\} \log[p(c_k = 1 | x_i, \theta)]), \tag{9.2}$$

$$p(c_k = 1 | x_i, \theta) = \frac{e^{o_k}}{\sum_{j=1}^{K} e^{o_j}}, \quad k = 1, ..., K, \tag{9.3}$$

where $p(c_k = 1 | x_i, \theta)$ is the probability of classifying the training sample x_i as the kth class of the overall K classes, θ represents the set of internal parameters to be estimated, y_i denotes the class label of the training sample x_i, $1\{\cdot\}$ is the indicator function, that is, $1\{y_i = k\}$ equals 1 if y_i belongs to the kth class and 0 otherwise, and o_k can be taken as the high-level output feature for the input of the soft-max layer. The minimization of the cost function $\mathcal{L}(x_i, y_i; \theta)$ can be achieved by the technique of Stochastic Gradient Descending (SGD).

In [22], multiple DCNNs are deployed to learn the differentiation of nodules and nonnodules based on 2D views. Specifically, for each positive and negative sample, nine cut-plane views are acquired from the 3D cube context as the representative views. A DCNN is further trained for each individual view. Afterward, the classification result of each DCNN with respect to each cut-plane view is integrated by the fusion strategies of committee-fusion, late-fusion, and mixed-fusion. Each DCNN in [22] is ended with a soft-max layer that constitutes of two neural nodes, which correspond to the classes of nodule and nonnodule. Very promising performance can be achieved with the detection sensitivities of 85.4% and 90.1% with respect to the one and four false-positive findings per scan on the 888 CT scans of the public Lung Image Database Consortium (LIDC) database [27,28]. The work [22] also achieved very good results on other datasets of ANODE09 challenge [29] and Danish Lung Cancer Screening Trial [30]. Different from using only 2D cues, the work in [11] developed three 3D DCNNs to consider 3D multiscale context for the false-positive reduction from the detected candidates provided by the LUNA16 challenge [31]. Specifically, the dimensions of the 3D image inputs of the three 3D DCNNs are $20 \times 20 \times 6$, $30 \times 30 \times 10$, and $40 \times 40 \times 26$, respectively, to include the anatomical contexts for better false-positive reduction effect. Similarly to [22], the topmost layers of 3D DCNNs of [11] are soft-max layers with two neural nodes that represent the nodule and nonnodule classes, respectively. The final differentiation results can be reached with the simple linear combination of the predicted values from three networks. With the data provided by LUNA16 challenge, the

3D DCNNs scheme [11] can attain sensitivities of 67.7%, 73.7%, 81.5%, 84.8%, 87.9%, 90.7%, and 92.2% with respect to false-positive findings of 1/8, 1/4, 1/2, 1, 2, 4, and 8 per scan, respectively. In summary, the introduction of deep learning techniques into the context of computer-aided detection for pulmonary nodules in CT images has been shown successfully in the reduction of false-positive findings while still keeping reasonably good sensitivity. Future research on this topic may focus on the improvement of sensitivity at the candidate proposal step, domain transfer to low dose CT, and so on.

9.2.2 Other detection problems for CT scans

Unlike the problem of nodule detection, the deep learning techniques have also been applied to retrieve other objects of interest from CT images. For example, similarly to [22], a DCNN model was proposed in [10] to detect lymph nodes, colon polyps, and sclerotic metastases from CT images. The differences between the two network models of [22] and [10] lie in the training samples of different views are used to train the same model in [10], where different view sets are applied for distinctive network models. It has been shown that, with sufficient random sampled 2D views, satisfactory performance can be achieved without the need of direct usage of 3D samples. This random 2D view aggregation is also called 2.5 D [10]. On the other hand, since a DCNN architecture is commonly composed of many convolutional and fully connected layers, the number of parameters to be adjusted is formidably large in the training process. This issue may render the task of parameter tuning to be very arduous. To address this issue, the work in [9] gave a very thorough investigation of a transfer learning technique for the improvement of network learning for the problems of computer-aided detection. Two specific computer-aided detection applications studied in [9] are the detection of lymph nodes in thoraco-abdominal CT scans and also the interstitial lung disease classification.

A transfer learning, also called inductive transfer, is a machine learning technique that tries to employ the knowledge, maybe in the form of parameters, learned from one source problem domain to the target problem domain [32]. The transferred knowledge is supposed to assist the task of machine learning in the target domain. In the context of deep learning, a simple-to-implement transfer learning just uses the neural parameters from one domain, for example, natural images, for the model initialization of a specific medical image domain, like ultrasound, CT, and so on [9,24,25]. Specifically, three DCNN architectures trained with natural images were exploited in [9] to see if the knowledge parameters learned from natural image domain can be useful for the image analysis problems in medical domain. The three DCNN architectures are CifarNet [33], AlexNet [3], and GoogleNet [34]. The efficacy of the transfer learning had been corroborated in [9] with the performance of 85% sensitivity along with three false-positive findings per scan for the mediastinal lymph node detection. In summary, since the data collection and annotation for medical images are usually expensive and the number of data is relatively small, to boost the learning performance of deep learning models, the implementation option of transfer learning from natural image domain can be considered.

9.3 COMPUTER-AIDED DIAGNOSIS

Different from the goal of computer-aided detection, which aims to retrieve objects of interest from images, the purpose of computer-aided diagnosis is to assist the clinical differentiation diagnosis, which is mostly for the identification of malignancy. In the literature, several studies have suggested that the incorporation of the computer-aided diagnosis schemes into the image reading process can boost the performance of image diagnosis via lowering-down the interobserver variation [35,36] and providing quantitative reference to support the clinical decision, for example, biopsy recommendations [37] and so on. Specifically, the computer-aided diagnosis systems were shown to be effective in helping the diagnostic workup for the reduction of unnecessary false-positive biopsies [38] and thoracotomy [39]. The computer-aided diagnosis applications have been extensively exploited in the past two decades on the differential diagnosis contexts of lesions and nodules on the breast, lung, and other organs in various image modalities.

To achieve accurate identification of lesion/nodule malignancy, the conventional design of computer-aided diagnosis scheme is often composed of three main steps: feature extraction [38,40–46], feature selection [46–49], and classification. These three steps need to be properly addressed and then integrated together for the overall performance tuning of computer-aided diagnosis. The performance tuning is difficult since the three steps affect to each other tightly in the processing pipeline. In the most previous works, engineering on effective feature extraction step for each specific problem was taken as the one of the most important issues [38,40–46]. Extraction of discriminative features can potentially relieve the latter steps of feature selection and classification. Nevertheless, the engineering of effective features is problem-oriented and, to achieve accurate lesion/nodule differentiation, still needs the guidance from the latter steps of feature selection and feature integration by classifier. Generally speaking, the diagnostic image features can be classified into features of morphology and texture.

The extraction of useful features is a complicated task that requires a series of image processing steps. These image processing steps include: 1) image segmentation [38,40, 43,44,50–52] for the computation of morphological features [37,40,43–45,50], which is unfortunately still an open problem [53], and 2) image decomposition [41,47] followed with statistical summarizations and representations [41,47,48,54] for the textural feature calculation [55]. Accordingly, the extraction of useful features highly depends on the quality of the intermediate result in each image processing step [38,40,43,50,55]. The feature extraction often needs many passes of trial-and-error design and engineering to find satisfactory intermediate results. Meanwhile, the computerized image segmentation results often require various case-by-case interventions by users, for example, manual refinement, solution selection, parameter adjustment, and so on, to improve the correctness of the generated contours [38,40,42,44,51]. However, with the user's intervention on the segmentation results, the differential diagnostic outcome from the computer-aided diagnosis system will be biased, and the objectiveness may no longer hold. In summary, the design and tuning of the overall performance of the conventional computer-aided diagnosis framework (with the above three steps) tend to be very arduous since many image processing issues need to be properly addressed.

On the other hand, experts' annotation on the training data is the main essential media that transfers experts' knowledge to the computer-aided diagnosis system. However, the annotation cost in the computer-aided diagnosis contexts is far more expensive than other pattern recognition applications because the annotation process requires professional medical knowledge. In particular, for both morphological and textural features, correct and reliable features can be more effectively derived from the annotations equipped with precise lesion/tumor boundaries. Nevertheless, the process of manual drawing for precise lesion/tumor boundaries is very laborious and costly for medical doctors. Meanwhile, the annotation of the training data should also consider a wide interobserver variation on the definition of lesion/tumor boundary and thus require as many manual lesion/tumor outlines as possible from distinctive medical doctors. Otherwise, the performance of the computer-aided diagnosis system could be biased by specific annotators. To address the issues of annotation cost and interobserver variation, training data with weak annotations could be a good compromise between the annotation cost and delicacy on the training data. One way of weak annotation can be realized is requesting medical experts to only define the Region Of Interest (ROI) with specification of the corresponding malignancy/benignancy. In this way, the process of meticulous manual drawing can be avoided, and hence the annotation cost can be significantly lowered. However, such weakly annotated training data will impose an extra challenge on the computer-aided diagnosis problem for the training of an efficient lesion/tumor classifier. An automatic feature extraction mechanism may be needed to explore potential useful features from weakly annotated data.

The advancement of deep learning techniques could potentially change the design paradigm of the computer-aided diagnosis framework for several advantages over the conventional frameworks. The deep learning techniques are the new artificial neural network techniques that can fulfill the processes of feature extraction, selection, and classification together based on the single network architecture. The advantages can be fourfold. First, deep learning can automatically and directly uncover features from the training data, and hence the effort of explicit elaboration on feature extraction can be significantly alleviated. Meanwhile, the neuron-crafted features from the deep learning techniques can carry high-level semantic representations to abstract the wide intraclass variation of lesion/tumor shapes and appearances. The neuron-crafted features may hence compensate and even to some extent surpass the discriminative power of the conventional feature extraction methods. Second, feature interaction and hierarchy can be exploited jointly within the intrinsic deep architecture of a neural network. Consequently, the feature selection process will be significantly simplified. In some cases a separate feature selection procedure may be unnecessary when satisfactory differentiation performance is achieved at the supervised fine-tuning stage of training. Third, the three steps of feature extraction, selection, and supervised classification can be realized within the optimization of the same deep architecture. With such a design, the performance can be tuned more easily in a systematic fashion. Fourth, the input to the deep learning architecture can be simply the ROI of the original image with the training labels of malignancy and benignancy. It provides an end-to-end fashion for training the system. Benefited from this, the training data for the deep-learning-based diagnosis scheme need not be annotated with the meticulous drawings of lesion/nodule boundaries [56]. In this case the annotation cost for the deep learning methods is significantly lower than the cost in the conventional computer-aided diagnosis scheme.

In this section, we will discuss the computer-aided diagnosis applications on breast lesions in ultrasound images and pulmonary nodules in CT scans.

9.3.1 Computer-aided diagnosis on breast lesions in ultrasound images

Ultrasound is a very important screening tool for breast lesions, particularly in many eastern Asian countries. The ultrasound imaging can be real time, relatively accessible, and cost effective. However, since the ultrasound image is reconstructed based on acoustic reflection signals, the image quality can be very noisy and presented with a posterior acoustic shadowing effect. Meanwhile, due to imaging physics, the quality of ultrasound image may also depend on the probe position. Since the ultrasound image quality is relatively low by comparison with other medical image modalities such as Computed Tomography (CT) and Magnetic Resonance Imaging (MRI), the image reading and diagnosis of ultrasound images are relatively difficult, and also the learning curve can be long. As a result, the screening and diagnostic outcomes of ultrasound breast images may highly depend on the experience and the condition of operator and medical doctors. With less reliable screening and diagnostic outcome, a further biopsy procedure is needed. In such a case, an unnecessary biopsy cannot be avoided, and the performance of biopsy may put more pressure on the subjects in terms of mental and physical aspects.

The purpose of computer-aided diagnosis for breast lesions depicted in ultrasound images is to provide a third party quantitative reference for diagnosis [37,38,40,45,57, 58] and potentially further alleviate the cases of unnecessary biopsy. In the convention design paradigm of computer-aided diagnosis for ultrasound breast lesions, the ultrasound images usually need a denoising step to alleviate the disturbance of speckle noise. Afterward, an automatic or semiautomatic image segmentation procedure is commonly

applied to demarcate the breast for feature extraction. As discussed earlier, the computational features for classification can be categorized as morphology and texture. With
the seeking of useful morphological and textural features, a classification technique is
applied to feature selection and final differentiation. One of the major computer-aided
diagnosis research lines focuses on the investigation of useful morphological features [40,
45,59,56], which are shown to be more robust to low image quality and less dependent
on the imaging parameter setting, for example, the adjustment of time gain compen-
sation [40,45]. As discussed earlier, the extraction of useful morphological features is
contingent on accurate lesion boundaries, which are expected to be obtained by the
computerized image segmentation method for objective analysis and clinical efficiency.
Because of the low ultrasound image quality and huge shape and appearance variations
of breast lesions, the results of image segmentation quite often require case-by-case
manual refinements [37,59,56,60] or user selection [40] to generate reliable morphological features. Meanwhile, image segmentation also usually requires parameter tuning
on other image processing steps such as speckle denoising to improve the segmentation results. The suitable parameter setting may vary from image to image and hence
increase the complexity of human intervention. In this regard, most ultrasound breast
computer-aided diagnosis systems equipped with image segmentation procedure are not
fully automatic and objective.

In terms of textural features [48,49,60] for the ultrasound breast computer-aided diagnosis problem, useful features can be computed from the Grey-Level Cooccurrence
Matrix (GLCM) [47,48]. To withstand the low image quality, wavelet related techniques [47] are commonly incorporated in the image preprocessing step to generate
plenty of textural features with the GLCM technique. A feature selection scheme is
further implemented to lower the feature number into a robust subset of features against
the low image quality. Prominently, the latest work of [47] employed the ranklet transform [61] to withstand the speckle noise and acoustic shadowing and yielded a better
performance than textural analysis with the standard wavelet technique. In summary,
the design of computer-aided diagnosis for breast lesions in ultrasound images needs to
well address the issues of low image quality and diverse phenotypes of breast lesions. Although morphological features can be more effective for this problem, the computation
of morphological features highly depends on the accuracy of lesion/tumor segmentation
and sometimes is not truly objective. The deep learning techniques can automatically
discover features from image content without the need of image preprocessing steps such
as denoising and segmentation and, therefore, can be very attractive for the computer-
aided differentiation diagnosis of breast lesions in ultrasound images. In the study [62]
the stacked denoising autoencoder [5] is applied for the differentiation of 520 breast
lesions in ultrasound images. The main idea is to solve the learning problem by greedily
constructing the neural network architecture in a layer-by-layer fashion [63]. For the
differentiation purpose, the trained architecture can be further adapted and fine-tuned

by the labeled data in the supervised training. Note that the unsupervised learning phase is often referred to as the pretraining step.

The stacked denoising autoencoder architecture can be constructed by the basic building block of autoencoder. An autoencoder is a two-layered neural network that attempts to seek the representative patterns (mostly sparse), denoted as s, of the original data d at the hidden (second) layer with the deterministic encoder mapping:

$$s = f_\theta(d) = \sigma(Wd + p), \tag{9.4}$$

where $\theta = \{W, p\}$ is the parameter set of synaptic weightings and bias, respectively, and the mapping σ is usually specified as the nonlinear sigmoid function. The hidden representations are sought by minimizing the difference between the original data d and the reconstructed data z from the hidden layer s, which can defined as

$$z = f_\vartheta(s) = \sigma(W's + 1), \tag{9.5}$$

where $\vartheta = \{W', q\}$ is the parameter set of synaptic weightings and bias for the reconstruction mapping, respectively. With the building block of autoencoder, a higher level of semantic representations for the object class can be further constructed by stacking up autoencoders into deeper layers.

To improve the robustness of the stacked autoencoder architecture against the noise and cluttered background, Vincent et al. [4] proposed to randomly corrupt the data in the training process as the stacked denoising autoencoder architecture. In such a framework, the desirable reconstructed data $\tilde{z}$ are sought with the minimization of

$$\underset{W, W', p, q}{\arg\min} \mathcal{L}(d, \sigma(W'\sigma(W\tilde{d} + p) + q)), \tag{9.6}$$

where $\mathcal{L}(\cdot, \cdot)$ is the loss function, and $\tilde{d}$ is the corrupted data d drawn from a stochastic mapping $q_D(\tilde{d}|d)$. The efficacy and robustness of the stacked denoising autoencoder architecture had also been demonstrated on *either* image *or* audio data with different degrees of corruption in [4]. With the capability of noise tolerance, the stacked denoising autoencoder can be possibly suitable for the analysis of noisy medical image data such as ultrasound images and so on.

The training of a stacked denoising autoencoder-based computer-aided diagnosis framework can be realized in two steps, the pretraining and supervised training. Fig. 9.2 illustrates the flowchart of the stacked denoising autoencoder computer-aided diagnosis framework. To facilitate the training and differentiation tasks on the stacked denoising autoencoder architecture, the ROIs are resized into smaller patches of the same size, where all pixels in each patch are treated as the input neurons. To preserve the original information, the resized scale factors of the two ROI dimensions and the aspect ratios of the original ROIs are later added into the input layer in the supervised training step.

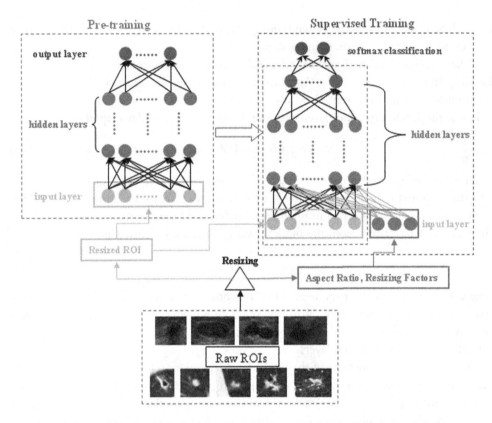

Figure 9.2 Flowchart of computer-aided diagnosis framework with a stacked autoencoder.

At the pretraining step, the input ROIs are degraded with random corruption, and then the noise-tolerance representative patterns for the lesion/tumor can be further discovered and encoded within the network architecture. Meanwhile, higher semantic levels of the representative patterns and the composite/interaction relations from the low semantic levels can be further sought by constructing autoencoders layer by layer. The constructed stacked denoising autoencoder architecture at the pretraining step can be served as reliable network initialization for the latter supervised training step.

The supervised training of the stacked denoising autoencoder computer-aided diagnosis framework performs the fine-tuning on the network architecture to yield the desirable discriminative performance. Since the size and aspect ratio of the ROI are omitted at the pretraining step, this information needs to be resumed at the supervised training to restore the completeness of the training data. Specifically, the scaling factors in the x and y dimensions and the aspect ratio are treated as three individual neurons as new inputs, whereas two extra output neurons of benign and malignant classes are also added on the top of the network for the supervised training. The modified network is then configured with the initialization from the pretrained architecture, three new input neurons, and the two outputs. Afterward, the supervised training is performed with the conventional back-propagation for the fine-tuning of the whole network. With such network architecture and two steps of training, the feature extraction and selection can be systematically and jointly realized with less need of *explicit* and *ad hoc* elaborations.

With the benefits of automatic feature extraction from the network of stacked autoencoder [62], the discrimination efficacy of breast lesions was shown to better than the compared conventional computer-aided diagnosis [47], which is a texture-based method that performed the ranklet transform to derive GLCM features to support the differentiation of benign and malignant breast lesions. In summary, the stacked denoising autoencoder can automatically discover useful textural, even morphological, features to support the difficult computer-aided diagnosis problem for the breast lesions depicted in ultrasound images.

Another work [64] proposed to employ the deep polynomial network for the differentiation of ultrasound breast lesions in a small dataset of 200 cases. The involved 200 cases include 100 benign and 100 malignant cases. To boost the differentiation performance, several deep polynomial networks are stacked up, similar to the mechanism of stacked denoising autoencoder, for more comprehensive feature learning. The stacked deep polynomial networks are compared with other deep learning methods of stacked autoencoder and deep belief network [65]. It was shown that the stacked deep polynomial networks can achieve better performance than other comparison methods.

9.3.2 Computer-aided diagnosis on pulmonary nodules in CT scans

Pulmonary nodules are small masses developed in the lung and appear as a relatively white round structure in the CT scan. A pulmonary nodule can possibly be spotted isolatedly within the lung parenchyma or as an attachment to the lung chest wall, airway, pulmonary vessel, and fissure. Recent studies have suggested that the phenotype features depicted in the CT images play an important role in the prediction of malignancy with the underlying mathematic models [39]. In clinical practice, the differential diagnosis of a pulmonary nodule based on the CT scan is a complicated decision process [66] but very crucial for the further diagnostic confirmation (with biopsy), management, follow-up, and treatment. Although the diagnostic guideline for the CT pulmonary nodule has been continuously modified and improved over the years [66,67], the diagnostic accuracy still remains varied for physicians [66,68] and also highly depends on experience [69]. The problems of the interphysician variation and experience dependency thus motivate the research and development of the computer-aided diagnosis system for the identification of nodule malignancy in CT images. The computer-aided diagnosis system is expected to provide objective and quantitative diagnostic evaluation and ultimately improve the diagnostic outcome [69] and avoid unnecessary biopsy, thoracotomy, and surgery [39].

Similarly to the computer-aided diagnosis scheme for ultrasound breast lesions, the differentiation of pulmonary nodules in CT also counts on the morphological [42,44, 69] and textural [41,69,70] features. The computation of the morphological features also mostly depends on the result of image segmentation for the automatic and quantitative analysis. The challenges of pulmonary nodule segmentation in CT images lie in the definition of the septal boundary between the nodule and its adherent tissue and diverse shape and appearance of nodules. The computed nodule boundaries still need manual or computerized refinement [69,71] for better quality of morphological feature computation.

Distinctly from the problem of breast lesion differentiation, the challenges for the classification of CT lung nodules can be twofold. First, as stated earlier, the pulmonary nodule can appear anywhere within the lung parenchyma. Therefore, the complicated anatomical background of pulmonary nodules can impose more difficulties on the extraction of useful features. Second, in some cases, the differentiation of a pulmonary nodule cannot be very certain since the phenotype in CT images can be ambiguous. In such a case, the deep learning techniques may offer an alternative method to find out better features for more discriminative power.

In the work [62], the stacked denoising autoencoder was also applied to tackle the lung nodule classification problem in CT images. Since a pulmonary nodule can be depicted within several slices in a CT scan, the image features of lung nodules can be much richer than 2D ultrasound breast images. However, since the slice thickness (z direction) may vary significantly from scan to scan, direct computing 3D image features may be problematic. The computation of a fully 3D representative form or feature in the z direction is less reliable than in the other x and y directions (dimensionality in 2D transversal slice). In such case, there will be an issue of slice selection for the representation of a nodule in the design of the CADx. For a pulmonary nodule, because in some *member* CT slices only small portions of the nodule are depicted, not every *member* CT slice of a nodule can be useful for the differential diagnosis. Accordingly, the efficacy of the involvement of all CT slices for the lung CT CADx is unknown. Two slice selection strategies are implemented in the study [62] for the lung CT computer-aided diagnosis problem. In one strategy, called "SINGLE" throughout this study, the middle slice of the nodule is selected as a representative sample. Each ROI sample stands for a distinctive nodule. In the other strategy, called "ALL," all *member* CT slices of each nodule participate in the training and testing. Specifically, in the training stage, all ROIs of the *member* CT slices from training nodules are treated as the training data for both the pretraining and supervised training phases. In the application stage, for the prediction of a testing nodule, a major voting scheme is implemented to reach the final classification result. Specifically, if more than half of the *member* CT slices are identified as malignant by the computer-aided model, then the nodule is regarded as a malignant nodule.

The SINGLE and ALL strategies are implemented with the stacked autoencoder and the baseline conventional computer-aided diagnosis methods to illustrate the corresponding classification performance of the pulmonary nodules in the work [62]. Therefore the sample sizes of the training and testing ROI can be different. Tenfold cross-validation scheme was implemented in [62] for the robustness evaluation on data dependence. 1400 pulmonary nodules were randomly selected from the public Lung Image Database Consortium (LIDC) [28,27] as the evaluation data.

The LIDC dataset includes retrospectively collected lung CT image data from 1010 patients (with appropriate local IRB approvals) at seven academic institutions with various CT machines in the United States. The slice thicknesses of the CT scans ranged diversely from 0.6 mm to 5 mm. 12 radiologists from five sites were involved in the rigorous two phases of the blinded and unblinded image reading and the annotation process. In particular, nodules with diameters larger than 3 mm were further annotated with subjective ratings, including the ranking of malignancy, by the radiologists to facilitate the computer-aided diagnosis research. For the malignancy ranking, the score ranges from 1 to 5, suggesting the subjective evaluation possibility of being a malignant nodule. The study [62] investigated if the performance of nodule malignancy identification by the stacked denoising autoencoder-based computer-aided diagnosis scheme can approach closely the ratings from the experienced thoracic radiologists. To facilitate the classification scheme, nodules with scores of 1 and 2 were taken as benign, whereas the nodules scored as 4 and 5 were enlisted in the malignancy category. For the balance of classification, the 1400 randomly selected nodules were composed of 700 benign and 700 malignant nodules.

Back to the slice selection strategies, for the SINGLE slice selection strategy, there were totally 1400 training and testing 2D ROI samples for the computer-aided diagnosis models. Each 2D ROI sample stands for a distinctive nodule. In the ALL strategy, *member* slices of each nodule are involved in the training and testing. Specifically, the *member* slices of all 1360 training nodules in the tenfold cross-validation scheme are randomly permutated as the training samples for the training of the computer-aided diagnosis models. For the 140 testing nodules, the final differentiation of a testing nodule is determined with the majority voting from the *member* slices. In the ALL strategy, the total number of involved slices is 10133. Because slice thickness variation is very high (0.6 mm to 5 mm), the fully 3D representative forms and features of image data for the lung computer-aided diagnosis may not be applicable with the LIDC dataset. In the experiments, the ALL strategy was shown to be better than the SINGLE strategy, suggesting that more training data can be helpful for deep learning methods. Meanwhile, the stacked denoising autoencoder was also shown to be capable of attaining better performance than the conventional computer-aided diagnosis scheme, which was based on the curvelet transform [72] and GLCMs for feature computing.

On the other hand, the techniques of deep convolutional neural network are also explored in the context of computer-aided diagnosis for pulmonary nodules in CT images [73–75,23]. In [73] the multiscale scheme is explored by training convolutional neural networks that take different scales of ROIs for pulmonary nodules in CT images to attain the purpose of differential diagnosis. Specifically, three convolutional neural networks were implemented to take input ROIs of different scales. The three convolutional neural networks have distinctive convolutional and max-pooling layers for the extraction of image context features at different scales. The multiscale image context features are further concatenated as an integral feature vector for the latter classifier to reach the final classification results. Two classifiers of Support Vector Machines (SVM) and Random Forests (RF) were implemented in [73]. To further improve the effectiveness of feature learning, the deep supervision strategy [76] was implemented. The deep supervision strategy aims to consider the learning of the three convolutional neural networks and feature integration at the same time with a linear combination of the objective loss functions. A fivefold cross-validation scheme was implemented to evaluate performance of multiscale deep convolutional neural networks. In each fold of cross-validation, three are 1100 nodules for training and 275 nodules for testing. To further illustrate the efficacy, conventional textural features computed with the techniques of Histogram of Oriented Gradient (HOG) [77] and Local Binary Pattern (LBP) [78] are implemented as baseline features. The experimental results suggest that the performance of the multiscale convolutional neural networks can achieve more than 80% in accuracy, and the classifier of random forest was shown to be better than SVM. The best accuracy performances of either the HOG or LBP features were around 70%, suggesting that the multiscale convolutional neural networks can have better capability to learn useful textural and image context features for differentiation of nodule malignancy. Different from [62], where 2D ROIs were used as input, the ROI in [73] is 3D. The involved pulmonary nodules are randomly selected from the LIDC database [28,27].

The work in [74] further improved the multiscale networks in [73] with the so-called multicrop pooling strategy. Specifically, given the feature map R_0, derived from a previous convolutional layer, two smaller features were obtained by two sequential cropping operations R_1 and R_2. The original feature map R_0 was convolved with two passes of max-pooling, whereas the smaller feature map R_1 with one cropping operation was undergone with one pass of max-pooling. The smallest feature map R_2 on the other hand was not convolved with any max-pooling operation. Afterward, the resulting feature maps of the multicrop pooling are further concatenated as a new feature map for latter convolutional layer. The multicrop pooling strategy was motivated by the work of spatial pyramid pooling network [79], where the feature pyramid was computed and concatenated as the one final integral feature vector. The multicrop pooling strategy aimed to better extract multiscale features than the previous multiscale convolutional neural networks [73]. To show the performance improvement and advantage, the multicrop pooling strategy was extensively compared with single convolutional neural network, multiscale convolutional neural networks, and so on. Meanwhile, three sets of datasets of different sizes of 340, 1030, and 1375 nodules from the LIDC database [28,27] were applied to illustrate the sensitivity to data size. Similarly to [73], fivefold cross-validation scheme was also applied to illustrate the robustness of the developed neural network methods to the data partition. In the experimental results, the multicrop pooling strategy can boost the prediction performance in terms of accuracy than the performance achieved by the multiscale convolutional neural networks. On the other hand, for the experiment of three datasets of 340, 1030, and 1375 nodules, the convolutional neural network with multicrop pooling can attain the accuracies of 83.09%, 86.36%, and 87.14%, respectively. It may thus suggest that the neural network model can remain perform well with a relatively small dataset of 340 nodules.

In [75] the same team further extended the technique of deep convolutional neural network in a regression framework to approximate the network prediction results toward the annotated likelihood scores of nodules from the radiologists. The involved nodule data were also selected from the LIDC database [28,27]. The work specifically explored [75] the issue of domain transfer from the expert rating domain to the patient level prediction of nodule malignancy. To leverage the malignancy prediction from the nodule level to the patient level, the technique of multiple instance learning [80] is further explored. For the domain transfer, the data were divided into the discovery set and diagnosed set. The discovery set included the LIDC pulmonary nodules, which were scored in terms of the subjective assessment of malignant likelihood by the radiologists. The score range of the nodule malignant likelihood is from 1 to 5. It is worth noting that the malignancy of each nodule in the discovery set was not confirmed definitely because the relevant diagnostic information was missed. On the other hand, the diagnosed set enclosed the nodules with firm diagnosis from the histological image examination results. The discovery set included 2272 nodules, whereas 115 nodules were enclosed in the diagnosed set. With the discovery set, a deep convolutional neural network was employed to learn the annotated scores from the radiologists with the formulation of regression. To specifically realize the regression norm, the loss function of the deep convolutional neural network is set as Euclidean distance, that is, the ℓ_2 distance, to approach the prediction scores close to the annotation ratings. The learned neural parameters in the discovery convolutional neural network were further applied as the network initialization of the deep convolutional neural network for the diagnosed set. To leverage the nodule level toward the patient level, the multiple instance learning paradigm was further applied to learn a convolutional neural network for the diagnosed set. Specifically, for the prediction of malignancy at the patient level, the feature of each nodule was treated as an instance in either positive or negative bags. A positive bag was supposed to contain at least one malignant nodule instance, whereas the nonmalignant nodule feature should be included in the negative bag. The effectiveness of the domain transfer and the multiple instance learning paradigm and of the feature learning with convolutional neural network was shown in the ablation experiment.

Instead of general lung nodule differentiation, the work [23] focused on the identification of perifissural nodules in CT images. The perifissural nodules are approximately 30% nodule findings in smokers [81] and attached to pulmonary fissures. It was suggested that the identification of perifissural nodules can possibly reduce the number of follow-up CT examinations for the workup of suspicious nodules [82]. Therefore, the automatic identification of perifissural nodules matters in the clinical image reading. The work [23] trained deep convolutional neural networks with 2D image inputs and considered the 3D information with multiple 2D views. Specifically, three convolutional neural networks were employed to learn the image context and textural cues from the three axial, coronal, and sagittal views. The off-the-shelf pretrained OverFeat network model [83] was employed as model initialization for the axial, coronal, and sagittal convolutional neural networks. With the three trained network models, the final identification result can be reached with the technique of ensemble learning. 4026 pulmonary nodules (3107 nonperifissural nodules and 919 perifissural nodules) were involved in the study [23]. The multiview convolutional neural networks were shown to be able to attain good classification performance with Area Under Receiver Operating Characteristic (ROC) Curve (AUC) values more than 0.80.

In summary, based on the experimental results of [62,73–75], promising nodule malignancy prediction results can be achieved with the learning of nodule textural and the image context features by deep learning techniques such as the stacked denoising autoencoder and deep convolutional neural network.

9.3.3 Other computer-aided diagnosis problems in pulmonary CT scans

Another prominent study that applied the deep learning technique for the analysis of pulmonary CT scans is [84]. The specific deep learning technique used is the deep restricted Boltzmann machine [85,86] for the textural classification and airway detection in the chest CT scans. Two types of restricted Boltzmann machine, the standard [86] and classification [85] restricted Boltzmann machines, were implemented to learn generative and discriminative features. Similarly to the stacked autoencoder, the standard restricted Boltzmann machine [86] can unsupervisedly learn image context features from the training data without the guidance of ground-truth labels. On the other hand, the classification restricted Boltzmann machine adds label nodes in the visible layer to consider the factor of object class membership for the feature learning. Since the object class is considered in the classification restricted Boltzmann machine, the learned features can be more discriminative. The work [84] combined the generative and discriminative restricted Boltzmann machine features for the tasks of lung tissue classification and airway centerline detection. For the problem of lung tissue classification, it aimed to find out useful textural descriptors that can effectively classify the CT image contents into the categories of healthy tissue, emphysema, ground glass, fibrosis, and micronodule. Referring to experimental results, the combination of generative and discriminative features for the random forest classifier can improve the classification accuracy with 1 to 8 percentage(s). For the application of airway centerline detection, 40 lung CT scans acquired from 20 subjects of the Danish Lung Cancer Screening Trial [30] were adopted. The accuracy of the airway centerline detection can be around 0.9 if the number of hidden nodes is increased to 64. In summary, the major hyperparameters of the restricted Boltzmann machine to determine the textural analysis results in the lung CT images are the filter kernel size, the number of hidden nodes, and so on. For further medical image analysis study considering the usage of restricted Boltzmann machine, two types of hyperparameters may be regarded to tune first.

9.4 AUTOMATIC MAPPING FROM IMAGE CONTENT TO THE SEMANTIC TERMS

Compared to the applications of computer-aided detection and computer-aided diagnosis, the topic of automatic mapping from the image content to the semantic terms used in the medical documents is a relatively new problem. The semantic mapping may provide richer quantitative features as the reference of clinical diagnosis and evaluation and assist the realization of automatic annotation, which may potentially alleviate the significant annotation shortage of clinical data for training machine learning models. Meanwhile, the semantic mapping may also support the application of Content-Based Image Retrieval (CBIR) or medical search engine as alternative similarity measures. By and large, the semantic mapping needs to find useful computational image features to describe each medical semantic term with satisfactory accuracy. In the conventional pattern recognition and machine learning paradigm, the feature engineering is the most crucial step, and the sought computational features for one semantic term may not be easily applied to the other semantic terms [87]. Since the medical semantic terms defined in several medical standard ontology, for example, Radiology Lexicon (RadLex) [88], can be very diverse, the feature engineering for several semantic terms on the same object of interest can be very problematic. Because the standard feature computing techniques, like HOGs, Scale Invariant Feature Transform (SIFT) [89], Haar-like features [90], LBPs [78], and so on, were shown to be domain agnostic [91], these handcrafted features are combined with proper classifier to find a suitable setting of hyperparameters and a subset of handcrafted features for the specific task and semantic mapping. As discussed earlier, the deep learning techniques can automatically discover textural and image context features from training data and also fulfill the classification and regression tasks with the same network architecture. Therefore the deep learning techniques may offer an easier way to achieve feature engineering for better performance of semantic mapping. In this section, we first review the relevant semantic mapping with the conventional pattern recognition paradigm and the relevant applications. Following that, we discuss a new study that utilizes the deep learning techniques to attain the goal of semantic mapping.

9.4.1 Semantic mapping with the conventional pattern recognition paradigm

There are less works about the semantic mapping between medical image content and medical terms. One of prominent works is described in [92], which aimed to predict the visual semantic descriptive terms directly from the medical image content. The visual semantic descriptive terms in [92] characterized the liver lesions. The involved semantic terms were the instantiated descriptions for the term categories of "lesion margin," "lesion substance," "perilesional tissue characterization," "lesion focality," "lesion attenuation," "overall lesion enhancement," "spatial pattern of enhancement," "temporal enhancement," "lesion uniformity," "overall lesion shape," and "lesion effect on liver." As can be found, these categories include the semantic textural and morphological features. Since the work [92] employed the technique of Riesz wavelet transform for the computing of textural features, the categories regarding to lesion morphology were excluded, and the remaining categories are "lesion margin," "lesion substance," "perilesional tissue characterization," "lesion attenuation," "overall lesion enhancement," "spatial pattern of enhancement," and "lesion uniformity." The category of "lesion margin" is instantiated with the terms of "circumscribed margin," "irregular margin," "lobulated margin," "poorly-defined margin," and "smooth margin." The category of "lesion substance" has only one term "internal nodules," whereas the category "perilesional tissue characterization" also has only one term of "normal perilesional tissue." The category of "lesion attenuation" has three descriptive terms of "hypodense," "soft tissue density," and "water density," whereas the "overall lesion enhancement" category is instantiated with the terms of "enhancing," "hypervascular," and "nonenhancing." The category of "spatial pattern of enhancement" are described with "heterogeneous enh.," "homogeneous enh.," and "peripheral discont. nodular enh." The last category of "lesion uniformity" is termed with "heterogeneous" and "homogeneous." In total, there were 18 instantiated visual semantic descriptive terms in [92]. To automatically profile the liver lesions depicted in the CT scans, the mapping of each visual semantic descriptive term was realized with the textural features computed from the Riesz wavelet transform and the SVMs classifier for the selection of term-oriented textural features. The mapping of each visual semantic descriptive term was performed separately since the textural features selected for one term may not be easily generalized to the other. The semantic mapping with Riesz wavelet transform and SVM was evaluated with 74 liver lesions imaged with 74 contrast-enhanced CT images. There were eight types of lesions: metastasis, cyst, hemangioma, hepatocellular carcinoma, focalnodular hyperplasia, abscess, laceration, and fat deposition. The semantic mapping was realized on the 2D ROI patches of liver lesions. The mapping results were assessed with the metrics of AUC, and the evaluation experiments were conducted with leave-on-patient-out cross-validation. There was performance variation for the 18 visual semantic descriptive terms. In particular, for the terms of "homogeneous," "heterogeneous," "homogeneous enhancement," "heterogeneous enhancement," "hypodense," "peripheral discont. nodular enh.," "nonenhancing," "smooth margin," and "irregular margin," the AUC values were more than 0.8. Other terms such as "normal perilesional tissue" and "lobulated margin" were relatively not stable in terms of AUC assessment metrics. Since the visual semantic descriptive terms like "circumscribed margin," "irregular margin," "lobulated margin," "poorly-defined margin," and "smooth margin" are of the same semantic category, there may exist some sorts of exclusive relation among these terms. The explicit consideration of the exclusive relation may be realized with the cotraining or multitask learning paradigm [93].

The work [94] further employed the semantic mapping technique developed in [92] into the context of CBIR of liver lesion depicted CT images. Specifically, a liver lesion was automatically profiled with the 18 visual semantic descriptive terms explored in [92]. Each semantic term was quantitatively described with a probability value. Therefore the probability values of the 18 semantic terms can then concatenated as an attribute vector to be served as the basis for the computation of similarity measure. The definition of similarity measure is very crucial for a CBIR system since it can quantitatively describe the association between the two images. The conventional handcrafted features do not consider the semantic meaning in the images and hence may be limited to address the cases of the same diagnosis but having significantly different image appearances, as well as cases with distinctive disease but having similar phenotype shown in the medical images. The attributes of the 18 visual semantic descriptive terms may leverage the computation of similarity measure from the low image context domain to the semantic domain for better consideration of the cases discussed before. In [94] the similarity measure was defined based on the combination of the differences between the visual semantic descriptive terms of the two comparing cases. Meanwhile, the ontological relation can also be easily incorporated by finding the path distances of major semantic terms in the ontological tree. With the works [92,94], the applicability of the semantic mapping on the content-based medical image retrieval can be well corroborated.

In the context of pulmonary nodules, recently, there were also studies that attempted to seek the mapping of the CT phenotypes of lung nodules toward the semantic descriptors [95,96] instead of simple prediction of the nodule malignancy. In the study [95], a computerized scheme was developed to automatically assess the nodule solidity with textural analysis. The computerized nodule solidity assessment was realized in a three-class classification scheme. The three classes were solid, part-solid, and nonsolid. The nonsolid is also called Ground Glass Opacity (GGO). Recent clinical studies have suggested that nonsolid and part-solid nodules may be more likely confirmed as malignancy since these types of nodules may stand for the spectrum of peripheral adeno-carcinomas or the subtypes of adenocarcinomas [97,98]. Therefore the classification and rating of pulmonary nodules into the categories of solid, part-solid, and nonsolid nodules has also been recommended to be implemented in the CT diagnostic procedure by the Fleischner Society [66] and other clinical studies [99,100]. The computerized nodule solidity assessment was based on the thresholding of Hounsfield Units (HU) on the CT image to segment the pulmonary nodules. Afterward, the histogram-based features like entropy, standard deviation, quantiles of 5%, 25%, 75%, and 95%, and so on were computed for the k-nearest neighbor classifier to identify the nodule solidity as solid, part-solid, and nonsolid. 150 pulmonary nodules including 50 solid, 50 part-solid, and 50 nonsolid nodules were involved in [95]. The performance of the computerized nodule solidity assessment was compared to four radiologists. It was shown that the computerized classification results achieved agreement with radiologists' classifications with the Cohen kappa values [101] in the range of 0.56 to 0.81, suggesting moderate agreement between the computerized results and expert annotations. Meanwhile, the interobserver variation was also investigated. The Cohen kappa values for the classification annotations from different radiologists were in the range of 0.54 to 0.72, suggesting moderate agreement among the four radiologists.

The work [96] employed the bag of visual words [102,103] technique on the frequency domain of the shape to characterize the semantic morphology feature of spiculation for pulmonary nodules. Specifically, an image segmentation procedure was incorporated to demarcate nodules in CT images and was constituted of steps of nodule diameter estimation, profile sampling, codebook generation, and so on. For nodule with complicated intensity and subtle or GGO nodules, the segmentation problem is relatively difficult. The bag-of-frequency technique was shown to successfully differentiate 51 spiculated and 201 nonspiculated nodules. In summary, the works [95,96] were formulated in the discrete trichotomy and dichotomy framework for semantic attributing. The feature engineering may require several intermediate steps to find out useful features for each semantic characteristic. The ambiguity and exclusive relation for one semantic category, for example, margin, solidity, and so on, were less explicitly explored.

9.4.2 Deep learning for semantic mapping

For the context of semantic mapping, the deep learning techniques may possibly circumvent the need of several intermediate processing steps including image segmentation. To consider the relation among multiple semantic features, multitask framework may help. The studies [104,105] employed two types of deep learning techniques for feature learning in the application of semantic mapping for the pulmonary nodules. Specific deep learning techniques were the stacked denoising autoencoder and the convolutional neural network. The specific goal of the semantic mapping for the pulmonary nodules in [104,105] was not only to perform automatic attributing of semantic terms but also to conduct the degree assessment on the semantic terms with quantitative scoring. The specifically exploited semantic terms are "texture," "subtlety," "spiculation," "sphericity," "margin," "malignancy," "lobulation," "internal structure," and "calcification." The semantic terms and their instantiated degrees can be found in Fig. 9.3. Except the feature "calcification," which can be rated from score 1 to score 6, and the feature "internal structure," which was scored from 1 to 4, the score range of the remainder features is 1–5. Specifically, the term "texture" is the nodule solidity as defined in [95]. Differently from the three-class formulation in [95], the degree score range of the "texture" term is from 1 to 5 for finer description and may have better consideration of the annotation ambiguity; and in [95] the interobserver agreement was shown to be moderate. The term "subtlety" indicates the degree of detection difficulty by human, and "calcification" suggests the degree of detection difficulty and calcification pattern of a nodule, respectively. The term "sphericity" describes the degree of roundness for a nodule, whereas the "margin" feature attempts to quantify how well defined are the margins of the nodule. The "spiculation" terms represent the degree of spiculation for a nodule shape. Similarly to the "spiculation" term, the "lobulation" term indicates the saliency degree of lobulation shape for a nodule. The "internal structure" term suggests that a nodule can be constituted of soft tissue, fluid, fat, or air, and the "malignancy" feature indicates the malignancy likelihood subjectively assessed by a radiologist. The nine semantic terms were defined in the annotation records of the LIDC database [28, 27], which is a resourceful public database to thrust the research of computer-aided diagnosis and detection on the application of pulmonary nodules. All involved nodules in the studies [104,105] were from the LIDC database.

The flowchart of the lung nodule semantic mapping and scoring is shown in Fig. 9.4, which is mainly constituted of two major parts, feature learning and multitask linear regression. As can be seen in Fig. 9.3, the characteristics described by the nine semantic terms can be very diverse, and the variation for the nodule morphology and appearance can be very significant. Meanwhile, since a nodule can be profile with several semantic terms, there may exist certain relation among these semantic terms. The relation may be automatically processed when a radiologist is doing the annotation ratings. On the other hand, because a nodule can have multiple attributes, the learning of computational features for one semantic term may be affected by other semantic terms. Accordingly, these issues impose more difficulty on learning useful features for each term. To address the aforementioned issues, the feature to be learned will be as diverse as possible to reserve for flexibility of latter feature selection and regression step. Since the nine semantic terms were described with degree scores, it will be more suitable to formulate the semantic mapping with a regression framework. The regression framework can possibly accommodate the scoring ambiguity between the consecutive scores by relaxing the output scores from the annotated integer level to the estimated floating point

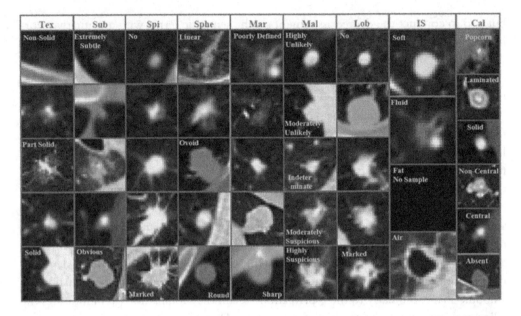

Figure 9.3 Illustration of nine semantic terms.

level. In the studies [104,105], the regression framework was formulated in the multitask paradigm to further explore the intertask relation.

In the feature learning step of [104,105], two deep learning techniques and the conventional feature extraction techniques of HOGs [77] and Haar-like [90] features were explored. Two deep learning techniques employed in [104,105] automatically extracted nodule features to support the semantic mapping. As shown earlier, the training of the stacked denoising autoencoder can be realized with the unsupervised and supervised phases. The unsupervised phase of training does not require the ground truth labels of data, and the nodule features can be automatically learned with the minimization of reconstruction errors between the input data and the output from the decoder. The network of the stacked denoising autoencoder is greedily constructed by stacking the shallow encoders. The higher hidden neurons may encode higher level of nodule features. At the supervised training phase, the network is added with soft-max layer and further fine-tuned with the guidance of ground truth labels of data. Since the purpose of using stacked denoising autoencoder was for feature learning, only the unsupervised training phase was carried out. A stacked denoising autoencoder with two hidden layers of neurons was used in [47,18]. Since the training of the stacked denoising autoencoder needs no ground-truth labels of data, the learnt features from the autoencoder can be more general nodule features. In [104,105] the overall number of stacked autoencoder features was 100.

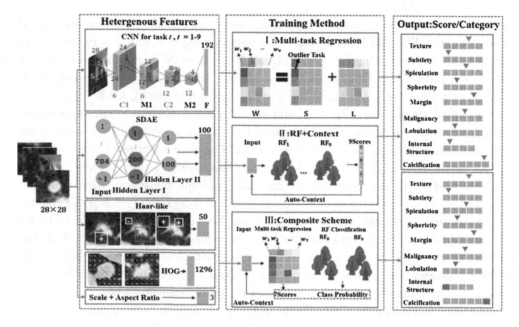

Figure 9.4 Flowchart of multitask regression for the semantic term mapping and scoring.

On the other hand, since the training of the convolutional neural networks requires the ground-truth labels, the corresponding learned features are more problem specific. In [104,105] the convolutional neural network features for each semantic term was trained with an independent convolutional neural network. Specifically, for one semantic term, each degree score was treated as an individual output class for the soft-max layer for the training of neural network. After the training, the soft-max layer of each convolutional neural network was removed, and the neurons of the final fully connected layer were treated as the feature neurons. Since the training of the convolutional neural network was trained with the guidance of ground-truth labels, the learned feature is more semantic term specific. For each convolutional neural network, there were 192 neurons in the fully connected layer, and therefore the total feature dimensionality of term-specific features was 1728 (192×9) [105]. On the other hand, the dimensionality of HOG features and Haar-like features were 1296 and 50, respectively.

Two multitask regression schemes were explored in [105]. The first multitask regression scheme is linear and exploits the intertask relations with the inclusion of regularization terms in the linear regression cost function. The cost function of the multitask linear regression was defined as

$$\sum_{i=1}^{n}\sum_{t=1}^{9}\|(l_t + s_t^T)x_i - y_i^t\|_F^2 + \lambda_L\|L\|_* + \lambda_S\|L\|_{1,2}, \tag{9.7}$$

where x_i is the feature vector of a stacked denoising autoencoder, convolutional neural network, Haar–like, and HoG features of a training sample i, and y_i^t is its annotation scores with respect to the task t; $y_i^t \in \{1, \dots, N^t\}$, where N^t is the maximal score of the task t, and d is the dimensionality of total computational features. The matrices L and S are defined as $L = [l_1, \dots, l_9]$ and $S = [s_1, \dots, s_9]$, respectively, $l_t \in \mathbb{R}^{d \times 1}$, $s_t \in \mathbb{R}^{d \times 1}$, n is the total sample, and $\|\cdot\|_F$, $\|\cdot\|_*$, and $\|\cdot\|_{1,2}$ are the Frobenius norm, trace norm, and $\ell_{1,2}$-norm, respectively. The columns of the matrices L and S are the coefficients of computational features of each semantic term. The trace norm regularization of the second term $\|L\|_*$ in Eq. (9.7) encourages to find low-rank structure [106], whereas the $\ell_{1,2}$-norm of the third regularization term $\|L\|_{1,2}$ prompts the group-sparse structure. The $\ell_{1,2}$-norm of S can be sought as $\sum_{t=1}^{9} \|s_t\|_2$, where $\|s_t\|_2$ is the ℓ_2-norm of a column vector. In such a case, the group-sparse structure has many zero columns in a matrix; λ_L and λ_S are the the second and third regularization terms, respectively. With the minimization of Eq. (9.7), the low-rank matrix L will pick up the shareable features from the diverse feature pool across the nine semantic terms. On the other hand, the nonzero columns in the matrix S may suggest the outlier semantic terms.

Another multitask regression scheme in [105] is based on the random forest [107] regression and autocontext [108] technique. The job of regression of the nine semantic terms was realized by nine random forests, whereas the intertask relations were explored with the autocontext. It was suggested that the best regression performance may be attained with one or two recursive iterations of autocontext.

Totally, 1400 and 2400 nodules were involved in the studies [104] and [105], respectively. The evaluation metrics to assess the regression performance is the difference between the computerized prediction and expert ratings. To illustrate the effectiveness of each type of features, the experiments of employing each feature type were conducted. The corresponding experimental results suggest that the features learned from convolutional neural networks can attain best performance, whereas the regression performance with features of stacked denoising autoencoder and HOG were relatively comparable. The usage of Haar-like features was not able to outperform other types of features. On the other hand, to illustrate the efficacy of multitask learning, the single-task linear regression methods of Elastic Net [109] and LASSO [110] were implemented for comparison. In the experimental results, the multitask linear regression can achieve the best regression performance, whereas random forest with autocontext regression can also attained reasonably good results. The single-task regression schemes were generally not able to outperform the multitask methods. Therefore the effectiveness of the multitask methods was corroborated. On the other hand, since the features from the convolutional neural networks were shown to be useful, the direct regression with the convolutional neural network architecture was also implemented. The loss function was set as the Euclidean distance to guide the network optimization and also approach the prediction scores toward the annotation ratings. It was shown that the direct regression results from the convolutional neural network were close to the results from autocontext + random forest. More details can be found in [104,105].

In summary, the computerized semantic mapping and attributing on medical images is a relatively new topic. It may serve as an intermediate technical component to support medical big data analysis, medical search engine, and research of precision medicine (see Chapter 8). Meanwhile, the deep learning techniques were shown to be beneficial in learning the textural and image context features of the objects of interest for e realization of computerized semantic mapping.

9.5 CONCLUSION

In this chapter the deep learning techniques of stacked denoising autoencoder, deep belief net, deep convolutional neural networks on the applications of computer-aided detection, computer-aided diagnosis, and automatic semantic mapping were discussed. With proper training and sufficient training data, promising medical image analysis performance can be attained. The current open issues of the deep learning techniques on the medical image analysis applications may lie in the limited training and annotation ground truth. Therefore recent deep learnings exploited the techniques of domain transfer [9] and multitask learning [105,104] to leverage the knowledge learned from other image domain and explore the relation of different image analysis tasks. On the other hand, the issues of medical image collection and ground truth annotation remain very crucial to support the research of big data analysis and precision medicine. Since the annotation of medical data are very expensive and difficult to realize, the future research trend with deep learning techniques may consist in how to learn from weakly labeled annotation data with the framework of weakly supervised learning [111,112].

ACKNOWLEDGMENTS

The work was supported by the National Natural Science Funds of China (Nos. 61501305, 61571304, and 81571758), the Shenzhen Basic Research Project (Nos. JCYJ20150525092940982 and JCYJ20140509172609164), and the Natural Science Foundation of SZU (No. 2016089).

REFERENCES

[1] Y. LeCun, Y. Bengio, G. Hinton, Deep learning, Nature 521 (7553) (2015) 436–444.

[2] J. Schmidhuber, Deep learning in neural networks: an overview, Neural Netw. 61 (2015) 85–117.

[3] A. Krizhevsky, I. Sutskever, G.E. Hinton, Imagenet classification with deep convolutional neural networks, in: Advances in Neural Information Processing Systems, 2012, pp. 1097–1105.

[4] P. Vincent, H. Larochelle, I. Lajoie, Y. Bengio, P.-A. Manzagol, Stacked denoising autoencoders: learning useful representations in a deep network with a local denoising criterion, J. Mach. Learn. Res. 11 (Dec) (2010) 3371–3408.

[5] P. Vincent, H. Larochelle, Y. Bengio, P.-A. Manzagol, Extracting and composing robust features with denoising autoencoders, in: Proceedings of the 25th International Conference on Machine Learning, ACM, 2008, pp. 1096–1103.

[6] H.-P. Chan, K. Doi, C.J. Vybrony, R.A. Schmidt, C.E. Metz, K.L. Lam, T. Ogura, Y. Wu, H. MacMahon, Improvement in radiologists' detection of clustered microcalcifications on mammograms: the potential of computer-aided diagnosis, Invest. Radiol. 25 (10) (1990) 1102–1110.

[7] T.W. Freer, M.J. Ulissey, Screening mammography with computer-aided detection: prospective study of 12,860 patients in a community breast center, Radiology 220 (3) (2001) 781–786.

[8] M. Das, G. Mühlenbruch, A.H. Mahnken, T.G. Flohr, L. Gündel, S. Stanzel, T. Kraus, R.W. Günther, J.E. Wildberger, Small pulmonary nodules: effect of two computer-aided detection systems on radiologist performance, Radiology 241 (2) (2006) 564–571.

[9] H.-C. Shin, H.R. Roth, M. Gao, L. Lu, Z. Xu, I. Nogues, J. Yao, D. Mollura, R.M. Summers, Deep convolutional neural networks for computer-aided detection: CNN architectures, dataset characteristics and transfer learning, IEEE Trans. Med. Imaging 35 (5) (2016) 1285–1298.

[10] H.R. Roth, L. Lu, J. Liu, J. Yao, A. Seff, K. Cherry, L. Kim, R.M. Summers, Improving computer-aided detection using convolutional neural networks and random view aggregation, IEEE Trans. Med. Imaging 35 (5) (May 2016) 1170–1181.

[11] Q. Dou, H. Chen, L. Yu, J. Qin, P.A. Heng, Multilevel contextual 3D CNNs for false positive reduction in pulmonary nodule detection, IEEE Trans. Biomed. Eng. 64 (7) (2017) 1558–1567, http://dx.doi.org/10.1109/TBME.2016.2613502.

[12] R.L. Siegel, K.D. Miller, A. Jemal, Cancer statistics, 2015, CA Cancer J. Clin. 65 (1) (2015) 5–29.

[13] C.I. Henschke, D.I. McCauley, D.F. Yankelevitz, D.P. Naidich, G. McGuinness, O.S. Miettinen, D. Libby, M. Pasmantier, J. Koizumi, N. Altorki, et al., Early lung cancer action project: a summary of the findings on baseline screening, The Oncologist 6 (2) (2001) 147–152.

[14] R. Yuan, P.M. Vos, P.L. Cooperberg, Computer-aided detection in screening CT for pulmonary nodules, Am. J. Roentgenol. 186 (5) (2006) 1280–1287.

[15] D.S. Gierada, T.K. Pilgram, M. Ford, R.M. Fagerstrom, T.R. Church, H. Nath, K. Garg, D.C. Strollo, Lung cancer: interobserver agreement on interpretation of pulmonary findings at low-dose CT screening, Radiology 246 (1) (2008) 265–272.

[16] T. Messay, R.C. Hardie, S.K. Rogers, A new computationally efficient CAD system for pulmonary nodule detection in CT imagery, Med. Image Anal. 14 (3) (2010) 390–406.

[17] A. Riccardi, T.S. Petkov, G. Ferri, M. Masotti, R. Campanini, Computer-aided detection of lung nodules via 3D fast radial transform, scale space representation, and Zernike MIP classification, Med. Phys. 38 (4) (2011) 1962–1971.

[18] X. Ye, X. Lin, J. Dehmeshki, G. Slabaugh, G. Beddoe, Shape-based computer-aided detection of lung nodules in thoracic CT images, IEEE Trans. Biomed. Eng. 56 (7) (2009) 1810–1820.

[19] K. Murphy, B. van Ginneken, A.M.R. Schilham, B.J. De Hoop, H.A. Gietema, M. Prokop, A large-scale evaluation of automatic pulmonary nodule detection in chest CT using local image features and k-nearest-neighbour classification, Med. Image Anal. 13 (5) (2009) 757–770.

[20] M. Tan, R. Deklerck, B. Jansen, M. Bister, J. Cornelis, A novel computer-aided lung nodule detection system for CT images, Med. Phys. 38 (10) (2011) 5630–5645.

[21] C. Jacobs, E.M. van Rikxoort, T. Twellmann, E.Th. Scholten, P.A. de Jong, J.-M. Kuhnigk, M. Oudkerk, H.J. de Koning, M. Prokop, C. Schaefer-Prokop, et al., Automatic detection of subsolid pulmonary nodules in thoracic computed tomography images, Med. Image Anal. 18 (2) (2014) 374–384.

[22] A.A.A. Setio, F. Ciompi, G. Litjens, P. Gerke, C. Jacobs, S.J. van Riel, M.M.W. Wille, M. Naqibullah, C.I. Sánchez, B. van Ginneken, Pulmonary nodule detection in CT images: false positive reduction using multi-view convolutional networks, IEEE Trans. Med. Imaging 35 (5) (2016) 1160–1169.

[23] F. Ciompi, B. de Hoop, S.J. van Riel, K. Chung, E.Th. Scholten, M. Oudkerk, P.A. de Jong, M. Prokop, B. van Ginneken, Automatic classification of pulmonary peri-fissural nodules in computed tomography using an ensemble of 2D views and a convolutional neural network out-of-the-box, Med. Image Anal. 26 (1) (2015) 195–202.

[24] H. Chen, D. Ni, J. Qin, S. Li, X. Yang, T. Wang, P.A. Heng, Standard plane localization in fetal ultrasound via domain transferred deep neural networks, IEEE J. Biomed. Health Inform. 19 (5) (2015) 1627–1636.

[25] H. Chen, Q. Dou, D. Ni, J.-Z. Cheng, J. Qin, S. Li, P.-A. Heng, Automatic fetal ultrasound standard plane detection using knowledge transferred recurrent neural networks, in: International Conference on Medical Image Computing and Computer-Assisted Intervention, Springer, 2015, pp. 507–514.

[26] X. Glorot, A. Bordes, Y. Bengio, Deep sparse rectifier neural networks, Aistats 15 (2011) 275.

[27] S.G. Armato III, G. McLennan, L. Bidaut, M.F. McNitt-Gray, C.R. Meyer, A.P. Reeves, B. Zhao, D.R. Aberle, C.I. Henschke, E.A. Hoffman, et al., The lung image database consortium (LIDC) and image database resource initiative (IDRI): a completed reference database of lung nodules on CT scans, Med. Phys. 38 (2) (2011) 915–931.

[28] S.G. Armato III, G. McLennan, M.F. McNitt-Gray, C.R. Meyer, D. Yankelevitz, D.R. Aberle, C.I. Henschke, E.A. Hoffman, E.A. Kazerooni, H. MacMahon, et al., Lung image database consortium: developing a resource for the medical imaging research community, Radiology 232 (3) (2004) 739–748.

[29] Anode09 challenge, https://anode09.grand-challenge.org/.

[30] J.H. Pedersen, H. Ashraf, A. Dirksen, K. Bach, H. Hansen, P. Toennesen, H. Thorsen, J. Brodersen, B.G. Skov, M. Døssing, et al., The Danish randomized lung cancer CT screening trial—overall design and results of the prevalence round, J. Thorac. Oncol. 4 (5) (2009) 608–614.

[31] Luna16 challenge, https://luna16.grand-challenge.org/.

[32] S.J. Pan, Q. Yang, A survey on transfer learning, IEEE Trans. Knowl. Data Eng. 22 (10) (2010) 1345–1359.

[33] A. Krizhevsky, G. Hinton, Learning multiple layers of features from tiny images, Technical Report and Master Thesis, University of Toronto, 2009, http://www.cs.utoronto.ca/~kriz/learning-features-2009-TR.pdf.

[34] C. Szegedy, W. Liu, Y. Jia, P. Sermanet, S. Reed, D. Anguelov, D. Erhan, V. Vanhoucke, A. Rabinovich, Going deeper with convolutions, in: Proceedings of the IEEE Conference on Computer Vision and Pattern Recognition, 2015, pp. 1–9.

[35] S. Singh, J. Maxwell, J.A. Baker, J.L. Nicholas, J.Y. Lo, Computer-aided classification of breast masses: performance and interobserver variability of expert radiologists versus residents, Radiology 258 (1) (2011) 73–80.

[36] B. Sahiner, H.-P. Chan, M.A. Roubidoux, L.M. Hadjiiski, M.A. Helvie, C. Paramagul, J. Bailey, A.V. Nees, C. Blane, Malignant and benign breast masses on 3D US volumetric images: effect of computer-aided diagnosis on radiologist accuracy, Radiology 242 (3) (2007) 716–724.

[37] M.L. Giger, N. Karssemeijer, J.A. Schnabel, Breast image analysis for risk assessment, detection, diagnosis, and treatment of cancer, Annu. Rev. Biomed. Eng. 15 (2013) 327–357.

[38] S. Joo, Y.S. Yang, W. Kyung Moon, H.C. Kim, Computer-aided diagnosis of solid breast nodules: use of an artificial neural network based on multiple sonographic features, IEEE Trans. Med. Imaging 23 (10) (2004) 1292–1300.

[39] M.B. McCarville, H.M. Lederman, V.M. Santana, N.C. Daw, S.J. Shochat, C.-S. Li, R.A. Kaufman, Distinguishing benign from malignant pulmonary nodules with helical chest CT in children with malignant solid tumors, Radiology 239 (2) (2006) 514–520.

[40] J.-Z. Cheng, Y.-H. Chou, C.-S. Huang, Y.-C. Chang, C.-M. Tiu, K.-W. Chen, C.-M. Chen, Computer-aided US diagnosis of breast lesions by using cell-based contour grouping, Radiology 255 (3) (2010) 746–754.

[41] T. Sun, R. Zhang, J. Wang, X. Li, X. Guo, Computer-aided diagnosis for early-stage lung cancer based on longitudinal and balanced data, PLoS ONE 8 (5) (2013) e63559.

[42] T.W. Way, B. Sahiner, H.-P. Chan, L. Hadjiiski, P.N. Cascade, A. Chughtai, N. Bogot, E. Kazerooni, Computer-aided diagnosis of pulmonary nodules on CT scans: improvement of classification performance with nodule surface features, Med. Phys. 36 (7) (2009) 3086–3098.

[43] S.G. Armato, W.F. Sensakovic, Automated lung segmentation for thoracic CT: impact on computer-aided diagnosis, Acad. Radiol. 11 (9) (2004) 1011–1021.

[44] T.W. Way, L.M. Hadjiiski, B. Sahiner, H.-P. Chan, P.N. Cascade, E.A. Kazerooni, N. Bogot, C. Zhou, Computer-aided diagnosis of pulmonary nodules on CT scans: segmentation and classification using 3D active contours, Med. Phys. 33 (7) (2006) 2323–2337.

[45] C.-M. Chen, Y.-H. Chou, K.-C. Han, G.-S. Hung, C.-M. Tiu, H.-J. Chiou, S.-Y. Chiou, Breast lesions on sonograms: computer-aided diagnosis with nearly setting-independent features and artificial neural networks, Radiology 226 (2) (2003) 504–514.

[46] D. Newell, K. Nie, J.-H. Chen, C.-C. Hsu, J.Y. Hon, O. Nalcioglu, M.-Y. Su, Selection of diagnostic features on breast MRI to differentiate between malignant and benign lesions using computer-aided diagnosis: differences in lesions presenting as mass and non-mass-like enhancement, Eur. Radiol. 20 (4) (2010) 771–781.

[47] M.-C. Yang, W. Kyung Moon, Y.-C.F. Wang, M.S. Bae, C.-S. Huang, J.-H. Chen, R.-F. Chang, Robust texture analysis using multi-resolution gray-scale invariant features for breast sonographic tumor diagnosis, IEEE Trans. Med. Imaging 32 (12) (2013) 2262–2273.

[48] W. Gómez, W.C.A. Pereira, A.F.C. Infantosi, Analysis of co-occurrence texture statistics as a function of gray-level quantization for classifying breast ultrasound, IEEE Trans. Med. Imaging 31 (10) (2012) 1889–1899.

[49] G.D. Tourassi, E.D. Frederick, M.K. Markey, C.E. Floyd Jr., Application of the mutual information criterion for feature selection in computer-aided diagnosis, Med. Phys. 28 (12) (2001) 2394–2402.

[50] B. Sahiner, N. Petrick, H.-P. Chan, L.M. Hadjiiski, C. Paramagul, M.A. Helvie, M.N. Gurcan, Computer-aided characterization of mammographic masses: accuracy of mass segmentation and its effects on characterization, IEEE Trans. Med. Imaging 20 (12) (2001) 1275–1284.

[51] J.-Z. Cheng, Y.-H. Chou, C.-S. Huang, Y.-C. Chang, C.-M. Tiu, F.-C. Yeh, K.-W. Chen, C.-H. Tsou, C.-M. Chen, ACCOMP: augmented cell competition algorithm for breast lesion demarcation in sonography, Med. Phys. 37 (12) (2010) 6240–6252.

[52] C.-M. Chen, Y.-H. Chou, C.S.K. Chen, J.-Z. Cheng, Y.-F. Ou, F.-C. Yeh, K.-W. Chen, Cell-competition algorithm: a new segmentation algorithm for multiple objects with irregular boundaries in ultrasound images, Ultrasound Med. Biol. 31 (12) (2005) 1647–1664.

[53] P. Arbelaez, M. Maire, C. Fowlkes, J. Malik, Contour detection and hierarchical image segmentation, IEEE Trans. Pattern Anal. Mach. Intell. 33 (5) (2011) 898–916.

[54] L. Sorensen, S.B. Shaker, M. De Bruijne, Quantitative analysis of pulmonary emphysema using local binary patterns, IEEE Trans. Med. Imaging 29 (2) (2010) 559–569.

[55] G.D. Tourassi, Journey toward computer-aided diagnosis: role of image texture analysis, Radiology 213 (2) (1999) 317–320.

[56] Y.-L. Huang, D.-R. Chen, Y.-R. Jiang, S.-J. Kuo, H.-K. Wu, W.K. Moon, Computer-aided diagnosis using morphological features for classifying breast lesions on ultrasound, Ultrasound Obstet. Gynecol. 32 (4) (2008) 565–572.

[57] K. Drukker, C.A. Sennett, M.L. Giger, Automated method for improving system performance of computer-aided diagnosis in breast ultrasound, IEEE Trans. Med. Imaging 28 (1) (2009) 122–128.

[58] M.L. Giger, H.-P. Chan, J. Boone, Anniversary paper: history and status of CAD and quantitative image analysis: the role of medical physics and AAPM, Med. Phys. 35 (12) (2008) 5799–5820.

[59] R.-F. Chang, W.-J. Wu, W.K. Moon, D.-R. Chen, Automatic ultrasound segmentation and morphology based diagnosis of solid breast tumors, Breast Cancer Res. Treat. 89 (2) (2005) 179–185.

[60] A.V. Alvarenga, W.C.A. Pereira, A.F.C. Infantosi, C.M. Azevedo, Complexity curve and grey level co-occurrence matrix in the texture evaluation of breast tumor on ultrasound images, Med. Phys. 34 (2) (2007) 379–387.

[61] M. Masotti, N. Lanconelli, R. Campanini, Computer-aided mass detection in mammography: false positive reduction via gray-scale invariant ranklet texture features, Med. Phys. 36 (2) (2009) 311–316.

[62] J.-Z. Cheng, D. Ni, Y.-H. Chou, J. Qin, C.-M. Tiu, Y.-C. Chang, C.-S. Huang, D. Shen, C.-M. Chen, Computer-aided diagnosis with deep learning architecture: applications to breast lesions in us images and pulmonary nodules in CT scans, Sci. Rep. 6 (2016).

[63] Y. Bengio, P. Lamblin, D. Popovici, H. Larochelle, et al., Greedy layer-wise training of deep networks, Adv. Neural Inf. Process. Syst. 19 (2007) 153.

[64] J. Shi, S. Zhou, X. Liu, Q. Zhang, M. Lu, T. Wang, Stacked deep polynomial network based representation learning for tumor classification with small ultrasound image dataset, Neurocomputing 194 (2016) 87–94.

[65] G.E. Hinton, S. Osindero, Y.-W. Teh, A fast learning algorithm for deep belief nets, Neural Comput. 18 (7) (2006) 1527–1554.

[66] D.P. Naidich, A.A. Bankier, H. MacMahon, C.M. Schaefer-Prokop, M. Pistolesi, J.M. Goo, P. Macchiarini, J.D. Crapo, C.J. Herold, J.H. Austin, et al., Recommendations for the management of subsolid pulmonary nodules detected at CT: a statement from the Fleischner society, Radiology 266 (1) (2013) 304–317.

[67] H. MacMahon, J.H.M. Austin, G. Gamsu, C.J. Herold, J.R. Jett, D.P. Naidich, E.F. Patz Jr., S.J. Swensen, Guidelines for management of small pulmonary nodules detected on CT scans: a statement from the Fleischner society, Radiology 237 (2) (2005) 395–400.

[68] A.A. Balekian, G.A. Silvestri, S.M. Simkovich, P.J. Mestaz, G.D. Sanders, J. Daniel, J. Porcel, M.K. Gould, Accuracy of clinicians and models for estimating the probability that a pulmonary nodule is malignant, Ann. Am. Thorac. Soc. 10 (6) (2013) 629–635.

[69] K. Awai, K. Murao, A. Ozawa, Y. Nakayama, T. Nakaura, D. Liu, K. Kawanaka, Y. Funama, S. Morishita, Y. Yamashita, Pulmonary nodules: estimation of malignancy at thin-section helical CT—effect of computer-aided diagnosis on performance of radiologists, Radiology 239 (1) (2006) 276–284.

[70] T. Sun, J. Wang, X. Li, P. Lv, F. Liu, Y. Luo, Q. Gao, H. Zhu, X. Guo, Comparative evaluation of support vector machines for computer aided diagnosis of lung cancer in CT based on a multidimensional data set, Comput. Methods Programs Biomed. 111 (2) (2013) 519–524.

[71] T. Kubota, A.K. Jerebko, M. Dewan, M. Salganicoff, A. Krishnan, Segmentation of pulmonary nodules of various densities with morphological approaches and convexity models, Med. Image Anal. 15 (1) (2011) 133–154.

[72] J.-L. Starck, E.J. Candès, D.L. Donoho, The curvelet transform for image denoising, IEEE Trans. Image Process. 11 (6) (2002) 670–684.

[73] W. Shen, M. Zhou, F. Yang, C. Yang, J. Tian, Multi-scale convolutional neural networks for lung nodule classification, in: International Conference on Information Processing in Medical Imaging, Springer, 2015, pp. 588–599.

[74] W. Shen, M. Zhou, F. Yang, D. Yu, D. Dong, C. Yang, Y. Zang, J. Tian, Multi-crop convolutional neural networks for lung nodule malignancy suspiciousness classification, Pattern Recognit. 61 (2017) 663–673.

[75] W. Shen, M. Zhou, F. Yang, D. Dong, C. Yang, Y. Zang, J. Tian, Learning from experts: developing transferable deep features for patient-level lung cancer prediction, in: International Conference on Medical Image Computing and Computer-Assisted Intervention, Springer, 2016, pp. 124–131.

[76] C.-Y. Lee, S. Xie, P. Gallagher, Z. Zhang, Z. Tu, Deeply-supervised nets, Aistats 2 (2015) 6.

[77] N. Dalal, B. Triggs, Histograms of oriented gradients for human detection, in: 2005 IEEE Computer Society Conference on Computer Vision and Pattern Recognition (CVPR'05), vol. 1, IEEE, 2005, pp. 886–893.

[78] Z. Guo, L. Zhang, D. Zhang, A completed modeling of local binary pattern operator for texture classification, IEEE Trans. Image Process. 19 (6) (2010) 1657–1663.

[79] K. He, X. Zhang, S. Ren, J. Sun, Spatial pyramid pooling in deep convolutional networks for visual recognition, in: European Conference on Computer Vision, Springer, 2014, pp. 346–361.

[80] O. Maron, T. Lozano-Pérez, A framework for multiple-instance learning, Adv. Neural Inf. Process. Syst. (1998) 570–576.

[81] National Lung Screening Trial Research Team, et al., The national lung screening trial: overview and study design, Radiology 258 (1) (2011) 243–253.

[82] B. de Hoop, B. van Ginneken, H. Gietema, M. Prokop, Pulmonary perifissural nodules on CT scans: rapid growth is not a predictor of malignancy, Radiology 265 (2) (2012) 611–616.

[83] P. Sermanet, D. Eigen, X. Zhang, M. Mathieu, R. Fergus, Y. LeCun, OverFeat: integrated recognition, localization and detection using convolutional networks, arXiv preprint, arXiv:1312.6229, 2013.

[84] G. van Tulder, M. de Bruijne, Combining generative and discriminative representation learning for lung CT analysis with convolutional restricted Boltzmann machines, IEEE Trans. Med. Imaging 35 (5) (2016) 1262–1272.

[85] H. Larochelle, M. Mandel, R. Pascanu, Y. Bengio, Learning algorithms for the classification restricted Boltzmann machine, J. Mach. Learn. Res. 13 (Mar) (2012) 643–669.

[86] G.E. Hinton, A practical guide to training restricted Boltzmann machines, in: Neural Networks: Tricks of the Trade, Springer, 2012, pp. 599–619.

[87] F. Zhang, Y. Song, W. Cai, S. Liu, S. Liu, S. Pujol, R. Kikinis, Y. Xia, M.J. Fulham, D.D. Feng, et al., Pairwise latent semantic association for similarity computation in medical imaging, IEEE Trans. Biomed. Eng. 63 (5) (2016) 1058–1069.

[88] C.P. Langlotz, RadLex: a new method for indexing online educational materials, Radiographics 26 (6) (2006) 1595–1597.

[89] D.G. Lowe, Object recognition from local scale-invariant features, in: The Proceedings of the Seventh IEEE International Conference on Computer Vision 1999, vol. 2, IEEE, 1999, pp. 1150–1157.

[90] Y. Gao, D. Shen, Collaborative regression-based anatomical landmark detection, Phys. Med. Biol. 60 (24) (2015) 9377.

[91] P. Prasanna, P. Tiwari, A. Madabhushi, Co-occurrence of local anisotropic gradient orientations (collage): distinguishing tumor confounders and molecular subtypes on MRI, in: International Conference on Medical Image Computing and Computer-Assisted Intervention, Springer, 2014, pp. 73–80.

[92] A. Depeursinge, C. Kurtz, C. Beaulieu, S. Napel, D. Rubin, Predicting visual semantic descriptive terms from radiological image data: preliminary results with liver lesions in CT, IEEE Trans. Med. Imaging 33 (8) (2014) 1669–1676.

[93] R. Caruana, Multitask learning, in: Learning to Learn, Springer, 1998, pp. 95–133.

[94] C. Kurtz, A. Depeursinge, S. Napel, C.F. Beaulieu, D.L. Rubin, On combining image-based and ontological semantic dissimilarities for medical image retrieval applications, Med. Image Anal. 18 (7) (2014) 1082–1100.

[95] C. Jacobs, E.M. van Rikxoort, E.Th. Scholten, P.A. de Jong, M. Prokop, C. Schaefer-Prokop, B. van Ginneken, Solid, part-solid, or non-solid?: classification of pulmonary nodules in low-dose chest computed tomography by a computer-aided diagnosis system, Invest. Radiol. 50 (3) (2015) 168–173.

[96] F. Ciompi, C. Jacobs, E.Th. Scholten, M.M.W. Wille, P.A. de Jong, M. Prokop, B. van Ginneken, Bag-of-frequencies: a descriptor of pulmonary nodules in computed tomography images, IEEE Trans. Med. Imaging 34 (4) (2015) 962–973.

[97] M.C.B. Godoy, D.P. Naidich, Subsolid pulmonary nodules and the spectrum of peripheral adenocarcinomas of the lung: recommended interim guidelines for assessment and management, Radiology 253 (3) (2009) 606–622.

[98] E.J. Hwang, C.M. Park, Y. Ryu, S.M. Lee, Y.T. Kim, Y.W. Kim, J.M. Goo, Pulmonary adenocarcinomas appearing as part-solid ground-glass nodules: is measuring solid component size a better prognostic indicator?, Eur. Radiol. 25 (2) (2015) 558–567.

[99] A. McWilliams, M.C. Tammemagi, J.R. Mayo, H. Roberts, G. Liu, K. Soghrati, K. Yasufuku, S. Martel, F. Laberge, M. Gingras, et al., Probability of cancer in pulmonary nodules detected on first screening CT, N. Engl. J. Med. 369 (10) (2013) 910–919.

[100] W.D. Travis, E. Brambilla, M. Noguchi, A.G. Nicholson, K.R. Geisinger, Y. Yatabe, D.G. Beer, C.A. Powell, G.J. Riely, P.E. Van Schil, et al., International Association for the Study of Lung Cancer/American Thoracic Society/European Respiratory Society international multidisciplinary classification of lung adenocarcinoma, J. Thorac. Oncol. 6 (2) (2011) 244–285.

[101] K.J. Berry, P.W. Mielke, A generalization of Cohen's kappa agreement measure to interval measurement and multiple raters, Educ. Psychol. Meas. 48 (4) (1988) 921–933.

[102] F.-F. Li, P. Perona, A Bayesian hierarchical model for learning natural scene categories, in: 2005 IEEE Computer Society Conference on Computer Vision and Pattern Recognition (CVPR'05), vol. 2, IEEE, 2005, pp. 524–531.

[103] J. Yang, Y.-G. Jiang, A.G. Hauptmann, C.-W. Ngo, Evaluating bag-of-visual-words representations in scene classification, in: Proceedings of the International Workshop on Workshop on Multimedia Information Retrieval, ACM, 2007, pp. 197–206.

[104] S. Chen, D. Ni, J. Qin, B. Lei, T. Wang, J.-Z. Cheng, Bridging computational features toward multiple semantic features with multi-task regression: a study of CT pulmonary nodules, in: International Conference on Medical Image Computing and Computer-Assisted Intervention, Springer, 2016, pp. 53–60.

[105] S. Chen, J. Qin, X. Ji, B. Lei, T. Wang, D. Ni, J.-Z. Cheng, Automatic scoring of multiple semantic attributes with multi-task feature leverage: a study on pulmonary nodules in CT images, IEEE Trans. Med. Imaging 36 (3) (2017) 802–814.

[106] J. Chen, J. Zhou, J. Ye, Integrating low-rank and group-sparse structures for robust multi-task learning, in: Proceedings of the 17th ACM SIGKDD International Conference on Knowledge Discovery and Data Mining, ACM, 2011, pp. 42–50.

[107] T.K. Ho, The random subspace method for constructing decision forests, IEEE Trans. Pattern Anal. Mach. Intell. 20 (8) (1998) 832–844.

[108] Z. Tu, X. Bai, Auto-context and its application to high-level vision tasks and 3D brain image segmentation, IEEE Trans. Pattern Anal. Mach. Intell. 32 (10) (2010) 1744–1757.

[109] H. Zou, T. Hastie, Regularization and variable selection via the elastic net, J. R. Stat. Soc., Ser. B, Stat. Methodol. 67 (2) (2005) 301–320.

[110] R. Tibshirani, Regression shrinkage and selection via the lasso, J. R. Stat. Soc., Ser. B, Methodol. (1996) 267–288.

[111] G. Papandreou, L.-C. Chen, K. Murphy, A.L. Yuille, Weakly- and semi-supervised learning of a DCNN for semantic image segmentation, arXiv preprint, arXiv:1502.02734, 2015.

[112] A. Vezhnevets, J.M. Buhmann, Towards weakly supervised semantic segmentation by means of multiple instance and multitask learning, in: 2010 IEEE Conference on Computer Vision and Pattern Recognition (CVPR), IEEE, 2010, pp. 3249–3256.

CHAPTER 10

Analysis of Histopathology Images
From Traditional Machine Learning to Deep Learning

Oscar Jimenez-del-Toro*,†, Sebastian Otálora*, Mats Andersson‡,
Kristian Eurén‡, Martin Hedlund‡, Mikael Rousson‡, Henning Müller*,†,
Manfredo Atzori*

*University of Applied Sciences Western Switzerland (HES-SO), Institute of Information Systems, Sierre,
Switzerland
†University of Geneva, Geneva, Switzerland
‡ContextVision, Stockholm, Sweden

Abstract

Digitizing pathology is a current trend that makes large amounts of visual data available for automatic analysis. It allows to visualize and interpret pathologic cell and tissue samples in high-resolution
images and with the help of computer tools. This opens the possibility to develop image analysis
methods that help pathologists and support their image descriptions (*i.e.*, staging, grading) with
objective quantification of image features. Numerous detection, classification and segmentation algorithms of the underlying tissue primitives in histopathology images have been proposed in this
respect. To better select the most suitable algorithms for histopathology tasks, biomedical image
analysis challenges have evaluated and compared both traditional feature extraction with machine
learning and deep learning techniques. This chapter provides an overview of methods addressing the
analysis of histopathology images, as well as a brief description of the tasks they aim to solve. It is
focused on histopathology images containing textured areas of different types.

Keywords

Histopathology, Deep learning, Biomedical texture analysis, Digital pathology

10.1 HISTOPATHOLOGY IMAGING: A CHALLENGE FOR TEXTURE ANALYSIS

Histopathology is the examination of a biopsy or surgical tissue specimen by a pathologist. The image analysis of histopathological samples is used to provide the final detailed
diagnosis of several diseases, including most cancers. Current histopathology practice
has several constraints, as it is highly time-consuming and often shows low agreement between pathologists. In this context, Computer-Assisted Diagnosis (CAD) of
histopathology images is a novel challenging field for biomedical image analysis. CAD
of histopathology images can help to solve these constraints, because histopathology
images are characterized by repetitive patterns at several scales that can be particularly
suited for texture analysis and automated recognition. The fusion of traditional diagno-

Biomedical Texture Analysis
DOI: 10.1016/B978-0-12-812133-7.00010-7

281

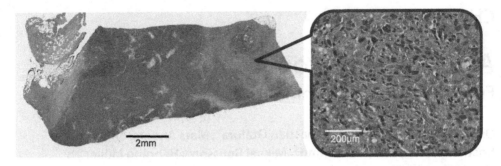

Figure 10.1 Example of a histopathology whole slide image, to the left, and a high power field (image zoomed at a high resolution) to the right.

sis methods with computational data analysis represents an opportunity to reduce the workload of pathologists while also harmonizing performance.

The standard process of histopathology image preparation passes through several phases that highlight specific structures in the images. First, the tissue is fixated and put onto glass slides with chemicals or by freezing the sections. Then the sections can be stained with pigments (*e.g.*, Haematoxylin-Eosin, H&E), using antibodies (Immuno-HistoChemistry, IHC) or with other methods (*e.g.*, immuno-fluorescence labeling). Afterward, the tissue slides are examined under a microscope by the pathologist. The pathologist's visual inspection of tissue samples is currently the standard method to diagnose a considerable number of diseases and to grade/stage most types of cancer [1,2]. When CAD is considered, there are two additional phases in the work flow. First, the image is scanned (generally using whole slide imaging, see Fig. 10.1). Then the digital images are processed and analyzed using computer-based methods such as visual feature extraction and machine learning.

The clinical analysis of histopathology images can be laborious for pathologists and prone to inconsistencies. The staging and grading of cancer is getting increasingly complex due to cancer incidence and patient-specific treatment options. The detailed analysis of a single case could require several slides with multiple stainings. Moreover, quantitative parameters are increasingly required (such as mitoses counting) [3]. Specific protocols were created to analyze biopsies or resected tissue specimens for the most common cancer types (such as lung, breast, or prostate). These protocols have led to precise and widely accepted prognostic grading strategies, for example, the Gleason grading system for prostate cancer [4]. However, the diagnostic practice is increasing the pressure for pathologists to handle large volumes of cases while providing a larger amount of information in the pathology reports [5]. Pathologists now spend much time on benign biopsies that represent approximately 80% of all biopsies [1]. The inter-observer agreement of pathologists for grading the same slide can be low with some cases reported between 0.52 and 0.73 and between 0.35 and 0.65 pairwise κ statistics representing levels of diagnostic agreement between subjects when viewing WSIs [6]. A second inspection of the same specimens can lead to 2.3% of total cases with major disagreements [7]. This disagreement between pathologists can lead to substantial changes in patient treatment [8,7].

Digital pathology has its origins in the 1980s, but many factors prevented it from being used in clinical practice including slow scanning speed, poor quality on screen, high costs, required memory, and limited network bandwidth. The first high-resolution, automated, Whole-Slide Imaging (WSI) system was developed in 1999 [9]. Since then the interest in WSI for pathology has continuously grown [10]. Whole slide imaging has been a strong advance for pathology because it overcomes the limitations of previous image acquisition methods, such as poor image quality and image navigation [10]. After whole slide imaging, the shift to a fully digital environment is currently expected for pathology, just as it previously happened for radiology. Digital pathology holds tremendous opportunities for histopathology practice: (1) it can allow online consultations; (2) it can provide access to pathology services in remote locations with limited pathology support; (3) it allows improving the productivity of pathologists by easily accessing and searching digital image archives (an activity representing up to 15% of the pathologists' work) [11]. Nevertheless, the advantages of digital pathology are not yet sufficient to overcome its limitations: WSI requires considerably large storage volumes (each WSI can require 2–3 GB) [12], scanning is an additional step in the prediagnostic work flow (thus representing an increase in the workload of pathologists or lab technicians) and the quality needs to be certified and accepted by pathologists. The number of scientific researchers and companies developing CAD algorithms for pathology has strongly increased in the last 10 years, further indicating that pathology is becoming digital. Computer-assisted diagnosis of WSI can help overcome the current limitations in digital pathology, promoting its use while reducing the workload of pathologists and rater disagreement.

Texture is generally characterized by homogenous areas with properties related to scale and regular patterns that can occur in 2D or 3D (see Section 1.2 of Chapter 1) [13]. Biomedical tissue samples contain a complex mixture of repetitive visual patterns at different scales, revealing that organ systems are composed of a few groups of tissue types (*e.g.*, connective, epithelial, muscle and nervous tissue) built from specific cells. The acquisition process (cutting bidimensional sections out of three dimensional structures), results in a partial representation of the organs and systems. It allows the visualization of alterations in the morphology of the tissue architecture that are associated with types and grades of disease. Several approaches have been presented to provide CAD of histopathology images but the field still faces several challenges. CAD methods can be subdivided into traditional machine learning approaches (*e.g.*, based on the detection of specific structures, texture analysis) and deep learning approaches, as will be discussed in Sections 10.2 and 10.3, respectively. The current challenges are presented with concrete

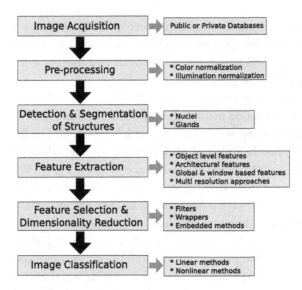

Figure 10.2 The traditional machine learning approach to histopathology CAD.

applications, such as the analysis of specific pathologies, the detection of mitosis, the classification of regions or WSI, and the segmentation of specific structures, providing a description of the most recent trends and achievements.

10.2 TRADITIONAL MACHINE LEARNING APPROACHES

Traditional machine learning approaches often include several phases to deal with histopathology images, as represented in Fig. 10.2. Each phase is described in the following sections.

10.2.1 Preprocessing

Preprocessing can compensate for differences between images that are diverse in color, illumination, and other defects, such as noise or artifacts that are often due to the scanning process.

10.2.1.1 Staining normalization

Digital pathology images can have strong color differences due to factors such as the use of different scanners, different stains or staining procedures, section thickness, and sample age. Despite standardization, a perfect color calibration among samples is hard to achieve [14], thus requiring color normalization. Examples of color normalization include histogram-based approaches and color deconvolution-based approaches [15, 16].

10.2.1.2 Illumination normalization

Shading correction uses empty images (recorded under each considered magnification and illumination condition) to correct the images [17]. Gaussian smoothing can also reveal the intrinsic illumination properties of the image [18]. Background estimation can be performed directly on the specimen [19]. Finally, the image can be modeled as a function of the excitation and emission patterns [20].

10.2.2 Detection and segmentation of structures

The presence as well as the number and the morphological characteristics of specific structures (such as nuclei and glands) are fundamental parameters to evaluate the presence and severity of a pathology such as prostate [4], breast [21], and colorectal [22] cancer.

10.2.2.1 Nuclei and cells

The nuclei are the central organelles of eukaryotic cells containing most of the cell DNA. Nuclei analysis usually involves detection, segmentation, and separation (*i.e.*, dividing overlaps). The identification of seed points in the nuclei is required by most nuclei segmentation and counting methods [23]. Many approaches have been proposed in the literature for nuclei detection, including methods based on Euclidean distance map peaks [24], H-maxima transform [25,26], Hough transform (detecting seed points for circular-shaped structures, requiring heavy computation) [27], multiscale Laplacian of Gaussian (LoG) filters [28], Radial Symmetry Transform (RST) [29], and (recently) deep learning based methods [30,31].

Many approaches have also been proposed to perform precise nuclei segmentation. Despite their simplicity, methods based on thresholding and morphological operations can perform well on uniform backgrounds [32,33] but they are not robust to size, shape, and texture changes. The watershed transform requires no tuning but the prior detection of seed points [26,25]. Active contour models can combine image properties with nuclei shape models [34,27] but they rely on seed points. Other methods are based on Gradients in Polar Space (GiPS) [24], graph-cuts [35,28], and machine learning approaches [36]. In H&E stained images, nuclei segmentation accuracy above 90% has been reported [36,35,27].

10.2.2.2 Glands

Glands are organs formed by an ingrowth from an epithelial surface. They synthesize and release substances (such as hormones or mucus) into the bloodstream (endocrine glands) or onto an outer surface of the body, such as the skin or the gastrointestinal tract (exocrine glands). Glands have been segmented automatically, although reliable gland segmentation for diverse cancer grades is still a challenge. Methods based on thresholding and region growing can identify the nuclei and the lumen that are used to initialize seed points for region growing (*e.g.*, [37]). Despite their simplicity, these methods perform well in segmenting healthy and benign glands but they have shown poor results in cancer cases where the gland morphology is deformed. Graph-based methods have been proposed as well (*e.g.*, [38]). Usually, tissue components (such as the nuclei) are identified, represented as vertices and properties of a graph are used to segment the glands. Segmentation based on polar coordinates (the center inside the gland) were proposed on benign and malign glands [39]. Approaches based on Bayesian inference allow to take into account prior knowledge of the structural properties and of the arrangement of glandular components (such as central lumen, surrounding cytoplasm, and nuclear periphery) [40,41].

10.2.3 Feature extraction

Histopathology images have been analyzed using many descriptors based on the knowledge of domain experts. Diagnosis criteria are mainly expressed with cytologic terms representing objects (*i.e.*, nuclei, cells, glands) and their presence in malign and benign areas. Thus several articles approached the problem at the object-level (using segmented object features) and the object relationship level (using architectural features). Multiresolution global and window-based features have also been used.

10.2.3.1 Object-level features

Object-level features depend strongly on the considered objects (usually nuclei or glands) and on the segmentation algorithms [1]. These features are relevant for any resolution, but in most cases they are extracted from high-resolution images. Object-level features are usually extracted for each color channel and can be grouped into size and shape features (*e.g.*, area, eccentricity, reflection symmetry), radiometric, and densitometric features (*e.g.*, image bands, intensity, hue), texture features (*e.g.*, co-occurrence matrix, run-length, and wavelet features) and chromatin-specific features (mean and integrated optical density) [42].

10.2.3.2 Architectural features

Architectural features are mainly based on graphs (mathematical structures used to model pairwise relations between objects). They are made up of vertices (also called nodes or points) that are connected by edges (arcs or lines). Graphs are an effective method to represent architectural information using topological features in histopathology images. Graph structures used in histopathology include: the Voronoi tesselation, the Delaunay triangulation, minimum spanning trees, O'Callaghan neighborhood graphs, connected graphs, relative neighbor graphs, k-NN graphs [1]. Each of these structures allows the computation of several descriptive features used for tissue classification, such as number of nodes, number of edges, edge length, number of triangles, area, and eccentricity. A detailed description of graph structures and features can be found in [42]. In the last decade, graph-based features have been investigated regularly as they are well suited to characterize tumor architecture. Different graphs can be built, usually to represent the spatial organization of epithelial nuclei. Demir et al. [43] performed a 3-class classification (healthy, malignant, inflammation) in brain cancer biopsies using a complete weighted graph. In Altunbay et al. [38], Delaunay Triangulation (DT) graphs included colon tissue components (nuclei, stroma, and lumen) as nodes. Chekkoury et al. [44] combined morphologic, network, and texton-based features for breast cancer diagnosis. In a fundamental study, Doyle et al. [45] presented a 90% accuracy in distinguishing between cancer and benign tissue for prostate cancer using a combination of features.

10.2.3.3 Global and window based features

These features can be related to color, texture (*e.g.*, cooccurrence matrices, run-length, and wavelet features). Average object and architecture features can be computed globally or in windows at multiple resolutions. At low resolution, color and texture analysis are often used to indirectly measure the properties of local constituents or of tissue architecture (*e.g.*, density of nuclei and the overall pattern of glands or stroma). General image statistics, histograms, gradients, and many other features can be extracted in order to perform measurements of the regions of interest and filters can also be used to extract local features. For instance, prostate cancer has been detected and graded with texture features such as Gabor filters [45–47], fractal dimension [48], wavelets [49], and morphological operations [50].

10.2.3.4 Multiresolution approaches

Multiresolution approaches were proposed to deal with the multiresolution characteristics of histopathology images and to mimic the analytic approach from pathologists. The Gaussian pyramid approach was used to represent the images in multiple resolutions [51]. Features can be extracted separately for each level and combined to classify image tiles. Color and texture features are commonly used at low resolutions. Architectural arrangement of glands and nuclei are used at medium scales. Nucleus and gland related features can be discriminatory at high resolutions [46].

10.2.4 Feature selection and dimensionality reduction

10.2.4.1 Feature selection

Feature selection is the process of selecting a subset of relevant features for model construction, thus reducing training times, simplifying the models (to make interpretation easier), and improving the chances of generalization, avoiding overfitting. There are three main feature selection strategies: filters (*e.g.*, information gain), wrappers (*e.g.*, search guided by accuracy), and embedded methods (in which the features are added or removed from the model depending on prediction errors). Well-known feature selection algorithms are represented by the sequential forward selection and sequential backward selection [52], sequential floating forward search – one of the most frequently used methods in pathology image analysis [1] – and sequential floating backward search [52]. Other available methods are genetic algorithms, simulated annealing, boosting [53], and grafting [54].

10.2.4.2 Dimensionality reduction

Dimensionality reduction has the following aims: first, to reduce storage space and computation time; second, to remove multicollinearity (thus improving the performance of machine learning models); third, to improve data visualization (reducing to low dimensions such as 2D or 3D). The data transformation may be linear, as in Principal Component Analysis (PCA), Independent Component Analysis (ICA), and Linear Discriminant Analysis (LDA) [55–57]. Several nonlinear dimensionality reduction algorithms (such as manifold learning and nonlinear embedding) were developed and shown to be useful in histopathology image analysis [45].

10.2.5 Classification

Classification methods aim at identifying the category of a new observation among a set of categories on the basis of a labeled training set. Depending on the task, anatomical structure, tissue preparation, and features the classification accuracy varies. Random forests with a set of 18 features (including, *e.g.*, histogram related features, autocorrelation, sum average, variance, entropy, contrast) obtained 83% median accuracy in the classification of prostate tissue into 7 classes (Gleason grade 3, 4, and 5; benign stroma; benign hyperplasia; intraepithelial neoplasia and inflammation) [58]. k-Nearest Neighbors (k-NN) with a set of 14 features (intensity, morphological, and texture) from segmented nuclei was tested on the classification of hepatocellular carcinoma [34]. Adaboost with a large set of features was used to segment nuclei with various shapes [36] and to detect suspicious areas on digital prostate histopathology [47]. Support Vector Machines (SVM) with radial basis function kernel and grid search were used to classify mitotic vs. nonmitotic regions [59], to distinguish between prostate tissue classes [45], to classify colon adenocarcinoma histopathology images vs. benign images [60], and to classify four subtypes of meningioma [61]. Multiclassifier ensembles [62] represent another solution to increase accuracy. Several classification methods can be integrated to increase accuracy because a fusion of methods can improve results [63]. For instance, in Huang et al. [34], an SVM-based decision graph classifier with feature subset selection on each decision node improved the classification of hepatocellular carcinoma.

The comparison of the methods is difficult, due to heterogeneity of datasets and evaluation metrics. Several studies compared the performance of classification procedures on prostate (*e.g.*, [48]). Alexandratou et al. [64] published a comparative study for prostate cancer diagnosis, showing a good performance for cancer detection, 80.8% accuracy for low–high Gleason grade discrimination, and 77.8% for cancer grading. Despite the existence of these studies, there are still only few images available for each pathology and obtaining good annotations of the images is expensive and time consuming. Databases and challenges have been proposed in the past few years (see Section 10.4).

10.3 DEEP LEARNING APPROACHES

Deep Learning (DL) methods are currently the most frequently studied and successful type of machine learning algorithms (see Chapter 4). In the last decade, DL outper-formed classical machine learning algorithms with handcrafted features in diverse fields such as computer vision [65], speech recognition [66], natural language processing [67], and also recently in biomedical fields (*e.g.*, functional genomics) [68]. The adoption of DL techniques in the biomedical image analysis community had a positive impact on several tasks, and automatic analysis of histopathology images is no exception [69,70]. Deep learning approaches in computer vision are based on the composition (*layers*) of nonlinear transformations over the raw input pixels (see Section 4.2 of Chapter 4). This composition builds increasingly abstract representations that are learned in a hierarchical fashion [71,72]. An *architecture* is the arrangement and the interconnection of parameters learned by an optimization algorithm for a given DL approach. An example of a com-monly used architecture, the convolutional neural network, for histopathology image analysis is depicted in Fig. 10.3.

One of the main characteristics of most DL architectures is that the output of the network is only based on the adjustable internal weights. These weights are learned by the network through iterative forward/backward propagation of the training samples and the errors, respectively. This is known as the backpropagation algorithm for neural networks [73]. The internal weights are updated using the error between the ground truth output of the samples (*i.e.*, the labels in the supervised case and the original representation in the unsupervised one) and the output of the network. This requires less explicit coding of domain knowledge than the traditional machine learning. For a more in depth review of the technical details of the deep learning methods, we suggest to the reader to go through the first sections of Chapter 4.

There are multiple considerations to take into account when applying a deep learning algorithm to a given histopathology task because the success of the algo-rithms is partially due to task-specific tuning. Histopathology images are large (*i.e.*, $100,000 \times 100,000$ pixels). One of the main characteristics of histopathology images is that the relevant patterns depend on the magnification level. The main considerations are then: the localization of the regions in the image where relevant histopathology primitives are located (*e.g.*, a ROI selection step); the size of the input image (or patch)

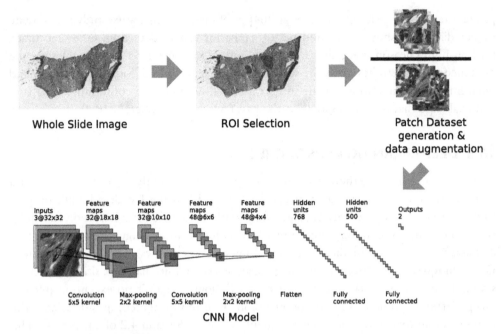

Figure 10.3 A framework for applying supervised deep learning (*e.g.*, CNN) to the classification of a WSI.

that is fed to the network and the homogeneity of the staining applied to the WSI (in cases with high variability, staining normalization is required). The architecture of the network plays an important role, although a considerable amount of articles decide to stay with predefined/pretrained network architectures (Fig. 10.4). We highlight the main characteristics and risks of DL approaches in histopathology, identifying trends in the biological problems and evaluation methods.

10.3.1 Supervised and unsupervised feature learning architectures

In machine learning and also in deep learning, there are two families of techniques that differ in supervision. Unsupervised Feature Learning (UFL) techniques do not depend on labeled samples to detect the inner structure of the data. UFL learns over-complete, sparse, or hierarchical unsupervised representations by reconstructing the original input with several constraints (see Fig. 10.5 for a basic UFL architecture).

On the other hand, we have supervised techniques that use the label information of the input and guide the model to a desired output. The problem of transitioning between the supervised toward unsupervised learning techniques is one of the biggest challenges for deep learning as discussed by [74,75]. In unsupervised feature learning, the input data are passed through a compressor to obtain a more compact representation.

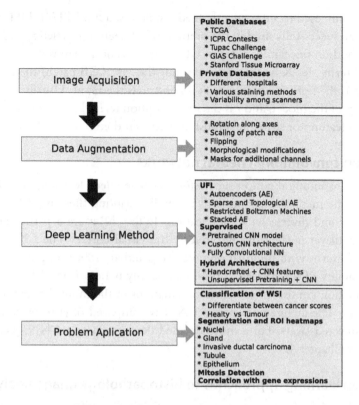

Figure 10.4 Steps involved when applying deep learning to histopathology images.

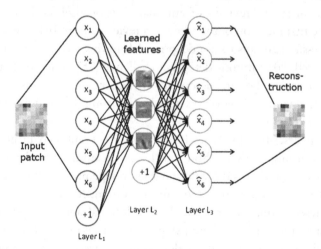

Figure 10.5 A shallow UFL architecture with only one hidden layer that encodes the features to represent a patch.

This is commonly used to subsequently feed a supervised model [74]. UFL representations have been successfully used in the pretraining of deep supervised models [76]. The multiple methods in this family differ in what regularization term is used.

In formal terms, let $\mathbf{X}$ be the original input (*i.e.*, an RGB patch) then $\mathbf{X} \approx \hat{f}(\mathbf{X}) + r(\theta)$, where $\hat{f}$ is the learned function that reconstructs the input. Usually $\hat{f} = \theta^T \theta$, θ are the parameters of the model, and r is the regularization term that can ensure the sparsity of the learned features or spatial relationships between them.

10.3.2 Deep convolutional neural networks

CNN models are among the most successful supervised deep learning models for computer vision. The medical imaging field is rapidly adapting these models to solve and improve a large and diverse set of applications [69]. CNNs are a particular kind of a supervised multilayer perceptrons inspired by the visual cortex. The CNN are able to detect visual patterns with minimal preprocessing and are robust in presence of distortion and variability of the pattern. CNNs can usually benefit from data augmentation. Data augmentation consists of subtle transformations in the input data, aimed at learning invariances. The architecture of the CNN is composed of convolutional, pooling and fully connected layers. For a more detailed description of a CNN, please refer to Section 4.2 of Chapter 4.

10.3.3 Deep learning approaches to histopathology image analysis

Most of the deep learning approaches for classification, segmentation, and localization in histopathology images are relatively new. The deep neural network methods are not even mentioned in recent reviews of automatic histopathology image analysis, such as [43,1,77]. The first mention of the topic in a review is from Irshad et al. [23]. One of the earliest successful attempts to use deep models in this scenario was from Cireşan et al. [31] at the ICPR 2012 mitosis detection challenge [78]. This work as well as other mitosis detection approaches are described in Section 10.5 of this chapter. A more detailed literature review and workflow for deep learning applications in breast cancer is described in Chapter 9.

The approach described by Cruz-Roa et al. [79] combined unsupervised (UFL) and supervised learning. This method first learns an unsupervised representation via sparse autoencoders and then a convolution and average pooling is performed over this representation to encourage both a translation invariant feature detection and a compact image representation. The proposed convolutional autoencoder neural network architecture is used for histopathology patch-based image representation learning. This representation allows automatic cancer detection and visually interpretable prediction results analogous to a digital stain. The method identifies image regions that are most relevant for diagnostic decisions, using the probabilities of the final softmax classifier layer of the model. The method was applied to 308 histopathology basal cell carcinoma images at 10X magnification, taking patches of 300 × 300 pixels. Arevalo et al. [80] computed a hybrid representation for basal cell carcinoma patches using a topographic unsupervised feature learning method and a bag of features representation. Their approach improved the classification performance by 6% with respect to classical texture based descriptor Discrete Cosine Transform (DCT). In Malon et al. [59], a novel combination of features is proposed. The authors build a feature set of basic statistical measures of the nucleus and cytoplasm pixels and combine them with a CNN classifier. The presented approach improves the performance in comparison with handcrafted features.

Currently, there is a diversification of architectures and applications of deep learning using WSIs. Nayak et al. [81], propose a variation of the unsupervised restricted Boltzmann machine method for learning image signatures. This method classifies patches of clear cell kidney carcinoma and glioblastoma multiforme images from the cancer genome atlas; with this signature the final stage is made using a multiclass regularized support vector classification. Xu et al. [82] developed a technique that deals with few labeled samples. In this work a multiple instance learning framework is introduced, where classification of colon histopathology images is performed. The authors propose a fully supervised approach and a weakly labeled one with similar accuracy (93.56% vs. 94.52%). In the work of Hou et al. [83] a similar idea is proposed. The authors use multiple instance learning to classify between glioblastoma and low-grade glioma images from the cancer genome atlas. The method uses three steps: first, it learns masks for discriminative areas using a CNN model with few selected discriminative patches; second, it makes a patch-level prediction using CNNs; third, the class count is performed. In Vanegas et al. [84] the authors do content-based image retrieval. The proposed technique is based on learning an unsupervised representation based on Topographic reconstructed Independent Component Analysis (TICA) and using it as an input for a multimodal semantic indexing method. The authors show that the Mean Average Precision (MAP) performance is greater when using the TICA representation in comparison with canonical texture descriptors such as DCT and Haar, showing that UFL based representations also have a good performance in retrieval tasks. In Cruz-Roa et al. [85], the authors propose a 3-layer CNN architecture to detect invasive ductal carcinoma using patches from 113 whole slide images as a training set. This approach obtains 10% higher balanced accuracy in comparison with Haralick and graph-based handcrafted features.

The recent use of unsupervised architectures includes the work of Arevalo et al. [86]. In this work, the authors describe a novel stacked model that shows the best performance when combining features from 2-layered TICA over patches for detecting basal cell carcinoma. The authors also introduce a *digital staining* method based on the weighting of the feature detectors by the classification probability to highlight the areas that are most related to the cancer. Their approach achieved ~0.99 in Area Under Receiver

Operating Characteristic (ROC) Curve (AUC) for a set of 100,000 patches. The approach described by Hang et al. [87] is based on learning a dictionary of 1024 features for classifying kidney clear cell carcinoma and glioblastoma multiforme. The authors used a curated version of the publicly available images from the Cancer Genome Atlas. The dictionary was built with a stacked UFL method called stacked predictive sparse decomposition, which is used in a spatial pyramid matching framework (with the last stage being a linear support vector machine classifier). Noel et al. [88] used a set of 3000 patches extracted from WSIs of the ICPR contests to go a step further in breast cancer detection, classifying each pixel using a CNN into stroma, nuclei, lymphocytes, mitosis, and fat. 90% accuracy was achieved, suggesting that a finer classification of the WSI primitives can help to improve the classification performance.

Besides the public challenges and contests, there have been some efforts to do systematic evaluations with the same datasets and experimental setups. Janowczyk and Madabhushi [70] recently published a tutorial on deep learning for digital pathology where they expose exemplar use cases of the technology on more than 1200 slides for seven tasks including mitosis detection, nuclei segmentation, epithelium segmentation, and lymphoma classification. Good results on the studied cases are shown and the source code is provided using the Caffe [89] deep learning framework. Interestingly, the authors did not design a different network architecture for each of the cases but decided to use Alexnet [65] for all of them. Questions such as the importance of manual annotation for ground truth generation are discussed when applying deep learning techniques and also some comparisons against pathologist annotators. It is shown that for some cases the DL system is doing a more refined job on detailed pixel-level; something a pathologist cannot really do. In Cruz-Roa et al. [90], the authors show an interesting comparison between unsupervised and supervised approaches for medulloblastoma tumor classification because in this case the proposed CNN architecture is not competitive against UFL methods such as TICA. A further improvement on this dataset was made by Otálora et al. in [91]. The authors combine features from unsupervised and supervised nature (the former from topographic independent component analysis and the latter using Riesz wavelets for two types of medulloblastoma brain cancer images) improving the use of the topographic unsupervised approach. It shows that a combination of a supervised texture representation and UFL contain complementary information about this particular pattern.

Some recent work goes beyond H&E patch classification. A line of work that is closer to the optical properties of the hardware involved in the tissue digitalization is the one presented by Chen et al. [92]. The authors present a new method that aims at detecting flowing colon cancer cells at high-throughput rates, extracting several biophysical features, and then building a deep fully connected network. Han et al. [93] present a novel deep UFL method for phenotypic characterization of glioblastoma multiforme. They are able to differentiate two major phenotypic subtypes with different survival curves using the extracted UFL features. Romo-Bucheli et al. [94] quantify tubule nuclei by feeding a CNN model with candidate patches to measure their tubule class membership probability that was then associated with high-low risk categories determined by an Oncotype DX test. To our knowledge, the application of deep learning stains different from H&E (such as immunohistochemistry) is still not well explored. An exception is the work of Chen et al. [95], in which the authors propose an immune cell detection with a 7-layer CNN model with patches of unmixed colors from the original RGB channels, highlighting immune cell markers. The authors compare their algorithm detection performance with the performance of pathologists achieving up to a 0.99 correlation coefficient. This shows promising results with other types of staining than H&E.

10.4 HISTOPATHOLOGY CHALLENGES

Open scientific challenges targeting tasks in the analysis of pathology images have been proposed in recent years similar to other medical imaging domains.[1] An advantage of having different methods tested on the same data and in the same tasks is the objective comparison of the strengths and limitations of state-of-the-art approaches. Particularly in the case of histopathology, the time-consuming and laborious tasks of searching whole slide images for relevant small tissue primitives (*e.g.*, mitosis, nuclei) can be improved. Selecting the most suitable approach to support and speed-up the visual interpretation of slides can help pathologists to focus on the most important questions when diagnosing these studies. An overview of histopathology related challenges is displayed in Table 10.1. In this section the datasets and tasks from some of these challenges are explained. The scope of this analysis includes only recent digital histopathology challenges, namely MITOS12, AMIDA13, MITOS-ATYPIA-14,[2] GlaS@MICCAI'2015,[3] CPM,[4] CAMELYON16,[5] and TUPAC16.[6]

10.4.1 Datasets

The size of the datasets provided to participants for training and testing their algorithms has increased significantly in the latest challenges. In the first challenges the algorithm's performance was evaluated on regions or High Power Fields (HPF) from a limited number of WSI (*e.g.*, MITOS12, 50 HPF from 5 slides). However, recent challenges

[1] http://grand-challenge.org/All_Challenges/, ★.
[2] https://mitos-atypia-14.grand-challenge.org/home/, ★.
[3] http://www2.warwick.ac.uk/fac/sci/dcs/research/combi/research/bic/glascontest/, ★.
[4] http://miccai.cloudapp.net/competitions/45, ★.
[5] https://camelyon16.grand-challenge.org, ★.
[6] http://tupac.tue-image.nl, ★.
    ★ as of January 1, 2017.

Table 10.1 Biomedical image analysis challenges related to histopathology. MD = Mitosis Detection, IC = Image Classification, SL = Structure Localization (segmentation)

Acronym	Challenge	Anatomy	Tasks
2012			
MITOS12 [78]	Mitosis detection in breast cancer histological images	breast	MD
2013			
AMIDA13 [96]	Assessment of mitosis detection algorithms	breast	MD
2014			
MITOS–ATYPIA14	Mitotic count and nuclear pleomorphism	breast	MD, IC
OCCISC[a]	Overlapping cervical cytology image segmentation	cervix	SL
2015			
2OCCISC[b]	2nd Overlapping cervical cytology image segmentation	cervix	SL
GlaS@MICCAI'2015	Gland segmentation challenge	bowel	SL
2016			
SLATMD [97]	Skin lesion analysis toward melanoma detection	skin	SL
CPM	Computational Precision Medicine	lung, neck, brain	IC, SL
CAMELYON16	Cancer metastasis detection in lymph nodes	lymph nodes	IC, SL
TUPAC16	Tumor proliferation assessment challenge	breast	MD, IC

[a] http://cs.adelaide.edu.au/~carneiro/isbi14_challenge/index.html, ★
[b] http://cs.adelaide.edu.au/~zhi/isbi15_challenge/index.html, ★

such as CAMELYON16 and TUPAC16 provide a large number (> 500) of WSIs, some of them with manual annotations by pathologists. This promotes the robustness of the participant algorithms as they have to analyze multiple images, sometimes from different scanners and in various disease stages. Table 10.2 shows an overview of the datasets from the digital pathology challenges.

10.4.2 Tasks

The complexity of the tasks in the challenges has increased in the past few years. The goals of the challenges can be grouped into three main tasks (Fig. 10.6):

- **Mitosis detection**: Identify the mitosis present in high power fields. This has a strong correlation to the aggressiveness of the cancer, *i.e.*, faster division of the cells will manifest in more mitosis.

Table 10.2 Histopathology challenge datasets. NA = Nonapplicable NM = Not mentioned

Challenge	Data	Staining	maxResol	Quantity	Scanners
MITOS12	High power fields	H&E	40X	50	2
AMIDA13	High power fields	H&E	40X	23	1
MITOS-ATYPIA14	High power fields	H&E	40X	1136	2
GlaS@MICCAI'2015	High power fields	H&E	20X	165	1
CPM	High power fields	H&E	NP	32	NP
CAMELYON16	Whole slide images	H&E	40X	400	NP
TUPAC16	Whole slide images	H&E	40X	821	1

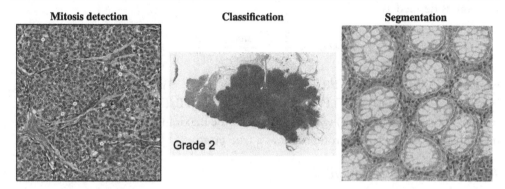

Figure 10.6 Tasks in anatomical pathology challenges.

- **Image classification**: Learn the characteristic set of features for a particular tissue class, potentially considering the underlying tissue primitives.
- **Structure localization/segmentation**: Localizing and delineating a boundary around specific tissue structures such as cell nuclei or glands.

Although these tasks are divided into different groups for their description, they can be combined or considered as an initial step for the other tasks, *e.g.*, a whole slide cancer grading algorithm can start by classifying an image as cancer tissue, then segment the tumor and finally grade the WSI based on the mitosis counting inside a tumor region of interest.

10.4.3 Evaluation metrics

An important element of a challenge is the selection of the right evaluation metric. A suitable metric should limit bias in the comparison of the algorithms. It is standard to provide multiple metrics in the challenge results, nevertheless algorithms are generally ranked according to a lead metric for each challenge task. The following metrics were selected for the evaluation of the digital pathology challenges included in Table 10.2.

10.4.3.1 Mitosis detection

In the evaluation of mitosis detection algorithms the mitotic events found by the algorithms are counted. Two sets of evaluation metrics have been used in these types of challenges: centroid-based, when each mitotic event is represented by a single location, and region-based when the mitoses are segmented and evaluated as structures with multiple pixels. Still, the centroid-based measures are more frequently used when ranking the participant algorithms in a benchmark. All 4 challenges with a mitosis detection task use the F_1-score as the preferred metric to rank the participant algorithms. The F_1-score is defined as

$$F_1\text{-score} = \frac{2 \cdot Precision \cdot Recall}{Precision + Recall},$$ (10.1)

where TP = True Positives, FP = False Positives, TN = True Negatives, FN = False Negatives, Precision = TP/(TP + FP), and Recall = TP/(TP + FN). A threshold of <7.5–8 μm (around 30 pixels) was determined as the maximum Euclidean distance from the centroid to consider a mitotic event as a true positive.

10.4.3.2 Image classification

Unlike mitosis detection, image classification challenges have used different evaluation approaches for ranking the participant algorithms. The classification can be binary, when the algorithms have to decide if an image contains a tumor/metastasis or not (e.g., CAMELYON16). Otherwise, the classification can be multiclass when the algorithms have to grade the pathologic stage of an image (e.g., TUPAC16). In some cases the referred challenges also request the probability with which the approaches grade each case (e.g., CAMELYON16) or measure the agreement between the algorithm classification and the pathologist-generated ground truth (e.g., TUPAC16). The proposed evaluation strategies include:

- A points based scheme for nuclear atypia scoring (MITOS-ATYPIA-14).
- Area under a receiver operating characteristic (ROC) curve for the discrimination of lymph node slides containing metastasis or not (CAMELYON16).
- Agreement with ground truth measured with quadratic weighted Cohen's kappa or Spearman's correlation coefficient for breast cancer grading (TUPAC16).

The quadratic weighted Cohen kappa is defined as

$$K_w = 1 - \frac{\sum_{i,j} w_{i,j} p_{i,j}}{\sum_{i,j} w_{i,j} e_{i,j}},$$ (10.2)

where p_{ij}: observed probabilities, $e_{ij} = p_{ij} q_{ij}$: expected probabilities and w_{ij}: weights (with $w_{ij} = w_{ji}$).

10.4.3.3 Structure localization and/or segmentation

Different types of tissue structures have been targeted for structure segmentation in histopathology challenges: cell nuclei (CPM), metastasis in lymph nodes (CAMELYON16), and colon and rectum glands (GlaS@MICCAI'2015). Three attributes of the output segmentations can be measured: number of structures detected, full structure overlap with the ground truth, and shape similarity. The favored segmentation metric (GlaS@MICCAI'2015 and CPM) is the Dice coefficient. Given a ground truth G and S as a set of segmented pixels, Dice is defined as

$$Dice(G, S) = \frac{2|G \cap S|}{|G| + |S|}.$$ (10.3)

In the GlaS@MICCAI'2015 challenge, both the Dice and Hausdorff distance are measured at object level, calculating the score over all the annotated objects included in the test set.

10.5 DETECTING MITOSES

Mitosis detection is an important problem in histopathology image analysis aiming at the identification and annotation of mitotic figures in high power fields. The mitotic figures can then be used as input for a grading system (*e.g.*, the Bloom–Richardson grading system for breast cancer [98]), to have an estimate of the patient's prognosis and to develop personalized treatments. Fig. 10.7 shows some examples of mitoses and highlights the difficulty to differentiate between mitotic and nonmitotic cells.[7] Most of the false positive detections arise when apoptotic nuclei (the nuclei going through the natural programmed cell death), are mistaken as mitoses because of their similar appearance. This problem was traditionally addressed with texture descriptors such as DCT, Haar, or Haralick [99]. However, traditional techniques are currently struggling to cope with the outstanding performance of deep learning algorithms [96,88,100, 101]. Several studies concluded that the performance of DL methods is close to the performance and interobserver agreement of expert pathologists [102,95,103].

There have been several challenges on the topic of mitosis detection (Table 10.1). The very first challenge for mitosis detection was the 2012 ICPR *Mitosis Detection in Breast Cancer Histological Images* challenge [78]. Seventeen teams submitted their results to the contest. The winner was the Swiss IDSIA team [31]. The approach of the team was based on a deep CNN with 13 sequential convolution-maxpooling layers trained over 50 whole slide images with 50 high power fields. This approach was computationally intense because it computed pixel-wise probabilities for mitosis. This method

---

[7] As of January 1, 2017, there is an online tool to compare the performance in mitosis detection: http://mitosis-detection.herokuapp.com/.

Mitosis

Non - mitosis

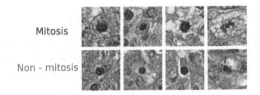

Figure 10.7 Example of hard to differentiate mitotic (top) and nonmitotic (bottom) images.

outperformed several classical machine learning classifiers trained with many features describing size, shape, color, and texture. Since then, several approaches focused on more clinically usable methods by taking larger areas as input, taking into account only the tumor regions in the WSI and building classifiers for difficult cases (ensemble of architectures among the others) as seen in Fig. 10.8.

In Veta et al. [96] the authors summarize the results of the AMIDA13 challenge. The AMIDA13 dataset consisted of 12 training and 11 testing subjects, with more than 1000 annotated mitotic figures by multiple observers. The accuracy of the best method (again, the one by IDSIA) is comparable with the interobserver variability. IDSIA's method is based on an efficient CNN without any candidate selection, but classifying each of the high power fields pixels.

In 2014, ICPR launched a second challenge for detecting mitosis: the MITOS-ATYPIA14 challenge. More data were made available and the tasks included also the differentiation between high and low grade nuclear atypia. Another major improvement from the first edition of the contest was that the annotations were provided by two senior pathologists and three junior pathologists. The winner of the nuclear atypia subtask

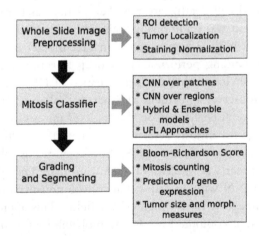

Figure 10.8 Common modules involved when applying DL to the mitosis detection problem; some of the elements overlap with those in Fig. 10.4. The main difference relies on the region of interest localization for patch extraction and the output of the DL classifiers.

was the team from University of Warwick, while the subtask of mitosis detection had two winners: the team from the Chinese University of Hong Kong for the Aperio scanner and the team from Institut Curie for the slides processed with Hamamatsu scanner. The Warwick team did not use an approach based on deep learning, but they used a generalized region covariance descriptor. The descriptor creates a geodesic measure of the manifold that represents the pixel-level feature space [104]. The approach described by Chen et al. [100] is currently the method with the best performance on the 2014 MITOS-ATYPIA dataset in the task of mitosis detection. This approach is based on two steps: (1) Collecting mitosis candidates by means of a fully convolutional neural network and (2) using transfer learning to classify the candidates. The parameters of the off-the-shelf CaffeNet model were used to train three architectures with a different number of neurons in three fully connected layers. The probability of a patch being mitosis or not was computed via the average of the model decisions.

In the 2016 *MICCAI Grand Challenge on Tumor Proliferation Assessment*-TUPAC16, there were two subtasks involving mitosis detection. The first consisted of predicting a proliferation score based on mitosis counting. The second consisted of the mitosis detection of tumor regions. The results of this challenge are available on its website.[8] The detailed descriptions of the tested methods have not been uploaded yet.

Besides the challenges in mitosis detection, many researchers have been tackling this problem using deep learning techniques [100,96,105,103,102,106]. The types of strategies can be differentiated into those involving ROI detection techniques and those that do not. Staining normalization between different microscopes/laboratories is a common preprocessing module. Further steps include tumor localization, data augmentation, and boosting classifiers with difficult cases. An approach of Xu et al. is based on stacked autoencoders that create high level feature representations for 34×34 pixel patches [106]. They evaluate the approach on 37 H&E stained breast histopathology images that were collected from a cohort of 17 patients. The result of the evaluation shows superior performance when compared with texture and morphological features. A hybrid approach is described by Wang et al. [105]. The authors propose a probabilistic fusion of two classifiers, one with handcrafted features as input and another with CNN derived features. For the difficult cases, they fuse both features to train and test a third classifier. The authors test their approach on the MITOS2012 and AMIDA13 challenge datasets showing that their method can obtain better performance than the other deep learning methods presented at the challenges.

Giusti et al. [103] make a man machine comparison. The authors measure the performance of human pathologists and algorithms presented in MITOS12 [78]. They show that the most accurate pathologists were outperformed by deep learning algorithms, specifically the CNN technique from IDSIA. More recently, Veta et al. [102]

---

[8] http://tupac.tue-image.nl/node/2 as of January 1, 2017.

go one step further in this direction and illustrate that the best measured agreement between the pathologists and an automatic DL method is comparable to the worst measured agreement among the pathologists. The authors also explain that the gap between the experts and the DL algorithm should be reduced if the researchers take into account the context of the regions in WSI and not only the patches centered at a given mitotic location.

10.6 FRAME AND WHOLE SLIDE IMAGE CLASSIFICATION

To confirm or discard a diagnostic hypothesis, pathologists visually inspect whole slide images at multiple resolutions, selecting Regions Of Interest (ROIs) and identifying structures. Image classification and disease grading in clinical routine are based on the architectural patterns in the histopathology tissue samples. In higher cancer grading, cells are poorly differentiated and prognosis is adverse with a higher risk of relapse. Pathologist grading can vary according to their personal experience and inherent subjectivity [107]. Novel approaches using deep learning are now able to classify a WSI or a region even without fully comprehending the relations between the underlying histological structures [85,70,108].

Different pipelines can be used to perform image classification in digital pathology. Some approaches first detect ROIs, then quantify histological primitives and finally decide the image classification. Other methods skip the middle steps entirely and decide the classification without targeting specific histological primitives. Depending on the pipeline selected, one or multiple classifiers can be trained at different resolutions and for separate histological structures. The sampling performed on the WSI can also vary according to the objectives of the proposed algorithms. Fig. 10.9 presents different sampling techniques used for WSI classification.

Image classification tasks were included in medical image analysis challenges only in recent years. In MITOS–ATYPIA–14, high power fields from breast cancer tissue slides were analyzed to generate a score for nuclear atypia (*i.e.*, nuclei shape variations compared to normal nuclei). The slides were divided into three grades according to

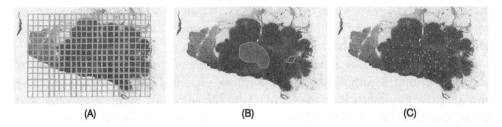

(A) (B) (C)

Figure 10.9 Sampling techniques performed in WSI classification. (A) Grid search. (B) Regions of interest. (C) Random sampling.

six criteria related to nuclear pleomorphism (such as size of the nuclei, size of the nucleoli, and thickness of nuclear membrane). The winning approach for this task was the one from Khan et al., both on the Aperio and Hamamatsu images [104]. This approach includes an image-level descriptor (implemented as the geodesic mean of region covariance descriptors) that enables a tractable geodesic-distance-based kNN classification using efficient kernels. This approach outperformed standard region covariance descriptors and textural descriptors for nuclear atypia.

CPM was a challenge on the classification of whole slide tissue images using two cohorts, one from patients with nonsmall cell lung cancer and one for head and neck squamous cell carcinoma. Participants were asked to classify patients from each cohort according to their molecular subtypes. No additional information regarding the challenge results is currently available on their website.[9]

CAMELYON16 was the first challenge aimed at classifying whole slide images: 270 for training and 130 for testing. The participant algorithms were implemented to discriminate between normal lymph node WSI and slides containing metastases. The probability of metastases present per image was also requested. The method by Wang et al. [109] obtained the highest score for both the WSI classification and tumor localization task. Their method consisted of a patch-based classification stage and a heatmap-based postprocessing. Positive and negative tumor samples were used for retraining a supervised classification model with the GoogLeNet architecture [110]. The predictions were embedded in a heatmap image from which 28 geometrical and morphological features were extracted to build a random forest classifier aimed at discriminating WSIs with metastases and normal WSIs.

Following the current trend in digital pathology, TUPAC16 distributed an even larger number of WSI to participants: 500 for training and 321 for testing. Unlike the binary classification from CAMELYON16, TUPAC16's main goals were to grade breast cancer WSIs into 3 classes (according to the mitosis counting among other components, *i.e.*, Bloom & Richardson grading system) and to predict a proliferation score based on molecular data. The method from Lunit Inc. Korea was the winner for both grading tasks among the automatic methods. For the moment, there is no other information available about the methods on the challenge website.[10]

10.7 STRUCTURE SEGMENTATION

Defining and measuring the boundaries of tissue primitives such as glands helps in disentangling the visual variability in the WSI. This is useful to identify malignant tumors

---

[9] http://miccai.cloudapp.net/competitions/46, as of January 1, 2017.
[10] As of January 1, 2017.

arising from glandular epithelium, which is a routinely performed task in histopathology. Structure segmentation in WSIs is a challenging task that is important for grading cancers like prostate and colon. The morphology of the glands has routinely been used by pathologists to assess the degree of malignancy of several adenocarcinomas, including prostate, breast, lung, and colon. The structure segmentation problem is now commonly addressed with many deep learning techniques such as Fully Convolutional Neural Networks (FCNN) [111], which helps to recover spatial information lost during downsampling of the usual classification architectures. Custom CNN architectures [79] have also been tested with promising results compared to the classical patch-based techniques [112].

In the 2015 MICCAI Gland Segmenting challenge (GlaS@MICCAI'2015, see Section 10.5), various deep learning approaches were promising regarding this direction [113]. A remarkable approach is the one of the winning group: CUMedVision[11] [114]. A modification of a fully convolutional neural network is used to segment colon tissue with a Dice score of 0.897 and an F1-score of 0.912. The approach uses *multilevel contextual features* that group features from several layers of the deep network, taking into account the intensities and varying sizes of receptive fields using a two–path FCNN.

An interesting approach for semantic segmentation in WSIs is proposed by Wang et al. [115]. They identify and tackle two main problems: the lack of training data and how to deal with the multiple sizes and shape patterns. They found that transfer learning is a reliable enhancer, both in training speed and segmentation performance. For the second problem, several FCNs are fused, varying the input size in order to be able to capture the different scales and shapes of the patterns. The approach outperforms FCN with a considerable margin in an inflammatory bowel disease dataset of H&E WSIs.

Patch-based techniques are a commonly used strategy in which the idea is that the structure to be segmented is fed in form of patches as input to the networks, for subsequently densely running of the model over the WSI. This creates a probability map for all pixels that can subsequently be used in the task of segmentation by setting a threshold in the map.

In [116], various patch-based deep learning settings are tested for both classification and segmentation of prostate cancer tissue showing promising results. Similarly, in [117], a pixel-wise classification architecture is proposed to segment prostate tissue reaching an error rate as low as 7.3% in patch classification. The problem of a better attempt in delineating the fine details in class boundaries is discussed, concluding that it is necessary to transfer these techniques to medical practice. This is an explicit trend seen in many recent methods for H&E segmentation.

---

[11] Department of Computer Science and Engineering, The Chinese University of Hong Kong.

In the paper of Sirinukunwattana et al. [118] a spatially constrained CNN is investigated to perform nucleus detection in colon cancer WSI. An F1 score of 0.78 is obtained classifying epithelial, inflammatory, fibroblast, and miscellaneous patches. After the model is trained, a comprehensive segmentation of a WSI is obtained. This is another example where the step of designing a particular segmentation architecture is skipped.

In the work of Drozdzal [119], the importance of the skip connections in very deep architectures for segmentation is investigated. Even though it is tested on electron microscopy images, the findings are also applicable to deep architectures that segment histopathology images. In Akram et al. [120], a two stage CNN-based method is proposed to accurately segment cells into three modalities: fluorescent imaging, phase contrast imaging, and H&E images. At the first stage, a CNN is build for detecting bounding boxes. In the second stage, a CNN predicts a segmentation mask. An interesting weight sharing scheme is used that puts the ROI information into the second CNN. Xu et al. [121] present a similar approach using three CNN architectures. One is used to remove the background, another for detecting the glands and a final network to refine the edges of the mask. An object Dice score of 0.908 is achieved with respect to a FCN baseline that achieves 0.742 on the same dataset.

Wenqi et al. [122] propose a hybrid approach to segment colon histology images using features from a CNN architecture and classical features like the Scale-Invariant Feature Transform (SIFT) and multiresolution local patterns among others. The performance improvement with respect to the CNN architecture alone was from 0.82 to 0.87 in Dice score. Xie et al. [123] proposed two fully convolutional regression networks to count cells in H&E images. A major component in the approach is that the training of the model is made over synthetically generated data. Nevertheless, the performance is comparable when trained on real data. Sadanandan et al. [124] also make use of data augmentation but in this case with specialized feature maps like the eigenvalue of Hessian and wavelet filtered images. Ben Taieb et al. [125] show how a CNN architecture can do joint segmentation and classification over colon glands, improving the usage of both approaches separately.

The approach of Kainz et al. [112] for the GlaS15 challenge consisted of two CNN architectures regularized by a total variation criteria. The first separated benign background, benign glands, malignant background, and malignant glands. A second CNN concentrated on separating very close gland objects. Tissue classification accuracies of 98% and 94% are obtained on two test sets. Ben Taieb et al. [126] encode the relationship between the different structures on the WSI with a topology aware FCNN that allows to encode geometric and topological priors of containment and detachment. This approach gained up to 10% in object-level Dice segmentation over the original FCNN approach.

Among all the work reviewed, the use of FCNN-based architectures is ubiquitous, showing good results in segmenting primitives like cells, glands, and epithelium. A current trend is to include additional channels for a finer segmentation and to encode spatial, geometric, and scale information in the loss functions of the models.

10.8 DISCUSSION AND CONCLUSIONS

Computer-assisted decision support of histopathology images is a challenging field that will meet the requirements of clinical practice in the near future. The fusion of traditional diagnosis with computational data analysis represents an opportunity to improve image interpretation by reducing the workload of pathologists for the most trivial tasks while also harmonizing agreement between pathologists. Since the advent of digital scanners, private companies and scientific researchers have been developing tools using machine learning that assist pathologists [1,2]. However, assisting pathologists is not trivial due to the many technical and clinical factors that surround histopathology images. Traditional machine learning approaches obtained good results for the detection, segmentation, and classification of tissue structures in the last 10–20 years. However, several technical challenges need to be solved before computer aided decision support can be applied to clinical practice. Deep learning obtained best results in most tasks in recent years and it has interesting perspectives for the future.

A first challenge in the field of histopathology CAD is represented by the high variability of tasks due to clinical characteristics of the images. Current CAD methods can obtain good results on highly specific tasks. The best performance is obtained when task-specific tuning is involved [70]. Thus current methods are strongly sensible to the fine tuning made for a specific, limited problem. To the best of our knowledge, there are currently no models that are capable to analyze WSI of several pathologies together.

A second challenge in the field is the development of methods that are capable of analyzing the many diverse parameters that are considered by a pathologist simultaneously. WSIs are now being used and distributed in challenges to develop algorithms that mimic the full workflow of a pathologist. However, the workflow of pathologists often follows almost instinctive paths. While looking at an image, a pathologist zooms to different areas, searching for properties that change depending on the image and on the clinical parameters of the subject. This diagnostic workflow is partially represented by the histopathology literature, but it has not been imitated in computer science literature. The workflow involves specific challenging tasks, such as the detection of mitosis or basal cells, the recognition of glands and the classification or regions. The development of methods that are capable to perform well in each of these specific tasks are important. Combining such methods can then lead to clinical applicability.

A third challenge is represented by analysis robustness. Recent results on specific tasks can be comparable to human performances. For instance, in the challenging task of mitosis detection, the performance of deep learning based methods is now approaching the interobserver agreement of expert pathologists [102,95,103]. Similarly, recent image analysis approaches can classify and grade WSI according to the tissue patterns resulting from different cancer types [85,70] with performances that are comparable with the interpretation of pathologists on the same images. However, the results are usually obtained on datasets that were usually acquired using up to 2–3 scanners, while metaanalysis comparing the same methods on datasets characterized by high heterogeneity have not been performed yet. Thus the challenge for the performance of each algorithm is to make it robust on different whole slide images.

Structure segmentation in WSI is a challenging task that is important for grading some cancers. The morphology of glands is routinely used by pathologists to assess the degree of malignancy of several adenocarcinomas, including prostate, breast, lung, and colon cancer. FCNN-based architectures are frequent and promising for cells, glands, and epithelium segmentation, while current trends include the use of additional channels to benefit of spatial, geometric, and scale information. However, also for this task, robust and generalizable results must be obtained before translating them into practice.

Future perspectives and trends for histopathology CAD systems include the capability to better interact with images from different scanners and across pathologies, as well as the creation of methods that can learn from unlabeled or weakly labeled data. During preprocessing, advanced normalization methods can increase image similarity, thus making the CAD methods more reliable even on images acquired with different scanners and with varying staining procedures. The creation of extended datasets that include numerous pathologies may allow the creation of algorithms that rely on a deep comprehension of the whole histopathology field, thus making CAD systems capable of working in several anatomical structures reliably. Finally, the recent advent and success of deep learning methods in digital pathology suggests that this quickly improve the performance of CAD systems and accelerate their use, possible via the development of unsupervised learning methods. Most of current learning algorithms require manually annotated training sets. This is particularly evident for deep learning methods that usually require large amounts of data to work well. It is often difficult to obtain a large amount of manually annotated WSIs because the task is tedious; it requires a medical background and pathologists are usually not very interested in doing it. Unsupervised learning methods (as in [72]) can help to skip the data annotation phase, thus allowing to exploit histopathology images automatically.

In conclusion, currently WSI and deep learning are revolutionizing histopathology CAD, and they may soon help to decrease the work load of pathologists for the most trivial tasks. This should allow pathologists to concentrate on the most difficult cases and result in an even deeper understanding of the pathologic processes via the machine learning approaches.

ACKNOWLEDGMENTS

This project has received funding from the Eurostars-2 Joint Program with cofunding from the European Union's Horizon 2020 research and innovation program.

REFERENCES

[1] M.N. Gurcan, L.E. Boucheron, A. Can, A. Madabhushi, N.M. Rajpoot, B. Yener, Histopathological image analysis: a review, IEEE Rev. Biomed. Eng. 2 (2009) 147–171.

[2] R. Rubin, D.S. Strayer, E. Rubin, et al., Rubin's Pathology: Clinicopathologic Foundations of Medicine, Lippincott Williams & Wilkins, 2008.

[3] G. Litjens, C.I. Sánchez, N. Timofeeva, M. Hermsen, I. Nagtegaal, I. Kovacs, C. Hulsbergen-van de Kaa, P. Bult, B. van Ginneken, J. van der Laak, Deep learning as a tool for increased accuracy and efficiency of histopathological diagnosis, Sci. Rep. 6 (2016).

[4] F.D. Gleason, Histologic grading of prostate cancer: a perspective, Hum. Pathol. 23 (3) (1992) 273–279.

[5] F. Ghaznavi, A. Evans, A. Madabhushi, M. Feldman, Digital imaging in pathology: whole-slide imaging and beyond, Annu. Rev. Pathol. Mech. Dis. 8 (2013) 331–359.

[6] J.L. Fine, D.M. Grzybicki, R. Silowash, J. Ho, J.R. Gilbertson, L. Anthony, R. Wilson, A.V. Parwani, S.I. Bastacky, J.I. Epstein, et al., Evaluation of whole slide image immunohistochemistry interpretation in challenging prostate needle biopsies, Hum. Pathol. 39 (4) (2008) 564–572.

[7] E. Manion, M.B. Cohen, J. Weydert, Mandatory second opinion in surgical pathology referral material: clinical consequences of major disagreements, Am. J. Surg. Pathol. 32 (5) (2008) 732–737.

[8] J.A. Woolgar, A. Ferlito, K.O. Devaney, A. Rinaldo, L. Barnes, How trustworthy is a diagnosis in head and neck surgical pathology? A consideration of diagnostic discrepancies (errors), Eur. Arch. Oto-Rhino-Laryngol. 268 (5) (2011) 643–651.

[9] J. Ho, A.V. Parwani, D.M. Jukic, Y. Yagi, L. Anthony, J.R. Gilbertson, Use of whole slide imaging in surgical pathology quality assurance: design and pilot validation studies, Hum. Pathol. 37 (3) (2006) 322–331.

[10] R.S. Weinstein, A.R. Graham, L.C. Richter, G.P. Barker, E.A. Krupinski, A.M. Lopez, K.A. Erps, A.K. Bhattacharyya, Y. Yagi, J.R. Gilbertson, Overview of telepathology, virtual microscopy, and whole slide imaging: prospects for the future, Hum. Pathol. 40 (8) (2009) 1057–1069.

[11] N. Velez, D. Jukic, J. Ho, Evaluation of 2 whole-slide imaging applications in dermatopathology, Hum. Pathol. 39 (9) (2008) 1341–1349.

[12] A. Huisman, A. Looijen, S.M. van den Brink, P.J. van Diest, Creation of a fully digital pathology slide archive by high-volume tissue slide scanning, Hum. Pathol. 41 (5) (2010) 751–757.

[13] A. Depeursinge, A. Foncubierta-Rodríguez, D. Van De Ville, H. Müller, Three-dimensional solid texture analysis and retrieval in biomedical imaging: review and opportunities, Med. Image Anal. 18 (1) (2014) 176–196.

[14] H.O. Lyon, A.P. De Leenheer, R.W. Horobin, W.E. Lambert, E.K.W. Schulte, B. Van Liedekerke, D.H. Wittekind, Standardization of reagents and methods used in cytological and histological practice with emphasis on dyes, stains and chromogenic reagents, Histochem. J. 26 (7) (1994) 533–544.

[15] A.M. Khan, N. Rajpoot, D. Treanor, D. Magee, A nonlinear mapping approach to stain normalization in digital histopathology images using image-specific color deconvolution, IEEE Trans. Biomed. Eng. 61 (6) (2014) 1729–1738.

[16] D. Magee, D. Treanor, D. Crellin, M. Shires, K. Smith, K. Mohee, P. Quirke, Colour normalisation in digital histopathology images, in: Proc Optical Tissue Image Analysis in Microscopy, Histopathology and Endoscopy (MICCAI Workshop), vol. 100, Citeseer, 2009.

[17] G.D. Marty, et al., Blank-field correction for achieving a uniform white background in brightfield digital photomicrographs, BioTechniques 42 (6) (2007) 716.

[18] F.J.W.M. Leong, M. Brady, J. O'D McGee, Correction of uneven illumination (vignetting) in digital microscopy images, J. Clin. Pathol. 56 (8) (2003) 619–621.

[19] F. Piccinini, E. Lucarelli, A. Gherardi, A. Bevilacqua, Multi-image based method to correct vignetting effect in light microscopy images, J. Microsc. 248 (1) (2012) 6–22.

[20] A. Can, M. Bello, H.E. Cline, X. Tao, F. Ginty, A. Sood, M. Gerdes, M. Montalto, Multi-modal imaging of histological tissue sections, in: 2008 5th IEEE International Symposium on Biomedical Imaging: From Nano to Macro, IEEE, 2008, pp. 288–291.

[21] J. Kong, O. Sertel, H. Shimada, K.L. Boyer, J.H. Saltz, M.N. Gurcan, Computer-aided evaluation of neuroblastoma on whole-slide histology images: classifying grade of neuroblastic differentiation, Pattern Recognit. 42 (6) (2009) 1080–1092.

[22] M.K. Washington, J. Berlin, P. Branton, L.J. Burgart, D.K. Carter, P.L. Fitzgibbons, K. Halling, W. Frankel, J. Jessup, S. Kakar, et al., Protocol for the examination of specimens from patients with primary carcinoma of the colon and rectum, Arch. Pathol. Lab. Med. 133 (10) (2009) 1539–1551.

[23] H. Irshad, A. Veillard, L. Roux, D. Racoceanu, Methods for nuclei detection, segmentation, and classification in digital histopathology: a review—current status and future potential, IEEE Rev. Biomed. Eng. 7 (2014) 97–114.

[24] J.-R. Dalle, H. Li, C.-H. Huang, W.K. Leow, D. Racoceanu, T.C. Putti, Nuclear pleomorphism scoring by selective cell nuclei detection, in: WACV, 2009.

[25] C. Wählby, I.-M. Sintorn, F. Erlandsson, G. Borgefors, E. Bengtsson, Combining intensity, edge and shape information for 2D and 3D segmentation of cell nuclei in tissue sections, J. Microsc. 215 (1) (2004) 67–76.

[26] C. Jung, C. Kim, Segmenting clustered nuclei using h-minima transform-based marker extraction and contour parameterization, IEEE Trans. Biomed. Eng. 57 (10) (2010) 2600–2604.

[27] E. Cosatto, M. Miller, H.P. Graf, J.S. Meyer, Grading nuclear pleomorphism on histological micrographs, in: 19th International Conference on Pattern Recognition, 2008. ICPR 2008, IEEE, 2008, pp. 1–4.

[28] Y. Al-Kofahi, W. Lassoued, W. Lee, B. Roysam, Improved automatic detection and segmentation of cell nuclei in histopathology images, IEEE Trans. Biomed. Eng. 57 (4) (2010) 841–852.

[29] M. Veta, A. Huisman, M.A. Viergever, P.J. van Diest, J.P.W. Pluim, Marker-controlled watershed segmentation of nuclei in H&E stained breast cancer biopsy images, in: 2011 IEEE International Symposium on Biomedical Imaging: from Nano to Macro, IEEE, 2011, pp. 618–621.

[30] E. Aptoula, N. Courty, S. Lefèvre, Mitosis detection in breast cancer histological images with mathematical morphology, in: Signal Processing and Communications Applications Conference (SIU), 2013 21st, IEEE, 2013, pp. 1–4.

[31] C.D. Cireşan, A. Giusti, L.M. Gambardella, J. Schmidhuber, Mitosis detection in breast cancer histology images with deep neural networks, in: International Conference on Medical Image Computing and Computer-Assisted Intervention, Springer, 2013, pp. 411–418.

[32] S. Petushi, F.U. Garcia, M.M. Haber, C. Katsinis, A. Tozeren, Large-scale computations on histology images reveal grade-differentiating parameters for breast cancer, BMC Med. Imaging 6 (1) (2006) 1.

[33] H. Irshad, et al., Automated mitosis detection in histopathology using morphological and multi-channel statistics features, J. Pathol. Inform. 4 (1) (2013) 10.

[34] P.-W. Huang, Y.-H. Lai, Effective segmentation and classification for HCC biopsy images, Pattern Recognit. 43 (4) (2010) 1550–1563.

[35] V.-T. Ta, O. Lézoray, A. Elmoataz, S. Schüpp, Graph-based tools for microscopic cellular image segmentation, Pattern Recognit. 42 (6) (2009) 1113–1125.

[36] J.P. Vink, M.B. Van Leeuwen, C.H.M. Van Deurzen, G. De Haan, Efficient nucleus detector in histopathology images, J. Microsc. 249 (2) (2013) 124–135.

[37] H.-S. Wu, R. Xu, N. Harpaz, D. Burstein, J. Gil, Segmentation of intestinal gland images with iterative region growing, J. Microsc. 220 (3) (2005) 190–204.

[38] D. Altunbay, C. Cigir, C. Sokmensuer, C. Gunduz-Demir, Color graphs for automated cancer diagnosis and grading, IEEE Trans. Biomed. Eng. 57 (3) (2010) 665–674.

[39] H. Fu, G. Qiu, J. Shu, M. Ilyas, A novel polar space random field model for the detection of glandular structures, IEEE Trans. Med. Imaging 33 (3) (2014) 764–776.

[40] K. Sirinukunwattana, D.R.J. Snead, N.M. Rajpoot, A stochastic polygons model for glandular structures in colon histology images, IEEE Trans. Med. Imaging 34 (11) (2015) 2366–2378.

[41] S. Naik, S. Doyle, M. Feldman, J. Tomaszewski, A. Madabhushi, Gland segmentation and computerized Gleason grading of prostate histology by integrating low-, high-level and domain specific information, in: MIAAB Workshop, Citeseer, 2007, pp. 1–8.

[42] L.E. Boucheron, Object-and Spatial-Level Quantitative Analysis of Multispectral Histopathology Images for Detection and Characterization of Cancer, University of California at Santa Barbara, 2008.

[43] C. Demir, B. Yener, Automated Cancer Diagnosis Based on Histopathological Images: A Systematic Survey, Tech. Rep., Rensselaer Polytechnic Institute, 2005.

[44] A. Chekkoury, P. Khurd, J. Ni, C. Bahlmann, A. Kamen, A. Patel, L. Grady, M. Singh, M. Groher, N. Navab, et al., Automated malignancy detection in breast histopathological images, in: SPIE Medical Imaging, International Society for Optics and Photonics, 2012, p. 831515.

[45] S. Doyle, M. Hwang, K. Shah, A. Madabhushi, M. Feldman, J. Tomaszewski, Automated grading of prostate cancer using architectural and textural image features, in: 2007 4th IEEE International Symposium on Biomedical Imaging: From Nano to Macro, IEEE, 2007, pp. 1284–1287.

[46] S. Doyle, A. Madabhushi, M. Feldman, J. Tomaszewski, A boosting cascade for automated detection of prostate cancer from digitized histology, in: International Conference on Medical Image Computing and Computer-Assisted Intervention, Springer, 2006, pp. 504–511.

[47] S. Doyle, C. Rodriguez, A. Madabhushi, J. Tomaszewski, M. Feldman, Detecting prostatic adenocarcinoma from digitized histology using a multi-scale hierarchical classification approach, in: Engineering in Medicine and Biology Society, 2006. EMBS'06. 28th Annual International Conference of the IEEE, IEEE, 2006, pp. 4759–4762.

[48] P.-W. Huang, C.-H. Lee, Automatic classification for pathological prostate images based on fractal analysis, IEEE Trans. Med. Imaging 28 (7) (2009) 1037–1050.

[49] K. Jafari-Khouzani, H. Soltanian-Zadeh, Multiwavelet grading of pathological images of prostate, IEEE Trans. Biomed. Eng. 50 (6) (2003) 697–704.

[50] J. Diamond, N.H. Anderson, P.H. Bartels, R. Montironi, P.W. Hamilton, The use of morphological characteristics and texture analysis in the identification of tissue composition in prostatic neoplasia, Hum. Pathol. 35 (9) (2004) 1121–1131.

[51] P. Burt, E. Adelson, The Laplacian pyramid as a compact image code, IEEE Trans. Commun. 31 (4) (1983) 532–540.

[52] P. Pudil, J. Novovičová, J. Kittler, Floating search methods in feature selection, Pattern Recognit. Lett. 15 (11) (1994) 1119–1125.

[53] Y. Freund, R.E. Schapire, A decision-theoretic generalization of on-line learning and an application to boosting, in: European Conference on Computational Learning Theory, Springer, 1995, pp. 23–37.

[54] S. Perkins, K. Lacker, J. Theiler, Grafting: fast, incremental feature selection by gradient descent in function space, J. Mach. Learn. Res. 3 (Mar) (2003) 1333–1356.

[55] K.L. Weind, C.F. Maier, B.K. Rutt, M. Moussa, Invasive carcinomas and fibroadenomas of the breast: comparison of microvessel distributions-implications for imaging modalities, Radiology 208 (2) (1998) 477–483.

[56] P.H. Bartels, D. Thompson, M. Bibbo, J.E. Weber, Bayesian belief networks in quantitative histopathology. Analytical and quantitative cytology and histology/the International Academy of Cytology [and] American Society of Cytology, Anal. Quant. Cytol. Histol. 14 (6) (1992) 459–473.

[57] I. Jolliffe, Principal Component Analysis, Wiley Online Library, 2002.

[58] M.D. DiFranco, G. O'Hurley, E.W. Kay, R. William, G. Watson, P. Cunningham, Ensemble based system for whole-slide prostate cancer probability mapping using color texture features, Comput. Med. Imaging Graph. 35 (7) (2011) 629–645.

[59] C.D. Malon, E. Cosatto, et al., Classification of mitotic figures with convolutional neural networks and seeded blob features, J. Pathol. Inform. 4 (1) (2013) 9.

[60] K. Rajpoot, N. Rajpoot, SVM optimization for hyperspectral colon tissue cell classification, in: International Conference on Medical Image Computing and Computer-Assisted Intervention, Springer, 2004, pp. 829–837.

[61] H. Qureshi, O. Sertel, N. Rajpoot, R. Wilson, M. Gurcan, Adaptive discriminant wavelet packet transform and local binary patterns for meningioma subtype classification, in: International Conference on Medical Image Computing and Computer-Assisted Intervention, Springer, 2008, pp. 196–204.

[62] J. Ghosh, Multiclassifier systems: back to the future, in: International Workshop on Multiple Classifier Systems, Springer, 2002, pp. 1–15.

[63] A. Depeursinge, H. Müller, Fusion techniques for combining textual and visual information retrieval, in: H. Müller, P. Clough, T. Deselaers, B. Caputo (Eds.), ImageCLEF, in: The Springer International Series on Information Retrieval, vol. 32, Springer, Berlin, Heidelberg, 2010, pp. 95–114.

[64] E. Alexandratou, V. Atlamazoglou, T. Thireou, G. Agrogiannis, D. Togas, N. Kavantzas, E. Patsouris, D. Yova, Evaluation of machine learning techniques for prostate cancer diagnosis and Gleason grading, Int. J. Comput. Intell. Bioinform. Syst. Biol. 1 (3) (2010) 297–315.

[65] A. Krizhevsky, I. Sutskever, G.E. Hinton, Imagenet classification with deep convolutional neural networks, in: Advances in Neural Information Processing Systems, 2012, pp. 1097–1105.

[66] L. Deng, G. Hinton, B. Kingsbury, New types of deep neural network learning for speech recognition and related applications: an overview, in: 2013 IEEE International Conference on Acoustics, Speech and Signal Processing, IEEE, 2013, pp. 8599–8603.

[67] X. Glorot, A. Bordes, Y. Bengio, Domain adaptation for large-scale sentiment classification: a deep learning approach, in: Proceedings of the 28th International Conference on Machine Learning (ICML-11), 2011, pp. 513–520.

[68] N. Pawlowski, J.C. Caicedo, S. Singh, A.E. Carpenter, A. Storkey, Automating morphological profiling with generic deep convolutional networks, bioRxiv (2016), http://dx.doi.org/10.1101/085118.

[69] H. Greenspan, B. van Ginneken, R.M. Summers, Guest editorial deep learning in medical imaging: overview and future promise of an exciting new technique, IEEE Trans. Med. Imaging 35 (5) (2016) 1153–1159.

[70] A. Janowczyk, A. Madabhushi, Deep learning for digital pathology image analysis: a comprehensive tutorial with selected use cases, J. Pathol. Inform. 7 (2016).

[71] Y. LeCun, Y. Bengio, G. Hinton, Deep learning, Nature 521 (7553) (2015) 436–444.

[72] I. Goodfellow, Y. Bengio, A. Courville, Deep Learning, MIT Press, 2016, http://www.deeplearningbook.org.

[73] Y. LeCun, L. Bottou, Y. Bengio, P. Haffner, Gradient-based learning applied to document recognition, Proc. IEEE 86 (11) (1998) 2278–2324.

[74] Y. Bengio, A.C. Courville, P. Vincent, Unsupervised feature learning and deep learning: a review and new perspectives, CoRR, arXiv:1206.5538, 2012, 1.

[75] J. Schmidhuber, Deep learning in neural networks: an overview, Neural Netw. 61 (2015) 85–117.

[76] D. Erhan, Y. Bengio, A. Courville, P.-A. Manzagol, P. Vincent, S. Bengio, Why does unsupervised pre-training help deep learning?, J. Mach. Learn. Res. 11 (Feb) (2010) 625–660.

[77] J. Arevalo, A. Cruz-Roa, et al., Histopathology image representation for automatic analysis: a state-of-the-art review, Revista Med 22 (2) (2014) 79–91.

[78] L. Roux, D. Racoceanu, N. Loménie, M. Kulikova, H. Irshad, J. Klossa, F. Capron, C. Genestie, G. Naour, M. Gurcan, Mitosis detection in breast cancer histological images an ICPR 2012 contest, J. Pathol. Inform. 4 (1) (2013) 8.

[79] A.A. Cruz-Roa, J.E.A. Ovalle, A. Madabhushi, F.A.G. Osorio, A deep learning architecture for image representation, visual interpretability and automated basal-cell carcinoma cancer detection, in: International Conference on Medical Image Computing and Computer-Assisted Intervention, Springer, 2013, pp. 403–410.

[80] J. Arevalo, A. Cruz-Roa, F.A. González, Hybrid image representation learning model with invariant features for basal cell carcinoma detection, in: IX International Seminar on Medical Information Processing and Analysis, International Society for Optics and Photonics, 2013, p. 89220M.

[81] N. Nayak, H. Chang, A. Borowsky, P. Spellman, B. Parvin, Classification of tumor histopathology via sparse feature learning, in: 2013 IEEE 10th International Symposium on Biomedical Imaging, IEEE, 2013, pp. 410–413.

[82] Y. Xu, T. Mo, Q. Feng, P. Zhong, M. Lai, I. Eric, C. Chang, Deep learning of feature representation with multiple instance learning for medical image analysis, in: 2014 IEEE International Conference on Acoustics, Speech and Signal Processing (ICASSP), IEEE, 2014, pp. 1626–1630.

[83] Le Hou, D. Samaras, T.M. Kurc, Y. Gao, J.E. Davis, J.H. Saltz, Efficient multiple instance convolutional neural networks for gigapixel resolution image classification, arXiv:1504.07947v1, 2015.

[84] J.A. Vanegas, J. Arevalo, F.A. González, Unsupervised feature learning for content-based histopathology image retrieval, in: 2014 12th International Workshop on Content-Based Multimedia Indexing (CBMI), IEEE, 2014, pp. 1–6.

[85] A. Cruz-Roa, A. Basavanhally, F. González, H. Gilmore, M. Feldman, S. Ganesan, N. Shih, J. Tomaszewski, A. Madabhushi, Automatic detection of invasive ductal carcinoma in whole slide images with convolutional neural networks, in: SPIE Medical Imaging, International Society for Optics and Photonics, 2014, p. 904103.

[86] J. Arevalo, A. Cruz-Roa, V. Arias, E. Romero, F.A. González, An unsupervised feature learning framework for basal cell carcinoma image analysis, Artif. Intell. Med. 64 (2) (2015) 131–145.

[87] H. Chang, Y. Zhou, A. Borowsky, K. Barner, P. Spellman, B. Parvin, Stacked predictive sparse decomposition for classification of histology sections, Int. J. Comput. Vis. 113 (1) (2015) 3–18.

[88] H. Noël, L. Roux, S. Lu, T. Boudier, Detection of high-grade atypia nuclei in breast cancer imaging, in: SPIE Medical Imaging, International Society for Optics and Photonics, 2015, p. 94200R.

[89] Y. Jia, E. Shelhamer, J. Donahue, S. Karayev, J. Long, R. Girshick, S. Guadarrama, T. Darrell, Caffe: convolutional architecture for fast feature embedding, arXiv preprint arXiv:1408.5093, 2014.

[90] A. Cruz-Roa, J. Arevalo, A. Basavanhally, A. Madabhushi, F. González, A comparative evaluation of supervised and unsupervised representation learning approaches for anaplastic medulloblastoma differentiation, in: Tenth International Symposium on Medical Information Processing and Analysis, International Society for Optics and Photonics, 2015, p. 92870G.

[91] S. Otálora, A. Cruz-Roa, J. Arevalo, M. Atzori, A. Madabhushi, A.R. Judkins, F. González, H. Müller, A. Depeursinge, Combining unsupervised feature learning and Riesz wavelets for histopathology image representation: application to identifying anaplastic medulloblastoma, in: International Conference on Medical Image Computing and Computer-Assisted Intervention, Springer, 2015, pp. 581–588.

[92] C. Lifan Chen, A. Mahjoubfar, L.-C. Tai, I.K. Blaby, A. Huang, K.R. Niazi, B. Jalali, Deep learning in label-free cell classification, Sci. Rep. 6 (2016).

[93] J. Han, G.V. Fontenay, Y. Wang, J.-H. Mao, H. Chang, Phenotypic characterization of breast invasive carcinoma via transferable tissue morphometric patterns learned from glioblastoma multiforme, in: 13th International Symposium on Biomedical Imaging (ISBI), IEEE, 2016, pp. 1025–1028.

[94] D. Romo-Bucheli, A. Janowczyk, H. Gilmore, E. Romero, A. Madabhushi, Automated tubule nuclei quantification and correlation with oncotype DX risk categories in ER+ breast cancer whole slide images, Sci. Rep. 6 (sep 2016) 32706.

[95] T. Chen, C. Chefd'hotel, Deep learning based automatic immune cell detection for immunohistochemistry images, in: International Workshop on Machine Learning in Medical Imaging, Springer, 2014, pp. 17–24.

[96] M. Veta, P.J. van Diest, S.M. Willems, H. Wang, A. Madabhushi, A. Cruz-Roa, F. Gonzalez, A.B.L. Larsen, J.S. Vestergaard, A.B. Dahl, D.C. Cireşan, J. Schmidhuber, A. Giusti, L.M. Gambardella, F.B. Tek, T. Walter, C.-W. Wang, S. Kondo, B.J. Matuszewski, F. Precioso, V. Snell, J. Kittler, T.E. de Campos, A.M. Khan, N.M. Rajpoot, E. Arkoumani, M.M. Lacle, M.A. Viergever, J.P.W. Pluim, Assessment of algorithms for mitosis detection in breast cancer histopathology images, Med. Image Anal. 20 (1) (2015) 237–248.

[97] D. Gutman, N.C.F. Codella, E. Celebi, B. Helba, M. Marchetti, N. Mishra, A. Halpern, Skin lesion analysis toward melanoma detection: a challenge at the international symposium on biomedical imaging (ISBI) 2016, hosted by the international skin imaging collaboration (ISIC), arXiv preprint arXiv:1605.01397, 2016.

[98] H.J.G. Bloom, W.W. Richardson, Histological grading and prognosis in breast cancer: a study of 1409 cases of which 359 have been followed for 15 years, Br. J. Cancer 11 (3) (1957) 359.

[99] M. Veta, J.P.W. Pluim, P.J. van Diest, M.A. Viergever, Breast cancer histopathology image analysis: a review, IEEE Trans. Biomed. Eng. 61 (5) (2014) 1400–1411.

[100] H. Chen, Q. Dou, X. Wang, J. Qin, P.A. Heng, Mitosis detection in breast cancer histology images via deep cascaded networks, in: Thirtieth AAAI Conference on Artificial Intelligence, 2016.

[101] M. Shah, C. Rubadue, D. Suster, D. Wang, Deep learning assessment of tumor proliferation in breast cancer histological images, arXiv preprint arXiv:1610.03467, 2016.

[102] M. Veta, P.J. van Diest, M. Jiwa, S. Al-Janabi, J.P.W. Pluim, Mitosis counting in breast cancer: object-level interobserver agreement and comparison to an automatic method, PLoS ONE 11 (8) (2016) e0161286.

[103] A. Giusti, C. Caccia, D.C. Cireşari, J. Schmidhuber, L.M. Gambardella, A comparison of algorithms and humans for mitosis detection, in: 2014 IEEE 11th International Symposium on Biomedical Imaging (ISBI), IEEE, 2014, pp. 1360–1363.

[104] A.M. Khan, K. Sirinukunwattana, N. Rajpoot, A global covariance descriptor for nuclear atypia scoring in breast histopathology images, IEEE J. Biomed. Health Inform. 19 (5) (2015) 1637–1647.

[105] H. Wang, A. Cruz-Roa, A. Basavanhally, H. Gilmore, N. Shih, M. Feldman, J. Tomaszewski, F. Gonzalez, A. Madabhushi, Mitosis detection in breast cancer pathology images by combining handcrafted and convolutional neural network features, J. Med. Imag. 1 (3) (2014) 034003.

[106] J. Xu, L. Xiang, Q. Liu, H. Gilmore, J. Wu, J. Tang, A. Madabhushi, Stacked sparse autoencoder (SSAE) for nuclei detection on breast cancer histopathology images, IEEE Trans. Med. Imaging 35 (1) (2016) 119–130.

[107] K. Trpkov, Contemporary Gleason grading system, in: Genitourinary Pathology, Springer, 2015, pp. 13–32.

[108] O. Jimenez-del Toro, M. Atzori, M. Andersson, K. Eurén, M. Hedlund, P. Rönnquist, H. Müller, Convolutional neural networks for an automatic classification of prostate tissue slides with high-grade Gleason score, in: SPIE Medical Imaging, International Society for Optics and Photonics, 2017.

[109] D. Wang, A. Khosla, R. Gargeya, H. Irshad, A.H. Beck, Deep learning for identifying metastatic breast cancer, arXiv preprint arXiv:1606.05718, 2016.

[110] C. Szegedy, W. Liu, Y. Jia, P. Sermanet, S. Reed, D. Anguelov, D. Erhan, V. Vanhoucke, A. Rabinovich, Going deeper with convolutions, in: Proceedings of the IEEE Conference on Computer Vision and Pattern Recognition, 2015, pp. 1–9.

[111] J. Long, E. Shelhamer, T. Darrell, Fully convolutional networks for semantic segmentation, in: Proceedings of the IEEE Conference on Computer Vision and Pattern Recognition, 2015, pp. 3431–3440.

[112] P. Kainz, M. Pfeiffer, M. Urschler, Semantic segmentation of colon glands with deep convolutional neural networks and total variation segmentation, arXiv preprint arXiv:1511.06919, 2015.

[113] K. Sirinukunwattana, J.P.W. Pluim, H. Chen, X. Qi, P.-A. Heng, Y.B. Guo, L.Y. Wang, B.J. Matuszewski, E. Bruni, U. Sanchez, et al., Gland segmentation in colon histology images: the glas challenge contest, arXiv preprint arXiv:1603.00275, 2016.

[114] H. Chen, X. Qi, L. Yu, P.-A. Heng, DCAN: deep contour-aware networks for accurate gland segmentation, arXiv preprint arXiv:1604.02677, 2016.

[115] J. Wang, J.D. MacKenzie, R. Ramachandran, D.Z. Chen, A Deep Learning Approach for Semantic Segmentation in Histology Tissue Images, Springer International Publishing, Cham, 2016, pp. 176–184.

[116] G. Flood, Deep Learning with a Dag Structure for Segmentation and Classification of Prostate Cancer, Master's Theses in Mathematical Sciences, 2016.

[117] A. Gummeson, Prostate Cancer Classification Using Convolutional Neural Networks, Master's Theses in Mathematical Sciences, 2016.

[118] K. Sirinukunwattana, S.E.A. Raza, Y.-W. Tsang, D.R.J. Snead, I.A. Cree, N.M. Rajpoot, Locality sensitive deep learning for detection and classification of nuclei in routine colon cancer histology images, IEEE Trans. Med. Imaging 35 (5) (2016) 1196–1206.

[119] M. Drozdzal, E. Vorontsov, G. Chartrand, S. Kadoury, C. Pal, The importance of skip connections in biomedical image segmentation, in: International Workshop on Large-Scale Annotation of Biomedical Data and Expert Label Synthesis, Springer, 2016, pp. 179–187.

[120] S.U. Akram, J. Kannala, L. Eklund, J. Heikkilä, Cell segmentation proposal network for microscopy image analysis, in: International Workshop on Large-Scale Annotation of Biomedical Data and Expert Label Synthesis, Springer, 2016, pp. 21–29.

[121] Y. Xu, Y. Li, M. Liu, Y. Wang, Y. Fan, M. Lai, E.I. Chang, et al., Gland instance segmentation by deep multichannel neural networks, arXiv preprint arXiv:1607.04889, 2016.

[122] W. Li, S. Manivannan, S. Akbar, J. Zhang, E. Trucco, S.J. McKenna, Gland segmentation in colon histology images using hand-crafted features and convolutional neural networks, in: 2016 IEEE 13th International Symposium on Biomedical Imaging (ISBI), 2016, pp. 1405–1408.

[123] W. Xie, J.A. Noble, A. Zisserman, Microscopy cell counting and detection with fully convolutional regression networks, Comput. Methods Biomech. Biomed. Eng. Imaging Vis. (2016) 1–10.

[124] S.K. Sadanandan, P. Ranefall, C. Wählby, Feature augmented deep neural networks for segmentation of cells, in: European Conference on Computer Vision, Springer, 2016, pp. 231–243.

[125] A. BenTaieb, J. Kawahara, G. Hamarneh, Multi-loss convolutional networks for gland analysis in microscopy, in: 2016 IEEE 13th International Symposium on Biomedical Imaging (ISBI), 2016, pp. 642–645.

[126] A. BenTaieb, G. Hamarneh, Topology aware fully convolutional networks for histology gland segmentation, in: International Conference on Medical Image Computing and Computer-Assisted Intervention, Springer, 2016, pp. 460–468.

CHAPTER 11

MaZda – A Framework for Biomedical Image Texture Analysis and Data Exploration

Piotr M. Szczypiński, Artur Klepaczko
Lodz University of Technology, Institute of Electronics, Lodz, Poland

Abstract

MaZda is a software package that provides a complete path for quantitative analysis of image texture and color, including image recognition, detection, and segmentation. The texture feature computation algorithms are generalized to three dimensions, to allow for analysis of 3D data from magnetic resonance imaging or computed tomography scanners. Moreover, the software can be used for computation of morphological features (shape properties) in segmented images. MaZda was utilized by numerous researchers in diverse applications. It has proven to be efficient and reliable for quantitative image analysis. It was successfully used in medical image analysis and food product quality assessment. This chapter presents the origins of MaZda, discusses the algorithms implemented in the software, shows selected applications, and recommends image analysis procedures. Finally, MaZda is compared to other software tools for image texture and color characterization and to image recognition. It is free and open-source software.

Keywords

Texture analysis software, Feature extraction, Feature selection, Biomedical texture classification, Image texture-based segmentation

11.1 INTRODUCTION

A texture perceived by humans is a visualization of complex patterns composed of spatially organized, repeated subpatterns, which have a characteristic, somewhat uniform appearance [1]. Examples of texture can be found in images of animal fur, tree canopies, or lawn. The human brain does not focus on individual elements that constitute textures, such as single hairs, leaves, or blades of grass. Instead, it considers the texture area as a single entity that displays specific characteristics. Thus the local subpatterns within an image demonstrate specific brightness, color, roughness, directionality, randomness, smoothness, granulation, and so on. Importantly, a texture may carry substantial information about the microstructure of physical objects. Consequently, texture analysis is an important tool for image processing and understanding.

Textures are easily perceived as somehow homogeneous image areas. Humans usually assess texture only qualitatively, although often quantitative texture analysis is required.

Biomedical Texture Analysis
DOI: 10.1016/B978-0-12-812133-7.00011-9

For instance, quantitative texture properties may be used to characterize properties of man-made materials or biological tissues. If an image is considered to be a mosaic of distinct textures, then texture analysis may be used for image segmentation (see Section 1.2 of Chapter 1). On the other hand, if an image is already segmented, then texture analysis may be used for classification of regions having specific properties. In both cases, texture quantification is required, a computation of numerical values corresponding to intrinsic texture properties. In brief, such computations utilize image transformations, image brightness, color distribution statistics, or image models. This step of quantitative image characterization is usually referred to as data extraction. MaZda [2] is a computer program for data extraction from images, which uses texture, color, and shape analysis.

The development of MaZda started in 1998 in the context of the European COST B11 (1998–2002) project. One of the project objectives was to develop methods for quantitative analysis of Magnetic Resonance Images (MRI) texture. At that time, there was no commercially available software capable of quantitative analysis of texture within freely selected Regions Of Interest (ROIs). MaZda was created to satisfy these objectives. The first algorithm for data extraction implemented in the software was gray-level cooccurrence matrix. The cooccurrence matrix translates into Polish as *Macierz Zdarzeń*. Consequently, the name of the software is an abbreviation of this term and has no intended connotation with the Japanese car manufacturer. Since its beginning, MaZda has been continuously improved and extended with new image analysis techniques, including image preprocessing, color-based feature computation, 3D image data extraction, image segmentation, machine learning, and data classification.

MaZda has been applied in many research areas including MRI protocol optimization [3], image-based medical diagnosis support [4–16], and food quality assessment [3, 17–19]. The studies have proven the existence of relationships between image texture properties and the presence of certain lesions. Thus they show the way for further development of computer vision-based methods to support medical diagnosis. The food quality assessment studies revealed that varieties of selected plants can be differentiated based on their texture properties, or the quality of some plants and meats can be assessed from such properties.

MaZda is written in C++. It was originally created in Borland C++ Builder environment exclusively for Windows systems. However, since 2012, it depends on the Qt cross-platform application framework and can be compiled for Linux, OS X, and Windows systems. The source codes of *qmazda*, as the currently maintained project is called, are available under GNU General Public License from GitLab repository.[1]

---

[1] The download page address is https://gitlab.com/qmazda/qmazda.

11.1.1 Related work

MaZda is not the only software capable of extraction and analysis of intensity-based texture and color attributes. There are other tools that can be applied for this task, some of them introducing different concepts for image characterization.

Recently, the MPEG-7 standard for "Multimedia Content Description Interface" gained interest of researchers working in the domains of image analysis and image recognition. The standard defines several feature groups (descriptors) of image color: Dominant Color Descriptor (DCD), Scalable Color Descriptor (SCD), Color Layout Descriptor (CLD), and Color Structure Descriptor (CSD). However, in computation of the DCD the color space has to be arbitrarily chosen beforehand, although the right choice of the color space is not obvious. Moreover, the SCD is computed by means of HSB space exclusively. The CLD is essentially a thumbnail of the whole image, and the CSD is an image histogram in Hue-Max-Min-Diff (HMMD) space. Despite the fact that algorithms of the standard were not particularly designed to characterize medical images, the features computed according to the standard were utilized in pathology detection, topological segmentation, or even motion analysis.

MaZda can be compared with the eXperimentation Model (XM) software, which is the simulation platform for MPEG-7 [20]. However, the two programs utilize different techniques for image features (or descriptors) computation. Moreover, the XM is focused on characterization and recognition of whole images, whereas MaZda can apply the texture computation within arbitrarily defined ROIs.

CellProfiler [21,22] is an open-source software for quantification of biological images. The program focuses on images of cells and on analysis automation of large amounts of images. The user can create computation pipelines starting from image segmentation, going through data extraction and ending with data categorization. Cell-Profiler can measure size, shape, intensity, and texture of cells or other similar objects. It extracts selected features Haralick's Gray Level Cooccurrence Matrices (GLCMs) and features derived from Gabor wavelets. The companion program CellProfiler Analyst implements tools for machine learning and data visualization. The texture extraction procedures of CellProfiler are not so versatile as these implemented in MaZda. Also, the program is not capable of 3D texture analysis or quantification of color features.

The feature selection and machine learning tools can be compared to other programs such as Weka [23]. However, tools implemented in Weka are more comprehensive and robust, whereas MaZda provides reliable solutions aimed at feature selection and design of simple classifiers, capable of quick data categorization and image segmentation. Since the data formats are exchangeable between the two programs, feature vectors computed in MaZda can be further analyzed by means of Weka.

It is worth noting that there are implementations of image analysis algorithms available as libraries or modules to be used in C/C++, Python, or Matlab programming.

In case of the latter the *Image Processing Toolbox* [24] contains functions for calculating properties of the gray-level cooccurrence matrix and features based on local image characteristics: entropy, intensity range, and standard deviation. Each feature can be determined within a selected ROI; however, selecting ROI of an arbitrary shape in an interactive manner is hampered due to the lack of a dedicated user interface. Moreover, analysis of 3D images is not directly supported. The *scikit-image feature* module for Python enables computation of Gabor wavelets and GLCM features for texture characterization [25]. The *OpenCV* [26] library enables computation of texture features derived from Haar wavelets or shape descriptive image moments. It implements a variety of image feature detection algorithms (among other algorithms, the scale-invariant feature transform and the speed up robust features) and tools for machine learning (*k*-nearest neighbors classifier, decision trees, support vector machines, multi-layer perceptrons, and others). The *InsightToolKit* (ITK) library implements a number of image filtration, segmentation, and feature extraction procedures for 2D and 3D images. It must be emphasized that all the three above-mentioned libraries are created as open-software projects. Thus they can be used not only in custom programs, but also the code can be easily examined and modified if required.

11.2 TEXTURE ANALYSIS WITH MAZDA

Image processing using texture analysis consists of a sequence of generic steps, such as visualization, identification of image regions designated to processing, feature engineering (including extraction, selection, or transformation), and pattern recognition. In the following sections, we present how these operations can be accomplished within the MaZda software, whereas the list of all implemented methods is summarized in the Appendix (see Table 11.3).

11.2.1 Overview of the image analysis workflows

MaZda (qmazda) consists of four modules: image editor, feature generator, report viewer, and maps viewer. The modules have different functions and goals. The first to be used is the image editor for image viewing and ROI editing. It provides tools for automatic image segmentation or free-hand region editing, both for two-dimensional images or three-dimensional data. This module also enables setting up options for the following feature extraction process.

The second module is the feature generator. It is usually invoked by the image viewer to compute the requested features. There are two modes for feature computation that the generator can handle. The first one is to compute feature vectors (descriptors) within user-defined ROIs. The second is to compute feature maps, images that represent feature values computed locally within small neighborhoods around every image pixel.

The feature vectors computed by the generator are displayed by the report viewer. This module presents data in a spreadsheet form, where every column represents a feature vector computed for an individual ROI, and every row presents values of a particular feature. Every column is labeled. The label should correspond with the class the particular ROI belongs to (e.g., "Healthy," "Bleeding," or "Ulcer"). With this additional information, the report module can be used as a supervised learning tool. The module implements data discrimination, feature selection, feature space reduction, and data classification algorithms. Users can also assess subspaces of selected features by viewing the vector distributions. Moreover, the module provides a full chain of procedures for designing and validation of classifiers.

The feature maps produced by the generator can be viewed in the maps viewer module. It displays a set of map images. Every map represents the distribution of a particular feature in the image (see Section 1.3.1 of Chapter 1). MaZda users can visually assess whether the maps enable image segmentation into regions of different texture characteristics. The map viewer can also use classifier specifications produced by the report viewer. Thus it is capable of automatic image segmentation by means of the classifiers.

There are several workflows or objectives of image analysis that are handled by MaZda (see Fig. 11.1). Starting with the input data, users have a choice between the analysis of 2D gray-scale, 2D color, or 3D gray-scale images. MaZda implements procedures for loading most popular image formats used in digital photography or by medical imaging techniques including Tagged Image File Format (TIFF), Neuroimaging Informatics Technology Initiative (NIFTI), or Digital Imaging and Communications in Medicine (DICOM). It loads color or gray-scale image files with pixel intensities encoded with 8 or 16 bits. Then the user is given a choice between data extraction from the whole image, analysis of freely defined ROIs or computation of feature distributions. Depending on the choice made, the results of the image texture analysis are feature maps or feature vectors. Feature maps may be useful for image segmentation, whereas the feature vectors can be used for classification of image content.

All the data produced by MaZda are saved in standard file formats that can be further processed by other programs. The feature vectors are saved in Comma-Separated Values (CSV) text format. Such files can be easily imported into most spreadsheet programs (e.g., Microsoft Excel or LibreOffice Calc) or data mining and statistical analysis programs (e.g., Weka [23], SPSS, or R). The maps are saved as floating-point number TIFF files and are compatible with several image processing programs, including ImageJ or Fiji [27].

The feature vectors computed by MaZda include up to several thousands elements per individual ROI. Such a large number of features creates a several-hundred-dimensional space, which is not easy to visualize, comprehend by humans, or to handle directly by statistical analysis. Thus MaZda implements techniques for reduction of

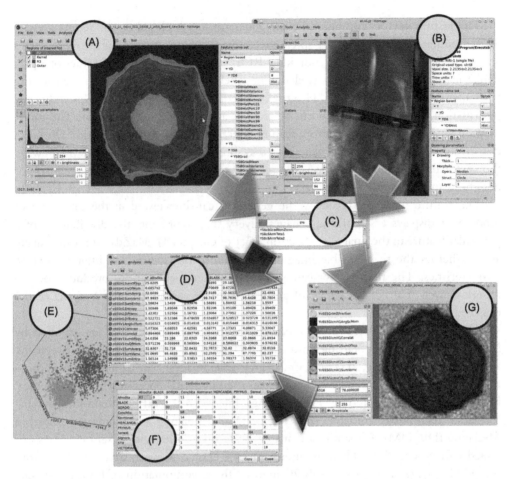

Figure 11.1 Graphical user interface and analysis workflow of MaZda: A), B) region of interest editor for 2D and 3D images, C) feature computation module, D) feature selection and machine learning module with E) vector distributions viewer and F) validation tools, and G) feature map viewer and image segmentation tool.

feature vector dimensionality by selecting the most discriminative features for further analysis. There are several methods for feature selection using various criteria, which can be chosen. Then the data (feature vectors) can be statistically analyzed to identify and visualize relations between features and image region categories. Also, there are several machine learning algorithms implemented to find classification rules. Finally, feature maps can be employed for image segmentation or for visual assessment of texture variability.

11.2.2 Regions of interest

The ROIs are created by the user to delimit the area in which the features are computed. The necessity for ROIs can be simply explained. In most cases, images present different entities or areas of varying characteristics (see Section 1.2.3.1 of Chapter 1). Features computed for the whole image area (or volume) would not correctly reflect characteristics of particular objects or meaningful areas. In such a case, feature values would expose averaged characteristics of both meaningful and irrelevant image regions. If the goal of feature computation is to characterize texture properties of some particular organ within a human body, then the features should not be computed in whole images but exclusively for image regions related to this organ.

Originally, MaZda was designed to analyze monochromatic images from medical devices, such as MRI, Computed Tomography (CT), or Ultrasonography Scanners (US). Therefore images loaded into the program were displayed in gray scale. Currently, even though the color image is loaded, it is always displayed in gray scale. However, to utilize and enable assessment of color information present in original data, the user is given a choice to monochromatically display various color channels.

For image rendering, MaZda uses color exclusively to indicate ROIs (Fig. 11.2). As the user creates regions, they appear as semitransparent color overlays on top of the gray-scale image. This way, both the image data and regions are visible simultaneously. In a single image, users can define as many regions as needed. Moreover, the regions may overlap if required. The regions are indicated with different colors, and therefore they can be easily distinguished. Every region can be assigned a name, either unique for a particular region or, more importantly, unique for the class the region belongs. Such region names are then used by the feature generator to label data vectors for supervised learning.

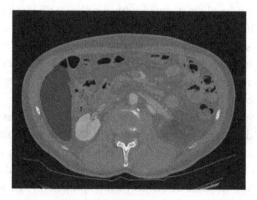

Figure 11.2 Cross-sectional view of the human abdomen CT image. Areas of selected organs are colored to indicate individual ROIs to be separately characterized and analyzed.

More formally, ROIs are subsets of pixels in 2D or voxels in 3D images. Defining a specific region of interest concentrates the computation effort on an image fragment that is relevant to the goal of analysis and thus enables one to avoid unnecessary processing. The regions are of great importance in biomedical image processing. For example, tomography images of the human body contain various kinds of organs or tissues. To analyze image properties in a selected organ and not in the surrounding tissue, the image fragment corresponding to the organ must be defined.

The ROIs in MaZda can be of arbitrary shape. They can be defined by free-hand drawing or by placing predefined shapes or blocks. To support editing, tools are available for image gray-level thresholding, flood-filling, identification of disjoint areas, active contour, and deformable surface matching [28,29]. Additionally, tools based on morphological transformations such as erosion, dilation, closing, or opening can also be used.

By default, MaZda uses the multipage TIFF format to store ROIs along with information on their colors and assigned names. The regions can also be saved individually as binary images or, in case of nonoverlapping regions, as a single image showing individual regions with unique colors. The files can be easily loaded into other image processing programs for editing. Moreover, the ROIs generated with other programs, stored as images with black background and uniformly colored areas, can be imported and used in MaZda.

11.2.3 Feature extraction

There are three categories of image descriptors calculated in MaZda: texture, color, and morphological features. Their detailed description is provided further.

11.2.3.1 Texture features

As already mentioned, texture can be understood as spatially organized, repeated patterns. Such patterns must be characterized within sufficiently large regions. In particular, it is not possible to characterize texture from a single pixel (or voxel). The region to characterize a texture should be large enough to contain several repeated patterns or a texture fragment that is representative for a whole texture. The texture feature computation algorithms examine brightness distributions within the region by searching for spatial relations between pixels having particular gray levels.

There are three categories of feature computation approaches utilized by MaZda: statistical, model-based, and image transform [29]. Statistical approaches represent the texture indirectly by the nondeterministic properties that govern the distributions and relationships between the gray levels of an image. Model-based texture analysis [1], using fractal or stochastic models, attempts to interpret an image texture by generative image models and stochastic models, respectively. Transform methods of texture analysis, such as Fourier, Gabor, or wavelet transforms [30–32], represent an image in a space with a coordinate system related to the characteristics of a texture. MaZda generates features based on image histogram, gradient, cooccurrence matrix, run-length matrix, autoregressive model [33], and Gabor and Haar wavelets [34].

The most common statistical method for image feature computation is based on the first-order histogram of an image. The histogram is computed from the intensity of pixels without taking into consideration any spatial relations between the pixels within the image. Features are simply statistical parameters of the histogram distribution such as mean brightness, variance, skewness, kurtosis, and percentiles. Another statistical method derives features from the gradient magnitude map of the image. MaZda computes features from the gradient image similarly to the way it computes histogram-based features from the image intensity distribution.

It should be noted that the statistical properties of the brightness first-order histogram do not carry any information about spatial relations of pixel (or voxel) intensities in the image. These features are referred to as intensity features and often regarded as nontextural attributes. In contrast, the gradient-based algorithm and extraction methods described in the following paragraphs do take into account the spatial relationships. Both the feature groups complement each other and allow for complete characterization of a region under study. In Mazda, both types of attributes are processed in the same manner. The discussion on the differences and properties of both types of features, characterizing the intensity exclusively and characterizing the spatial relations, can be found in Section 1.2.1 of Chapter 1.

The Gray-Level Cooccurrence Matrix (GLCM) is a second-order histogram computed from intensities of pairs of pixels, where the spatial relationship of the two pixels in a pair is defined. In MaZda, there are four directions between the pixels taken into account: horizontal, vertical, 45°, and 135°. For every direction, there are several interpixel distances that can be used. MaZda enables selection of distances from 1 to 8. The cooccurrence matrix can be computed for every combination of direction and distance. Therefore, using four directions and eight distances, 32 different matrices can be computed by MaZda. The GLCM-based features are derived from the matrix to demonstrate statistics, such as angular second moment, contrast, correlation, sum of squares, and variously defined averages, variances, inverse moments, and entropies [35]. There are 12 such features that can be computed for every individual matrix. Therefore, since there are 32 matrices and 12 features, the number of GLCM-based features computed by MaZda for a single region amounts to 384.

The Gray-level Run-Length Matrix (GRLM) holds counts of pixel runs with the specified gray-scale level and length. In MaZda, there are computed four various run-length matrices, for four directions of pixel runs: horizontal, vertical, at 45°, and at 135°. There are five run-length matrix-based features computed for each of the matrices: short-run emphasis inverse moment, long-run emphasis moment, gray-level nonuniformity, run-length nonuniformity, and fraction of image in runs [35]. MaZda can compute seven GRLM features using four different directions, which results in 28 GRLM features computed for a single region.

There are also model-based textural features computed by the software, which are based on a first-order autoregressive model of the image [33]. The model assumes that the pixel intensity, in reference to the mean value of image intensity, may be predicted as a weighted sum of four neighboring pixel intensities. These neighboring pixels are left, top, top-left, and top-right adjacent. Therefore the model has four parameters, which are weights associated with these pixels, plus a fifth parameter, the variance of the minimized prediction error [3,33].

One of the two transform-based methods implemented in MaZda makes use of the discrete Haar wavelet [36,37]. The wavelet images are scaled up to five times, both in the horizontal and vertical directions. It results in image transformation into twenty frequency channels. Energies computed within the channels provide data on texture frequency components and are used as texture-characterizing features. There are 15 Haar wavelet-based features, which can be computed this way. The second transform-based method implemented in MaZda computes Gabor wavelets. Users can select several options for Gabor wavelet computation including frequency, orientation, and standard deviation of the Gaussian envelope. Average wavelet magnitudes computed within the ROI are used as texture descriptors. There are eight orientations (changing with an increment of 22.5°), 48 frequencies, and several settings of envelope size. By default, MaZda calculates 30 Gabor-wavelet-based features for every individual ROI.

11.2.3.2 Color features

It is evident that, in images, color often carries essential information required for image differentiation or recognition. However, all the algorithms for textural features computation implemented in MaZda require a monochromatic (gray-scale) image as an input. To enable color analysis by means of gray-scale image-based algorithms, MaZda uses a specific image-processing procedure to convert color image into a set of monochromatic images.

It is known that every color image perceived by humans is represented by three color components or image channels. In the additive color model, color image is composed by Red, Green, and Blue (RGB) components. Every channel can be viewed as a separate image, each of them carrying somewhat different information. In MaZda, each of these color channels are inputs for the texture feature computation algorithms.

It is known that information carried by red, green, and blue channels may be highly correlated. Thus the feature values extracted from these channels may be also correlated. This is an unwanted situation in the discriminant analysis, the following analysis step applied by the MaZda package. To minimize such correlations, features are computed for color channels that separate brightness from color information (see Fig. 11.3). The example is YIQ model, which separates image luminance (brightness information) in

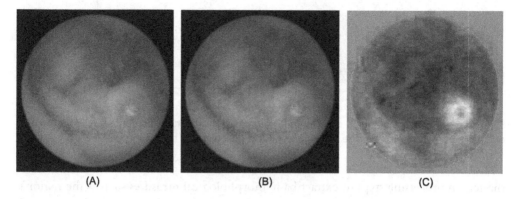

Figure 11.3 Example capsule endoscopy image of small ulcer [38]: A) original color image, B) the image gray-scale channel, and C) the image Q channel of YIQ color model. The Q component of color information shows the highest contrast between the lesion and the normal area, to at best differentiate or detect the abnormality.

Y channel from image chrominance (exclusive color information) carried by I and Q channels. The other color models MaZda uses to split luminance from chrominance are YUV (where U and V define chrominance) and HSB (where chrominance is given by the hue and saturation, and luminance by the brightness component).

Therefore MaZda can be used to analyze not only the image brightness. It computes textural descriptors based on color components of RGB, YIQ, as well as YUV, HSB, CIE XYZ, or CIE L⋆a⋆b⋆ color models. The textural features computed for various components can be combined to obtain a comprehensive characterization of a colored texture. Therefore feature vectors computed for color image regions may include more than several thousands of elements.

11.2.3.3 Morphological features

Regions of interest, used as masks for texture analysis, themselves can also be viewed as binary images. They represent silhouettes of objects and may carry key information for classification of such objects. Therefore another approach for feature computation is to measure characteristics of these regions, such as location, orientation, size, geometry, and topology. Numerical features characterizing region shapes are usually called morphological parameters.

MaZda computes parameters, such as areas, perimeters, various diameters, and radix, including Feret's and Martin's diameters [39], and parameters of the inscribed circle, circumscribed circle, ellipsis, and rectangle. Moreover, it computes various ratios of these parameters, like elongation, roundness, and compactness. Most of such ratios are size and rotation invariant. Other morphological feature computation algorithms implemented in the software are first- and second-order central moments [39]. Figure 11.4 illustrates

Figure 11.4 Illustration of region processing for morphological features computation. From the left: original region, the region's profile and convex profile, skeleton, and boundary pixels.

the region processing steps to extract basic morphological measures such as the region's area, area of the region's profile and the convex profile, skeleton-based features, and perimeter.

Another group of morphological parameters is based on transformations into a profile (holes are removed from a region), into a convex region (concavities are filled in), and into a skeleton. Skeletonization is implemented through a thinning algorithm that removes outer pixels of a region to find its medial axis. The result, a skeleton, preserves information on the region's topology. The skeletal descriptors computed by MaZda are length of the skeleton, number of branches, branching points and loops, and minimal and maximal thicknesses of the body region surrounding the skeleton.

11.2.4 Image preprocessing

In some situations the visual appearance of the same texture instance may vary in different images. In photography, it may be caused by different exposure setups or scene illumination, resulting in darker or brighter pictures. In magnetic resonance the images vary between different scanner models or imaging sequences. Nevertheless, the texture features should characterize texture patterns exclusively and should be unaffected by image acquisition-related alterations. Unfortunately, in practice, all the feature values somehow depend on image average brightness or overall contrast resulting from factors often unrelated to texture attributes.

MaZda copes with such situations by introducing an image preprocessing step prior to feature computation. To ensure that the features characterize an image texture exclusively and do not depend on any global image characteristics, such as the overall brightness or contrast, histogram normalization and standardization procedures have been implemented. These operations reduce dependency of higher-order parameters from first-order gray-level distribution [40].

There are two image histogram normalization approaches available. One of them remaps an image brightness range from minus to plus three standard deviations around mean brightness onto the black-to-white gray-scale range. The other approach remaps an image histogram in the range between the first and the ninety-ninth percentile onto the black-to-white range [2].

The GLCM and GRLM derived feature values strongly depend on yet another image property, which is not related to texture attributes. This property is the number of image gray levels related to the number of bits used for coding of pixel brightness. In the GLCM algorithm, the size of the matrix corresponds to the number of gray levels. In the GRLM algorithm, one of the two matrix dimensions also matches the number of gray levels, and what is more, the lengths of pixel chains increase as the number of gray levels is reduced. It was presented in [41] that, for natural images, GLCM and GRLM features discriminate texture at best if they are computed at 5 or 6 bits per pixel. Moreover, it is worth noting that in many algorithms also the computation time grows with the increase of image gray levels, and thus increasing this number is not recommended.

To enable selection of the most appropriate number of bits per pixel parameter, the histogram normalization step is followed by quantization of gray levels. The image histogram normalization step and gray-level quantization are performed prior to the computation of textural features and thus have substantial impact on their values.

11.2.5 Feature naming convention

Feature values computed in MaZda result from several steps of image processing. As already explained, the image processing involves selection of color channel, histogram normalization, and gray-level quantization. Moreover, most of the implemented algorithms produce results that depend on some parameters. These parameters may involve selection of specific frequencies, distances, or direction angles. Successively, this creates a problem with feature name uniqueness. Using feature names such as *mean, variance,* or *entropy* is insufficient to identify particular characteristics since the feature values identified by the same name could be computed after application of different image preprocessing procedures and with different algorithm parameters.

In particular, let us imagine a situation where there are several sets of feature vectors. In every set, features are computed with varying parameter setups or image preprocessing procedures. If the data sets are merged and used in discriminant analysis, its results would be unreliable. Feature values computed in different ways would vary not because of disparate texture characteristics but because of differences in the computation procedure. Successively, this would lead to false conclusions on high discriminative ability of such features.

To prevent this situation, in MaZda version 14 and higher, feature names uniquely identify the complete image processing procedure, feature extraction algorithm, and the algorithm parameters. The feature name is a combination of symbols, numbers, and names that uniquely identify the way in which the feature value is generated. The most left symbol identifies the first image processing procedure applied to the image.

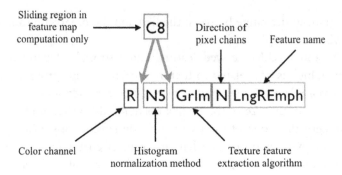

Figure 11.5 Explanation of MaZda feature naming convention.

Consecutive symbols, from the left to the right, code the following computation steps. The most right part of MaZda's feature name is an abbreviation of the particular feature produced by the extraction algorithm.

For example, the name *RN5GrlmNLngREmph* (see Fig. 11.5) means that the red (*R*) color channel intensity was normalized (*N*) and then quantized to use 5 bits per pixel. The *Grlm* identifies the gray-level run-length matrix algorithm. The letter *N* codes direction of pixel chains at a 135° angle. Finally, *LngREmph* is an abbreviation identifying long-run emphasis measure.

As in the example, the first letter of the feature name identifies the color channel: *Y* stands for brightness; *R*, *G*, and *B* identify respectively red, green, and blue channels of the RGB model; and *I*, *Q*, *U*, and *V* identify chroma channels of the YIQ and YUV models. There are also *x*, *y*, and *z* (lower case) to identify channels of the CIE XYZ color model, and *L*, *a*, *b* to identify channels of CIE L⋆a⋆b⋆ model. The image histogram normalization is coded by the letters *N* (first and ninety-ninth percentile), *S* (three standard deviations around mean value), or *D* (original unmodified image gray-scale range). If the algorithm requires direction to be specified, then it is usually coded by one of four letters: *V* (vertical), *H* (horizontal), *Z* (45°), or *N* (135°). If the number follows the direction code, then it defines the distance (in the GLCM algorithm) or directional frequency (Gabor transform). Table 11.2 summarizes in the Appendix all symbols used by MaZda in the proposed feature naming convention.

11.2.6 Feature maps and image segmentation

As already mentioned, MaZda computes feature distributions within image, feature maps (see Fig. 11.6). Each point of a map represents a particular textural feature value that corresponds to a given point of a textured image (see Section 1.3.1 of Chapter 1). The map is represented by a gray-scale image, in which a textural feature value is depicted by a relative gray level. The feature at a given image location is computed within a square or a semicircular region (mask) centered at this location. The mask slides over

(A)   (B)   (C)

Figure 11.6 Illustration of image texture segmentation. A) Artificially rendered image with two areas covered with similarly looking but different textures. It is difficult to segment the image using gray-scale-based image segmentation algorithms. B) The feature map to display differences between the two textures. C) Feature map value thresholding result.

the image surface by given vertical and horizontal steps in order to fill the whole output image with computed feature values. The user can select features for which maps will be computed, size of the mask, and the horizontal and vertical steps.

The names of feature maps slightly differ from the feature names computed within user defined regions. They code additional information on the shape and size of a sliding mask. The semicircular (represented with the raster precision) masks are coded with the letter C, and square regions are coded with S. The number following the code defines the region's radius. Thus $S15$ means that the sliding mask is a 31×31 pixel square. The example name may be $RC12N5GrlmNLngREmph$ or $RN5C12GrlmNLngREmph$. It must be noted that the two names are not identical. In the first case the code of the mask precedes the code of the image gray-scale normalization. This means the normalization is performed within the mask region after the mask is created. In the second case the $N5$ code comes first, which means that the normalization is performed within the whole image before the sliding mask is created.

Segmentation is an image-processing task that partitions the image into separate regions, which are in some way homogeneous. The most common segmentation routine is performed through image gray-level thresholding. In this method, image pixels of intensities higher than the threshold level fall into one region, and the others fall into the other. Unfortunately, if the goal is to segment a texture, then the image gray-level thresholding alone usually fails. Nevertheless, the thresholding can still be effective if preceded by a feature map computation. If the feature is able to discriminate between two different textures and the feature map is computed on an image containing such textures, then the result is an image showing one texture as a dark area and the other one as a bright area. Therefore thresholding the brightness of the feature map separates the two texture areas from each other. A visual inspection of maps produced by MaZda allows determination of feature maps, which can be used for texture segmentation.

Moreover, to study the feasibility of image texture segmentation based on multiple feature maps, a machine learning method can be used. In example images, users can indicate areas to be differentiated and then compute feature maps. The feature maps in the indicated areas are used as an input for machine learning procedures. Subsequently, the resulting classification rules can be applied for the segmentation of new images. The machine learning methods available in MaZda are addressed in the following section.

11.2.7 Machine learning

The last step in the image processing chain is the analysis of the collected data so that a diagnostically relevant information can inferred. It can be approached using two alternative scenarios. The first strategy, which is perhaps more popular within medical doctors community, involves statistical modeling. It exploits probability distribution formalism to build relationships between random variables (such image descriptors) and a response (such as a texture class). It further operates on statistical tests to verify a hypothesis, for example, of a significant correlation between a given texture feature and a certain lesion. In contrast, machine learning consists of a set of algorithms, very often with a heuristic flavor, capable of constructing rules for class label prediction. Although many of such methods disregard the underlying models of data sample distribution functions, they very often occur to be the only feasible technique to effectively find nonlinear, complex relations and class boundaries in multidimensional data space. MaZda provides a set of tools that fall into the machine learning category. However, owing to the data export option available in the report window, it is also possible to analyze the computed data vectors in any external program for classical statistics.

The number of features computed by MaZda is enormous, reaching a few thousands per region of a color image. Consequently, this number of features is difficult to handle. Several-hundred-dimensional feature vectors require analysis in a several-hundred-dimensional space. Such a space is difficult to conceive for humans. More importantly, it would be time consuming, inefficient, and in some cases not feasible to directly examine such spaces. Moreover, with many machine learning algorithms, the higher number of features (dimensions) increase the risk of classifier overfitting. If over-fitted, the classifier is not able to describe underlying relationships presumably present in data; it has poor predictive performance and is not able to correctly categorize new data, not used in learning.

Usually, only a limited number of features carry relevant information needed for texture discrimination. Finding these features and rejection of the others usually improve machine learning outcomes. Therefore there are several methods for feature selection implemented in MaZda. Given the a priori knowledge of which feature vector or sample belongs to which predefined class, these methods select a subset of features according to a given mathematical criterion.

11.2.7.1 Feature selection

There are two general approaches to attribute (feature) selection in data mining. In the filter strategy, attribute saliency is estimated individually, and the main advantage is low computational complexity. However, the discriminative power of a feature may be revealed only when it is accompanied by other attributes. Thus a subset of individually optimized features may prove to be inferior to the actual best solution. In the alternative wrapper approach the discriminative power of attribute subsets is evaluated. The efficiency of the process depends on the strategy of constructing subsequent feature subsets and the computational complexity of the evaluation routine.

The machine learning algorithms are implemented in the report module of MaZda. The feature vectors computed by the generator are presented in the report window in the form of a spreadsheet. Each column of the table, except the first one, represents a ROI for which the features were computed. The rows correspond to the particular features. Each column is assigned a name, or a label, that identifies the class that the feature vector belongs to. The user must assure that the columns are correctly labeled before running any supervised learning algorithms, including feature selection.

The feature selection algorithms used in MaZda combine the above mentioned two strategies. The selection begins with an estimation of discriminative power of individual features. Subsequently, all combinations of attribute pairs are examined. The attributes are then ordered according to their discrimination ability. For higher-dimensional feature subspaces, examination of all possible combinations is extremely time consuming. To overcome this obstacle, the search is limited in time, and the attributes that had been highly ranked so far are examined first.

There are three criteria MaZda uses in the feature filtration approach: Fisher discriminant, classification error combined with the correlation coefficient, and mutual information [40,42]. The features are ranked based on the chosen criterion. For the following analysis steps, several of the most discriminative features are chosen. The discriminative power of a feature subset is related to the performance of a given classification algorithm. Hence a set of features selected for a particular classifier may be unsuitable for another one. Therefore in the wrapper strategy, attributes are selected to maximize classification ability of a particular classifier. MaZda implements several classifiers employed in the feature selection procedure. These classifiers are described further in this section.

If the number of selected features is still unacceptably large, then it is possible to perform its further reduction by transformation of the original feature space into a new space of lower dimensionality. This yields a set of new features, reduced in number in comparison to the original feature set. This approach is called feature projection [43]. Procedures implemented in MaZda comprise Principal Component Analysis (PCA) [44] and Linear Discriminant Analysis (LDA) [45], which are aimed at feature space projection. It is not recommended to apply these methods directly to the original several-hundred-dimensional feature spaces since they are prone to overfitting. As a rule of thumb, the number of feature vectors (examples) used in this procedure should substantially exceed the number of dimensions (features) to be reduced.

11.2.7.2 Classification

The next step in machine learning is classifier training. The objective of this stage is to gain decision rules to categorize data vectors. The rules may be imagined as boundaries in the feature space. If such a boundary is a line, a plane, or a hyperplane, then the classifier is called linear. In this case, feature vectors appearing on one side of the boundary fall into the one class, and the vectors located on the other side fall into the other class. If the shape of the boundary is more complex, then the classifier is called nonlinear.

There are several classifiers available for training in MaZda. The simplest one is a linear classifier. It can be used if there are two classes of vectors to be separated, assuming that the vector distributions within the classes are normal. The training of the classifier employs linear discriminant analysis to find the orientation of the decision hyperplane and minimization of classification error to establish its actual location.

If there are more than two classes, then ensembles of linear classifiers can be used [46]. The classes are analyzed in pairs, and the goal of the analysis is to differentiate the two classes. This is a reasonable approach if the vectors of two different classes form two clusters separated by a hyperplane. The ensemble consists of as many linear classifiers as existing pairs of classes. During classification, feature vectors are presented to every linear classifier of the ensemble. A vector is identified as a member of a specific class if all the classifiers corresponding to the class confirm its membership.

MaZda implements a nonlinear Hyperellipsoidal Decision Boundary (HDB) classifier [46]. It finds the decision boundary between a single class of vectors and all the other classes. It is assumed that the vectors of the class are separated form a cluster surrounded by vectors belonging to other classes. The working principle of the algorithm is as follows: a PCA is performed on vectors of a class to be separated from other classes. The PCA transforms the feature space into a space of principal components, which are uncorrelated and ordered according to their standard deviations. The decision boundary is a hyperellipsoid with center located in the center of the cluster and radii aligned with the principal components and proportional to their standard deviations. The size of the hyperellipsoid is adjusted to maximize the balanced accuracy of classification and to expand the margin between the vectors of separated classes.

The other nonlinear classifier implemented by MaZda is the Vector-Supported Convex Hull (VSCH) [47]. The assumption on feature vector distribution is similar as in HDB. Feature vectors of a particular class form a compact cluster surrounded by scattered vectors belonging to other classes. The decision boundary is a convex hull built on the vectors forming the cluster (Fig. 11.7).

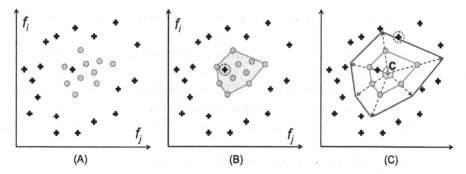

Figure 11.7 Illustration of convex hull decision boundary in 2D space [38]. A) Specific distribution of vectors, B) Estimation of the first criterion, and C) Estimation of the second criterion.

Both the algorithms, HDB and VSCH, can be used in the wrapper approach feature selection procedure. In HDB the goal is to select features for which the balanced accuracy of classification computed on the training set is the highest. If there are more feature subsets exhibiting the same balanced accuracy, then the second criterion is the size of the margin between the decision boundary and the nearest correctly categorized vector of the surrounding classes. In VSCH the count of vectors incorrectly included within the convex hull is used as the selection criterion. The feature subset is chosen for which this number is the smallest. If there are more such subsets, then the second criterion is used, which is the size of the margin between the convex hull and the closest vector located on its outside.

The sets of vectors of selected or projected features can be saved in the comma-separated vector format. This is a common data exchange format used in most data mining software. The classification rules are saved in text format. They can be used in the map viewer module for image segmentation.

11.2.7.3 Processing beyond MaZda – data mining using Weka

In its legacy versions, MaZda was accompanied by a dedicated program – called B11 – for performing basic tasks on the extracted texture feature vectors. The main advantage of B11 was the ability to launch it directly from MaZda without the need to parse the native MaZda report files where the feature vectors were stored. The program B11, however, had a series of limitations that hindered a deepened analysis of the image data. It concerned mainly the number of allowed features and data vectors and the catalogue of available machine learning algorithms. On the other hand, the development of MaZda and its growing number of image analysis options called for more advanced and diverse data exploration techniques. Therefore with the advent of Weka [23], a powerful and open-source data mining application suite, we decided to make benefit of this implemented toolkit instead of further supporting our own data analysis Application Programming Interface (API).

The advantages of this approach are manifold. Weka offers a bunch of feature selection algorithms. It is especially crucial in the case of texture analysis, where the abundance of extracted parameters often entails searching for their most discriminative and representative subsets. The selection schemes implemented in Weka embrace:

• feature ranking according to various criteria, such as information gain and chi-squared statistics;
• correlation, redundancy, entropy, or mutual information;
• sequential search strategies, in forward and backward variants;
• evolutionary and particle swarm optimization-based techniques.

The wrapper-based approach implemented in Weka allows evaluation of feature subsets using one of the rich collection of classifiers. The available options include such schemes as k-nearest neighbor, support vector machines, neural networks, Bayesian-based algorithms, decision trees, inference rules, logistic regression, classifier ensembles, and many others. In addition to attribute selection, the user can also transform features using principal component analysis or random projection.

A few modifications were required to enable MaZda and Weka cooperation. Most importantly, it concerned feature naming conventions and file formats of the data set exported from MaZda. To solve the latter problem, the CSV format was selected since it is one of the formats accepted by Weka. Moreover, it is a common file format, also adopted in various scientific software, such as Python, Matlab, R, or SPSS. Regarding feature names, the required modification consisted in the introduction of a unified naming scheme where only alphanumeric characters were allowed. These adjustments made it feasible to link MaZda with any of the Weka operating modules, including *Explorer*, *Knowledge Flow*, and *Experimenter*. Section 11.3.1 presents an example of using the *Explorer* module in the image texture processing chain.

11.3 APPLICATIONS

Since patterns or texture areas appear in almost all biomedical images, they can be investigated with textural analysis tools such as MaZda. The software was already utilized in many domains, including MRI measurement protocol optimization [3], various medical studies [4–16], food quality studies [17–19,46,48], and others. MaZda turned out to be useful in analysis of food images. It was applied to identify and discriminate biomedical image areas with different textural characteristics. It proved feasibility of discrimination between potato varieties, cooked and raw potatoes [19], analysis of the influence of apple ripening process on storage [18], assessment of soft cheese quality [17]. and barley grains quality assessment [48]. More examples related to MaZda applications can be found in [2,3].

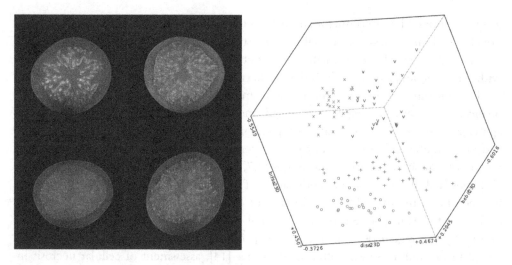

Figure 11.8 Example magnetic resonance images of four varieties of potatoes. The varieties differ by their internal starch structure resulting in differences in texture. Vector distribution, resulting from feature selection and dimensionality reduction, shows four separable clusters of vectors, each representing potato images of respective variety.

Figure 11.8 presents results of potato image analysis. Potato samples belonging to four potato varieties were examined with an MRI scanner [19]. The obtained images were analyzed with the MaZda package. The procedure of textural feature computation and feature selection led to the conclusion that the examined potato varieties have dissimilar structures of storage parenchyma, which results in significant interclass differences of a certain set of textural features. The sample distributions in Fig. 11.8 show that samples belonging to different potato varieties tend to group in specific areas of the LDA projected feature space. The study shows that features extracted from MR-images of potatoes enable classification of sensory properties in five potato varieties, thereby enabling categorization for desired quality attributes. It also demonstrates that the image acquisition techniques, originally meant for applications in medical imaging and diagnosis, can be successfully used in other domains.

MaZda was also applied for texture analysis of pork and poultry cold meats [49]. Color and texture features were extracted from images acquired from a flatbed scanner. The dry matter, protein, fat, ash, and collagen content of the analyzed products was determined by the reference methods. The results were statistically compared by correlation, discriminant, and canonical analysis. The results revealed correlations between the chemical composition and image texture of cold meats. Moreover, the developed statistical model discriminated cold meats with high accuracy.

Another study [46] evaluated effectiveness of identification of barley varieties based on image-derived shape, color, and texture attributes of individual kernels. Varieties were determined by discriminant analysis, including reduction of feature space dimensionality and linear classifier ensembles. The study demonstrated that color, shape, and texture features enable determination of varieties and that classification results improve with simultaneous analysis of both sides of kernels, dorsal and ventral.

The software was also used in studies of medical images from magnetic image resonance, computer tomography, ultrasound, and confocal microscopy. These studies, by applying methods of image analysis implemented in MaZda, prove relations between texture attributes and existence of particular lesions. Example studies involved early detection of osteoporotic changes in bones [4,7,13], detection of amygdale activation in rat brains [16], detection and quantification of hippocampal sclerosis [5], distinguishing between brain tumors [3], discrimination of healthy and cirrhotic livers [9], textural analysis of trabecular bone images targeted at osteoporosis detection [11], monitoring of atrophy and regeneration of muscles [12], monitoring of teeth implants [10], analysis of myocardium tissue in ultrasound images [14], assessment of cellular necrosis in epithelial cells [15], evaluation of antivascular therapy of mammary carcinomas in mice [6], and prediction of mezoscale vascular tree growth parameters, such as blood viscosity and outflow pressure and number of outlet branches, both in synthetic and real vessel systems [50,51].

In [52] a study aimed to determine whether texture or morphological features extracted from MRI maps are sensitive to the presence of abnormalities at the posterior aspect of lumbar intervertebral discs. T2-weighted fast spin-echo sequences and multiecho spin-echo sequences were used. ROIs indicating discs were manually defined and graded into normal, bulging, and herniation classes. Texture features (the GLCM's entropy and difference variance) and shape attributes (the Feret diameter and the second-order moment of inertia) were identified to distinguish between normal discs and discs with bulging or herniation. Figure 11.9 shows an example image of the lumbar spine with the three ROIs indicating lumbar intervertebral discs.

In [38] a study presented several approaches to exploratory analysis of wireless capsule endoscopy images. It demonstrated correspondence of texture- and color-based descriptors to various anomalies of the gastrointestinal tract. Image texture analysis enabled accurate detection of pathologies in endoscopic video. Moreover, feature maps segmentation was presented to partition images into abnormal- and healthy-looking regions (see Fig. 11.10). For both detection and segmentation tasks, the same procedures of feature extraction, selection, machine learning, and classification were applied.

The goal of the study was to develop a method for detection of bleeding and focal ulcer and excessive ulcer images. To quantify bleeding and excessive ulcerations, the image segmentation procedure based on color and texture attributes was applied. The experiment involved image classification of the three abnormality classes versus reference class of images displaying normal appearance of small intestine. The best classification measures were obtained for bleeding, with sensitivity of 0.98 and specificity

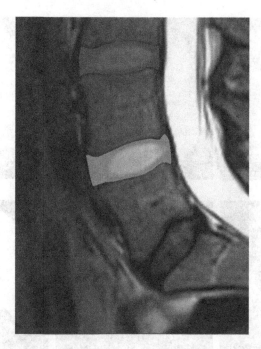

Figure 11.9 Sagittal T2 fast spin echo (FSE) image of the lumbar spine with ROIs indicating lumbar intervertebral discs – the blue one exhibits a posterior herniation (courtesy of Dr. Marius E. Mayerhoefer). (For interpretation of the references to color in this figure legend, the reader is referred to the web version of this chapter.)

of 1.00. The classification of excessive ulcerations gained sensitivity of 0.88 and specificity of 0.94. The Jaccard index computed for segmented ulceration images reached 0.62, with recall 0.92 and precision of 0.67. The results of automatic segmentation were compared with regions depicted by an expert.

11.3.1 Lesion detection with MaZda and Weka

The case study presented in this section [38] describes the above-mentioned analysis of the Wireless Capsule Endoscopy (WCE) images in greater detail. The WCE video sequence consists of thousands of frames, among which only a fraction is of clinical importance. Previewing the whole sequence, even in the increased frame rate, is a very time-consuming and fault-prone task. To aid in the WCE-based diagnosis, it is proposed to extract image texture descriptors that can be used to train a classifier to automatically recognize pathological structures characteristic to the gastrointestinal tract. Examples of such structures—bleeding, focal and excessive ulcers—are shown in Fig. 11.10.

Before constructing a classification rule, a set of training frames was collected. It contained 35 images of each pathology and 500 control frames acquired for different

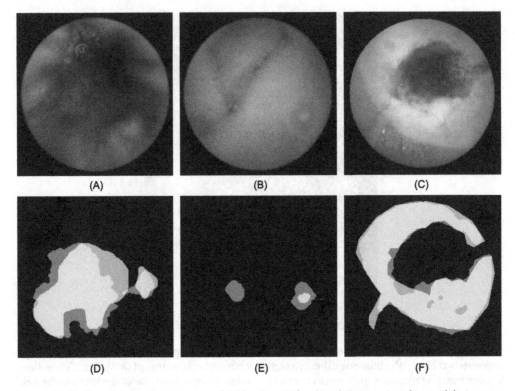

Figure 11.10 Capsule endoscopy images of A) bleeding, B) focal, and C) excessive ulcer and their corresponding segmentation results D), E), and F). Green and yellow areas denote regions found by manual delineation; red and yellow indicate areas automatically extracted by MaZda. (For interpretation of the references to color in this figure legend, the reader is referred to the web version of this chapter.)

patients. The pathological structures were annotated by two gastroenterologists, experts in capsule endoscopy diagnostics. Similarly, the selected nonpathology frames were reviewed by these experts to ensure that the reference class is indeed free from lesions. Every frame was then divided into 48 equally sized, overlapping circular ROIs covering the whole field of view (see Fig. 11.11). Those ROIs that collocated with the malformations were assigned a class label corresponding to a given lesion. Otherwise, a ROI was assigned to the reference class. For each region, MaZda computed a set of 2495 texture attributes of various texture models and color channels. Eventually, data sets composed of a specific pathology and reference classes were imported into the Weka Explorer module.

Figure 11.12 shows the *Preprocess* tab of the *Explorer* window after loading a data set created for the bleeding class. As it can be seen on the bottom-right panel, the proportion of reference (denoted by red bar) to pathological (blue bar) data vectors is highly unbalanced, 1032 to 101 instances for each class, respectively. It can bias the

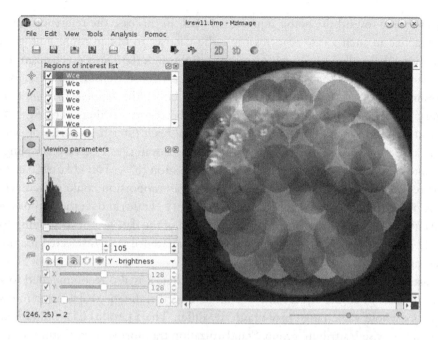

Figure 11.11 Circular ROIs covering the field of view of an image from wireless capsule endoscope.

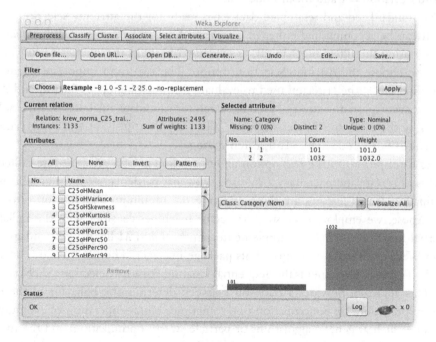

Figure 11.12 The *Preprocess* tab of the Weka *Explorer* module. (For interpretation of the references to color in this figure, the reader is referred to the web version of this chapter.)

classifier model toward the better represented class, in this case the reference one. The good practice in such a case is to resample the data set to obtain more uniform class distribution. Weka allows us to perform this step through the *Filters* mechanism. In the upper section of the *Preprocess* tab, we can select a desired transformation algorithm by pressing the `Choose` button, then picking up either the *supervised* or *unsupervised* category, and eventually the *attribute* or *instance* subcategory of filters. The `Resample` filter can be found in the *supervised/instance* group. Each filter has a series of options, which can be used to parameterize the result of processing. In case of resampling, we decided to enforce approximately equal class distribution (`biasToUniformClass=1`), disable replacement, in which case the desired classes proportion could be achieved by cloning of some of the data vectors (`noReplacement=True`) and set the output sample cardinality to 25% of the original data set, so that all of the data vectors belonging to the bleeding class are preserved, and only part of the reference class is selected for further processing.

Apart from the application of the `Resample` filter, it is also recommended to standardize features to ensure equal influence on data vectors distance calculation in case of parameters whose value range may drastically differ. The `Standardize` filter belongs to the unsupervised/attribute group. Standardization transforms every feature so that its standard deviation $= 1$ and mean value $= 0$.

The standardized and resampled data set is ready for the attribute selection procedure. It can be accomplished either in a separate *Explorer* tab, called *Select attributes*, or again by using the filters mechanism, as explained here. In the *supervised/attribute* category, we choose the AttributeSelection option. The two most important parameters of this filter are 1) the criterion used to evaluate features saliency (evaluator) and 2) the feature space exploration strategy used to search for candidate attributes or their subsets assessed by the evaluator. In this study, we select the `WrapperSubsetEval` criterion and `LinearForwardSelection` search strategy.

The `WrapperSubsetEval` option allows assessment of a given feature collection in terms of classification accuracy associated with that particular attribute combination. The finally selected features are those ensuring the minimum classification error. For that purpose, we employed the support vector machines classifier with the radial basis kernel function (`gamma=3`), implemented in the *LibSVM* library [53] and installed within Weka as an extension through its package manager. The `LinearForwardSelection` strategy implements the Sequential Floating Forward Selection (SFFS) scheme [54]. This approach provides a trade-off between deterministic behavior of the optimization procedure (in contradiction to—sometimes blind—operation of evolutionary techniques such as genetic algorithms or particle swarm optimization) and the moderate computational complexity. The number of selected features depends either on the user/application demands or the required accuracy.

Eventually, the reduced data set serves for creation of the final classification model, which can later be used to detect similar image patterns in a new WCE image sequence. This stage is performed through the *Classify tab*. The user must first choose a classification algorithm. It can be any of the available schemes; however, it is reasonable to select the one used as a feature evaluator since its operation has been proven optimal for the finally constructed reduced data space. Thus we present the results for the Support Vector Machines (SVM) classifier. There are four possible options for assessing the quality of a trained classifier:

1. Use training set: the classification error is calculated for the same set as the one used for building the classifier model. As such, it is an overoptimistic score.
2. Supplied test set: the accuracy is measured based on classification performance achieved on data vectors that were not included in the training stage. In most cases, it gives the best estimate of the true (unknown) error committed on an infinitely large data sample.
3. Cross-validation: this option allows us to approximate the true error when a separate test set is not available. It splits the training set into n folds, and then the folds are sequentially combined into subsets, each time leaving one fold apart as a testing one. The average of *n* errors is the final estimate of the classification quality.
4. Percentage split: if a training set is large enough, then a fraction of it can be separated to form the test set and evaluate the quality similarly to option 2.

In the experiments presented here, we calculated both the cross–validation and test set errors. A screen shot of the results obtained for the focal ulcer class in the cross-validation experiment is shown in Fig. 11.13. It contains not only the overall accuracy and inaccuracy ratios, but also a series of other statistics, such as TP and FP rate, precision, recalls, AUC, and so on. Moreover, the program presents the confusion matrix, where false positive and false negative errors are visualized on the off-diagonal entries. Similar results were calculated for the separate test sets and also for the two other classification problems formulated above, focal ulcer vs. reference class and excessive ulcer vs. reference class. The obtained estimates of the classification accuracy are shown in Table 11.1.

As it can be seen, the constructed SVM classifier is capable of detecting pathologies in image frames, which were not present during the training stage. The obtained results

Table 11.1 Sensitivity and specificity ratios obtained in the pathology detection experiment. Evaluation performed using 10-fold cross-validation scheme. In each case, size of the reference class = 133 data vectors

Pathology class	Sensitivity	Specificity	Size of the pathology class
Bleeding	0.980	1.000	101
Focal ulcer	0.804	0.963	131
Excessive ulcer	0.592	1.000	118

```
=== Stratified cross-validation ===
=== Summary ===

Correctly Classified Instances      270              98.5401 %
Incorrectly Classified Instances      4               1.4599 %
Kappa statistic                      0.9708
Mean absolute error                  0.0146
Root mean squared error              0.1208
Relative absolute error              2.9234 %
Root relative squared error         24.1795 %
Total Number of Instances           274

=== Detailed Accuracy By Class ===

              TP Rate  FP Rate  Precision  Recall  F-Measure  MCC    ROC Area  PRC Area  Class
              0,992    0,021    0,978      0,992   0,985      0,971  0,986     0,974     1
              0,979    0,008    0,993      0,979   0,986      0,971  0,986     0,983     2
Weighted Avg. 0,985    0,014    0,986      0,985   0,985      0,971  0,986     0,979

=== Confusion Matrix ===

   a   b   <-- classified as
 131   1 |   a = 1
   3 139 |   b = 2
```

Figure 11.13 The summary of classification results in the *Classify* tab.

allow us to conclude that combining texture and data exploratory analyses constitutes a potentially reliable framework for automation of the WCE image-based diagnosis. Most importantly, this conceptual processing path, starting from texture feature extraction, through preprocessing and ending at classifier testing, is embodied in the integration of two powerful software tools, MaZda and Weka.

11.4 SUMMARY

The MaZda package is an efficient and reliable set of tools for analysis of image textures. Its efficiency was confirmed by the participants of the COST B11 and COST B21 projects and other researchers who applied this software for many different texture analysis tasks. In biomedical applications the software was used for MRI measurement protocol optimization, detection and characterization of various lesions, and image segmentation based on texture and color characteristics. However, MaZda is not limited to biomedical studies only. It was successfully used for food quality assessment and for grain, fruit, and vegetable variety determination.

The methods implemented in MaZda were carefully selected to be useful for 2D and 3D image analysis applications. The features may be computed after several steps of image preprocessing, including color space transformation and image brightness normalization. In particular, combining the color analysis and the gray-scale texture attribute computation enables characterization of color textures. In addition, the introduced feature naming convention prevents confusion of feature values computed for different preprocessing steps. Moreover, MaZda applies feature extraction algorithms to locally characterize images and to compute feature maps. The maps can be used for image texture segmentation.

Additional information, which includes tutorials, the manual and executable codes, is available from the qmazda web page.[2]

Although MaZda has proven to be useful in many medical image studies, it must be clearly stated that it was never officially certified for usage in medical practice or in the food industry.

APPENDIX 11.A

11.A.1 List of feature name symbols

The list of symbols is presented in Table 11.2.

Table 11.2 Symbols used in MaZda feature names

Feature extraction component	Symbols used
Color channel	• Y – brightness • R, G, B – red, green, blue • I, Q, U, V – chroma channels • x, y, x – CIE XYZ channels • L, a, b – CIE L⋆a⋆b channels
Sliding region	• C – circular • S – square
Histogram normalization	• D – no normalization (default) • N – 1st–99th percentile normalization • S – $\pm 3\sigma$ standardization
Extraction algorithm	• Hist – histogram • Grad – gradient • Glcm – gray-level cooccurrence matrix • Grlm – gray-level run-length matrix • Arm – autoregressive model • Gab – Gabor wavelet
Direction of pixels	• H - horizontal • V – vertical • Z – 45° • N – 135°

---

[2] The qmazda page address is http://www.eletel.p.lodz.pl/pms/SoftwareQmazda.html.

11.A.2 List of implemented methods

The list of implemented methods is presented in Table 11.3.

Table 11.3 Summary of methods implemented in MaZda

Analysis category	List of implemented methods

Image preprocessing	• Regions of interest editing
	  – predefined and arbitrary shape drawing
	  – gray-level thresholding
	  – flood-filling
	  – active contour
	  – deformable surface
	  – morphological operations
	• Histogram normalization
	  – $\pm 3\sigma$
	  – 1st–99th percentile
	• Gray-level quantization

Feature extraction	• Image histogram (2D & 3D)
	• Texture features
	  – Image gradient (2D & 3D)
	  – Gray-level cooccurrence matrix (2D & 3D)
	  – Run-length matrix (2D)
	  – Autoregressive model (2D)
	  – Gabor wavelets (2D)
	• Color analysis, feature computation in channels:
	  – Brightness
	  – RGB
	  – YIQ
	  – YUV
	  – HSB
	  – CIE XYZ
	  – CIE L$\star$a$\star$b
	• Morphological features

Feature selection	• Filter approach
	  – Fisher discriminant
	  – Mutual information
	• Wrapper approach
	  – Linear Discriminant Analysis
	  – Nonlinear hyperellipsoid decision boundary
	  – Vector supported convex hull

Feature transformation	• Linear Discriminant projection
Classification	• Linear Discriminant Analysis • Nonlinear hyperellipsoid decision boundary • Vector-supported convex hull

REFERENCES

[1] A. Materka, M. Strzelecki, Texture Analysis Methods – A Review, COST B11 report, Technical University of Lodz, Institute of Electronics, Brussels, 1998.

[2] P.M. Szczypiński, M. Strzelecki, A. Materka, A. Klepaczko, MaZda – a software package for image texture analysis, Comput. Methods Programs Biomed. 94 (1) (2009) 66–76.

[3] M. Hajek, M. Dezortova, A. Materka, R. Lerski, Texture Analysis for Magnetic Resonance Imaging, Med4 publishing, Prague, 2006.

[4] S. Blouin, M.F. Moreau, M.F. Baslé, D. Chappard, Relations between radiograph texture analysis and microcomputed tomography in two rat models of bone metastases, Cells Tissues Organs 182 (3–4) (2006) 182–192.

[5] L. Bonilha, E. Kobayashi, G. Castellano, G. Coelho, E. Tinois, F. Cendes, L.M. Li, Texture analysis of hippocampal sclerosis, Epilepsia 44 (12) (2003) 1546–1550.

[6] G. Chen, S. Jespersen, M. Pedersen, Q. Pang, M.R. Horsman, H. Stødkilde-Jørgensen, Evaluation of anti-vascular therapy with texture analysis, Anticancer Res. 25 (5) (2005) 3399–3405.

[7] M. Cichy, A. Materka, J. Tuliszkiewicz, Computerised analysis of X-ray images for early detection of osteoporotic changes in the bone, in: Proc. Conf. Information Technology in Medicine TIM '97, Jaszowiec, Poland, 1997, pp. 53–61.

[8] S. Herlidou, R. Grebe, F. Grados, N. Leuyer, P. Fardellone, M.-E. Meyer, Influence of age and osteoporosis on calcaneus trabecular bone structure: a preliminary in vivo MRI study by quantitative texture analysis, Magn. Reson. Imaging 22 (2) (2004) 237–243.

[9] D. Jirák, M. Dezortová, M. Hájek, P. Taimr, Texture analysis of human liver, J. Magn. Reson. Imaging 15 (1) (2002) 68–74.

[10] M. Kozakiewicz, M. Stefanczyk, A. Materka, P. Arkuszewski, Selected characteristic features of radiotexture of bone of the dental alveolus in the human maxilla and mandible, Mag. Stomatol. 3 (2007) 40–43.

[11] E. Lerouxel, H. Libouban, M.F. Moreau, M.F. Baslé, M. Audran, D. Chappard, Mandibular bone loss in an animal model of male osteoporosis (orchidectomized rat): a radiographic and densitometric study, J. Osteoporos. Int. 15 (10) (2004) 814–819.

[12] D. Mahmoud-Ghoneim, Y. Cherel, L. Lemaire, J.D. de Certaines, A. Manier, Texture analysis of magnetic resonance images of rat muscles during atrophy and regeneration, Magn. Reson. Imaging 24 (2) (2006) 167–171.

[13] A. Materka, P. Cichy, J. Tuliszkiewicz, Texture analysis of X-ray images for detection of changes in bone mass and structure, in: Series in Machine Perception & Machine Intelligence, vol. 40, World Scientific, 2000, pp. 189–195.

[14] V. Punys, J. Puniene, R. Jurkevicius, J. Punys, Myocardium tissue analysis based on textures in ultrasound images, Stud. Health Technol. Inform. 116 (2005) 435–440.

[15] A. Santos, C. Ramiro, M. Desco, N. Malpica, A. Tejedor, A. Torres, M.J. Ledesma-Carbayo, M. Castilla, P. García-Barreno, Automatic detection of cellular necrosis in epithelial cell cultures, in: M. Sonka, K.M. Hanson (Eds.), Medical Imaging, Proc. SPIE 4322 (2001) 1836–1844.

[16] O. Yu, N. Parizel, L. Pain, B. Guignard, B. Eclancher, Y. Mauss, D. Grucker, Texture analysis of brain MRI evidences the amygdala activation by nociceptive stimuli under deep anesthesia in the propofol-formalin rat model, Magn. Reson. Imaging 25 (1) (2007) 144–146.

[17] G. Collewet, M. Strzelecki, F. Mariette, Influence of MRI acquisition protocols and image intensity normalization methods on texture classification, Magn. Reson. Imaging 22 (2004) 81–91.

[18] J. Letal, D. Jirak, L. Suderlova, M. Hajek, MRI texture analysis of MR images of apples during ripening and storage, Lebensm.-Wiss. Technol. 36 (7) (2003) 719–727.

[19] A.K. Thybo, P.M. Szczypinski, A.H. Karlsson, S. Donstrup, H. Stodkilde-Jorgensen, H.J. Andersen, Prediction of sensory texture quality attributes of cooked potatoes by NMR-imaging (MRI) of raw potatoes in combination with different image analysis methods, J. Food Eng. 61 (1) (2004) 91–100.

[20] J.M. Martinez, MPEG-7 Overview, Technical Report JTC1/SC29/WG11N5525, ISO/IEC, March 2003.

[21] A.E. Carpenter, T.R. Jones, M.R. Lamprecht, C. Clarke, I.H. Kang, O. Friman, D.A. Guertin, J. Han Chang, R.A. Lindquist, J. Moffat, P. Golland, D.M. Sabatini, CellProfiler: image analysis software for identifying and quantifying cell phenotypes, Genome Biol. 7 (10) (2006) R100.

[22] L. Kamentsky, T.R. Jones, A. Fraser, M.-A. Bray, D.J. Logan, K.L. Madden, V. Ljosa, C. Rueden, K.W. Eliceiri, A.E. Carpenter, Improved structure, function and compatibility for CellProfiler: modular high-throughput image analysis software, Bioinformatics 27 (8) (2011) 1179–1180.

[23] I.H. Witten, E. Frank, Data Mining: Practical Machine Learning Tools and Techniques, 2nd edition, Morgan Kaufmann, San Francisco, 2005.

[24] MATLAB, version 9.1, Image Processing Toolbox, The MathWorks Inc., Natick, Massachusetts, 2016.

[25] SciKit, scikit-image: image processing in Python, 2016.

[26] G. Bradski, A. Kaehler, Learning OpenCV: Computer Vision with the OpenCV Library, O'Reilly Media, Inc., 2008.

[27] J. Schindelin, I. Arganda-Carreras, E. Frise, V. Kaynig, M. Longair, T. Pietzsch, S. Preibisch, C. Rueden, S. Saalfeld, B. Schmid, J.-Y. Tinevez, D.J. White, V. Hartenstein, K. Eliceiri, P. Tomancak, A. Cardona, Fiji – an open source platform for biological image analysis, Nat. Methods 9 (7) (2012) 06, http://dx.doi.org/10.1038/nmeth.2019.

[28] P.M. Szczypiński, Center point model of deformable surface, in: K. Wojciechowski, B. Smolka, H. Palus, R.S. Kozera, W. Skarbek, L. Noakes (Eds.), Proceedings International Conference Computer Vision and Graphics ICCVG 2004, in: Computational Imaging and Vision, vol. 32, Springer-Verlag, 2006, pp. 343–348.

[29] P.M. Szczypiński, M. Strzelecki, A. Materka, Mazda – a software for texture analysis, in: Proc. of ISITC 2007, Jeonju, Korea, 2007, pp. 245–249.

[30] A.C. Bovik, M. Clark, W.S. Geisler, Multichannel texture analysis using localized spatial filters, IEEE Trans. Pattern Anal. Mach. Intell. 12 (1) (1990) 55–73.

[31] L. Cohen, Time-frequency distributions, Proc. IEEE 77 (1989) 941–981.

[32] J.G. Daugman, Complete discrete 2-D Gabor transforms by neural networks for image analysis and compression, IEEE Trans. Acoust. Speech Signal Process. 36 (7) (1988) 1169–1179.

[33] A. Jain, Fundamentals of Digital Image Processing, Prentice-Hall International, 1989.

[34] S.G. Mallat, A theory for multiresolution signal decomposition: the wavelet representation, IEEE Trans. Pattern Anal. Mach. Intell. 11 (7) (1989) 674–693.

[35] R.M. Haralick, Statistical and structural approaches to texture, Proc. IEEE 67 (5) (1979) 786–804.

[36] M. Kociolek, A. Materka, M. Strzelecki, P. Szczypinski, Discrete wavelet transform-derived features for digital image texture analysis, in: International Conference on Signals and Electronic Systems, Łódź-Poland, 2001, pp. 99–104.

[37] J. Racz, A. Dubrawski, B. Siemiatkowska, M. Nieniewski, Computer-aided diagnosis of breast cancer based on analysis of microcalcifications, in: Proceedings of 5th International Workshop on Digital Mammography, Toronto, Canada, Medical Physics Publishing, Madison, June 2000.

[38] P. Szczypiński, A. Klepaczko, M. Pazurek, P. Daniel, Texture and color based image segmentation and pathology detection in capsule endoscopy videos, Comput. Methods Programs Biomed. 113 (1) (2014) 396–411.

[39] D.W. Luerkens, Theory and Application of Morphological Analysis: Fine Particles and Surfaces, CRC Press, 1991.

[40] A. Materka, M. Strzelecki, P. Szczypiński, MaZda Manual, 2006.

[41] M. Kociołek, A. Materka, M. Strzelecki, P. Szczypiński, Badanie wpływu liczby poziomów jasności obrazu na zdolność dyskryminacji tekstur przy użyciu macierzy zdarzeń, Elektron.-Pr. Nauk. 5 (2000) 21–33.

[42] G.D. Tourassi, E.D. Frederick, M.K. Markey, C.E. Floyd Jr., Application of the mutual information criterion for feature selection in computer-aided diagnosis, Med. Phys. 28 (12) (2001) 2394–2402.

[43] A.K. Jain, R.P.W. Duin, J. Mao, Statistical pattern recognition: a review, IEEE Trans. Pattern Anal. Mach. Intell. 22 (1) (2000) 4–37.

[44] W. Krzanowski, Principles of Multivariable Data Analysis, Oxford University Press, 1988.

[45] K. Fukunaga, Introduction to Statistical Pattern Recognition, Academic Press, New York, 1991.

[46] P.M. Szczypiński, A. Klepaczko, P. Zapotoczny, Identifying barley varieties by computer vision, Comput. Electron. Agric. 110 (2015) 1–8.

[47] P. Szczypiński, A. Klepaczko, Convex hull-based feature selection in application to classification of wireless capsule endoscopic images, in: Proceedings ACIVS 2009 Advanced Concepts for Intelligent Vision Systems, in: Lecture Notes in Computer Science, vol. 5807, Springer-Verlag, Berlin, Heidelberg, 2009, pp. 664–675.

[48] P. Zapotoczny, M. Zielinska, Z. Nita, Application of image analysis for the varietal classification of barley, J. Cereal Sci. 48 (1) (2008) 104–110.

[49] P. Zapotoczny, P.M. Szczypiński, T. Daszkiewicz, Evaluation of the quality of cold meats by computer-assisted image analysis, LWT-Food Sci. Technol. 67 (2016) 37–49.

[50] A. Klepaczko, M. Kociński, A. Materka, Quantitative description of 3D vascularity images: texture-based approach and its verification through cluster analysis, Pattern Anal. Appl. 14 (4) (2011) 415–424.

[51] M. Kociński, A. Klepaczko, A. Materka, M. Chekenya, A. Lundervold, 3D image texture analysis of simulated and real-world vascular trees, Comput. Methods Programs Biomed. 107 (2) (2012) 140–154.

[52] E.M. Mayerhoefer, D. Stelzeneder, W. Bachbauer, G.H. Welsch, T.C. Mamisch, P. Szczypinski, M. Weber, N.H.G.M. Peters, J. Fruehwald-Pallamar, S. Puchner, et al., Quantitative analysis of lumbar intervertebral disc abnormalities at 3.0 Tesla: value of T_2 texture features and geometric parameters, NMR Biomed. 25 (6) (2012) 866–872.

[53] C.-C. Chang, C.-J. Lin, LIBSVM: a library for support vector machines, ACM Trans. Intell. Syst. Technol. 2 (3) (2011) 27:1–27:27.

[54] P. Pudil, J. Novovičová, J. Kittler, Floating search methods in feature selection, Pattern Recognit. Lett. 15 (11) (1994) 1119–1125.

CHAPTER 12

QuantImage: An Online Tool for High-Throughput 3D Radiomics Feature Extraction in PET-CT

Yashin Dicente Cid*,†, Joël Castelli‡,§,¶, Roger Schaer*, Nathaniel Scher‡,
Anastasia Pomoni‖, John O. Prior‖, Adrien Depeursinge*,**

*University of Applied Science Western Switzerland (HES-SO), Institute of Information Systems, Sierre, Switzerland
†University of Geneva, Computer Vision and Multimedia Laboratory, Geneva, Switzerland
‡Lausanne University Hospital (CHUV), Radiotherapy Department, Lausanne, Switzerland
§INSERM, U1099, Rennes, France
¶Université de Rennes 1, LTSI, Rennes, France
‖Lausanne University Hospital (CHUV), Nuclear Medicine and Molecular Imaging Department, Lausanne, Switzerland
**École Polytechnique Fédérale de Lausanne (EPFL), Biomedical Imaging Group, Lausanne, Switzerland

Abstract

The processes of radiomics consist of image-based personalized tumor phenotyping for precision medicine. They complement slow, costly, and invasive molecular analysis of tumoral tissue. Whereas the relevance of a large variety of quantitative imaging biomarkers has been demonstrated for various cancer types, most studies were based on 2D image analysis of relatively small patient cohorts. In this work, we propose an online tool for automatically extracting 3D state-of-the-art quantitative imaging features from large batches of patients. The developed platform is called QuantImage and can be accessed from any web browser. Its use is straightforward and can be further parameterized for refined analyses. It relies on a robust 3D processing pipeline allowing normalization across patients and imaging protocols. The user can simply drag-and-drop a large zip file containing all image data for a batch of patients and the platform returns a spreadsheet with the set of quantitative features extracted for each patient. It is expected to enable high-throughput reproducible research and the validation of radiomics imaging parameters to shape the future of noninvasive personalized medicine.

Keywords

Quantitative imaging biomarkers, Radiomics, Texture analysis, Medical imaging, Radiology, Web technologies, Precision medicine

12.1 INTRODUCTION

Cancer ecosystems are composed of microhabitats defining the cancer subtype, stage, response to treatment, and patient survival [1]. Estimating the composition of tumoral tissue can therefore provide valuable information for optimal personalized disease management. Current promising approaches for precision medicine are mostly based on molecular analyses of biopsied or resected tissue [2]. Nevertheless, the success of such

Biomedical Texture Analysis
DOI: 10.1016/B978-0-12-812133-7.00012-0

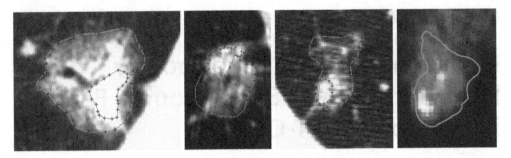

Figure 12.1 Axial views of heterogeneous NSCLC tumors. The three CT images on the left show tumors with solid (red) and ground glass (blue) regions delineated. The PET image on the right shows the metabolic heterogeneity of the tumor. (For interpretation of the references to color in this figure legend, the reader is referred to the web version of this chapter.)

approaches are hindered by their invasiveness as well as the requirement of slow, costly, and complex analysis of "-omics" data. In addition, the tissue samples used for molecular analyses are most often acquired from a highly localized tumor region or from the grinding of whole tumor mass, which does not allow to accurately capture molecular heterogeneity [3]. The computerized quantitative analysis of existing diagnostic, treatment planning and follow-up images enables reproducible and comprehensive analysis of tumoral regions as a whole, potentially allowing the exploration of tumor heterogeneity in a noninvasive fashion. The latter spawned the new research fields of *radiomics* [4] and *imaging genomics* [5] (see Chapter 8). The metabolism, density and structure of tumoral tissue observed in three-dimensional Positron Emission Tomography (PET), Magnetic Resonance (MR), and Computed Tomography (CT) images reflects their nature [6], including regions of active cancer cells, angiogenesis, necrosis [7], and even subsets of underlying cancer-related genomics [8,9]. Examples of PET and CT images of heterogeneous Nonsmall Cell Lung Cancer (NSCLC) tumors are depicted in Fig. 12.1.

Radiomics consists of image-based personalized tumor phenotyping, complementing slow, costly, and invasive molecular analysis [4,10]. Computerized quantitative image analysis yields a collection of variables, which are further linked to disease outcomes and subtypes using advanced statistical and machine learning methods. These processes are detailed in Fig. 12.2. The main categories of quantitative imaging biomarkers are intensity, shape, and texture, which are characterizing the distribution of voxel values, the contour of tumors and the spatial transitions between voxel values, respectively. Current state-of-the-art in radiomics provided initial evidence about the relevance of these quantitative imaging parameters for precision medicine [11–15]. However, several limitations are concerning texture-based image measures in particular. First, large-scale clinical studies are required to further validate texture-based features in diverse disease-specific contexts. This is of primary importance because texture parameters are mostly informative when used in large groups (*e.g.*, 50 to 100 attributes), requiring large cohorts

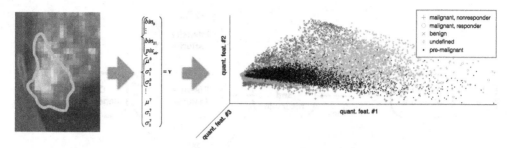

Figure 12.2 Illustration of the radiomics processes (toy example): Several quantitative imaging features are extracted from the tumoral region (green) in PET-CT. The resulting collection of variables spans a hyperspace where distinct cancer subtypes and treatment responders are ideally occupying well-defined regions. The latter can be further revealed by advanced statistical and machine learning methods.

of patients to respect acceptable ratios between the number of variables and instances. Second, most studies are based on suboptimal texture operators applied in 2D slices that are not able to capture the wealth of complex three-dimensional tissue architectures available in modern imaging protocols (see Chapter 3). 3D approaches are emerging [16]. However, the interpretation of such image measures is challenging since the human cannot fully visualize opaque 3D solid images. Moreover, few computer tools are available and they are not specific to PET and CT imaging. Exceptions include LifeX[1] and CGITA[2] [17], which allow extracting three-dimensional texture measures. However, they are both based on gray-level matrices only, which have shown limited abilities to mine rich textural patterns (see Section 3.3 of Chapter 3).

This work presents an online tool for extracting high-throughput, advanced 3D radiomics features in PET-CT images. No software installation is required and the platform can be accessed through any web browser. It allows submitting large batches of patient files and to download resulting patient-wise collections of radiomics features in a standard Comma-Separated Values (CSV) data structure. The radiomics attributes include PET- and CT-specific intensity values, novel distance measures characterizing metastatic spread, as well as three distinct groups of texture measurements combining advanced analysis and interpretability. All attributes can be parameterized in a simple web page.

The chapter is structured as follows. The methods of the proposed 3D quantitative image analysis pipeline are detailed in Section 12.2. Each subsection contains the necessary technical details to understand and reproduce every quantitative imaging biomarkers: intensity-, distance-, and texture-based features. Section 12.3 details the

---

[1] http://www.lifexsoft.org, as of 5 December 2016.
[2] https://code.google.com/archive/p/cgita/, as of December 5, 2016.

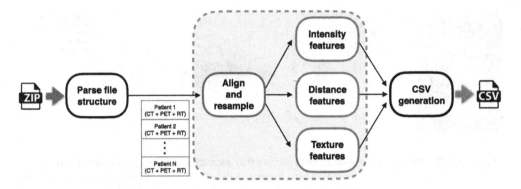

Figure 12.3 Architecture of the proposed 3D quantitative image analysis pipeline.

workflows and processes of the proposed online tool, which can be used as a reference manual for the end-user. Conclusions and perspectives are provided in Section 12.4.

12.2 METHODS

The pipeline of our system is shown in Fig. 12.3. The very first step of the pipeline consists in finding and regrouping image studies and series from a batch zip file uploaded to the server. For each patient, the latter should include Digital Imaging and Communications in Medicine (DICOM) files of both CT and PET scans, as well as a DICOM RT structure file containing the Region(s) Of Interest (ROI) to analyze. The structure of the zip file and its parsing is detailed in Section 12.3.2. Nonetheless, it is required to explain at this point that the DICOM RT file must contain the ROI corresponding to main tumor referred to as *GTV_T*. Optionally, it may contain secondary ROIs referring to nodes or metastases named *GTV_N*. The structure of the returned CSV file containing all 3D quantitative imaging parameters is described in Section 12.3.3. The following subsections detail the various steps of the pipeline contained inside the orange box in Fig. 12.3.

12.2.1 PET-CT alignment

The DICOM RT file is generated with respect to the CT scan and the ROI it is always in the same scale and alignment as the CT image. This is not the case for the PET scan. Most often, the resolution of the CT scan is higher than the PET and the volume covered might be unaligned between the two. The volume position and voxel dimensions specified in the headers of the DICOM files series from the PET and CT scans are used to resample and to perform a rigid alignment of the PET series on the CT. This transformation requires an interpolation procedure. The nearest neighbor method is used to preserve the uptake values of the PET. After this step, the PET and CT

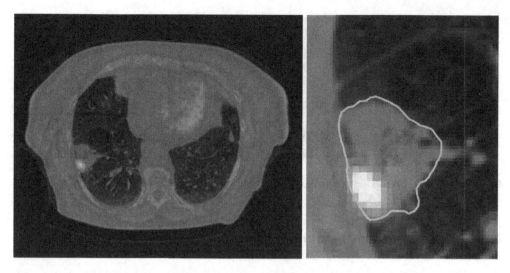

Figure 12.4 Left: axial CT slice and PET image overlay after the alignment and rescaling of the PET image. Right: closeup view of the ROI (green). (For interpretation of the references to color in this figure legend, the reader is referred to the web version of this chapter.)

volumes contain the same number of voxels and cover the same 3-dimensional region. Fig. 12.4 shows a full slice as well as a zoomed view of the around the ROI (marked in green) of fused PET and CT volumes after the alignment. This example will be further used to illustrate the various steps of the system.

12.2.2 Intensity-based features

This section details intensity-based features, which are based on the regional distributions of the voxel values. The latter correspond to the Hounsfield Units (HU) and the Standardized Uptake Values (SUV) in CT and PET images, respectively. Only the voxels inside GTV_T and GTV_N are used. The system allows to further refine the ROIs based on multiple user-defined metabolic thresholds on the SUV values. This is illustrated in Fig. 12.5, where the initial ROI shown in Fig. 12.4 shrinks with respect to increasing metabolic thresholds τ.

12.2.2.1 Statistics of voxels values

The first four statistical moments are used to characterize the intensity distribution inside the ROI. By intensity, we refer to the voxel values, *i.e.*, HU in CT and SUV in PET (see Section 1.2.1 of Chapter 1). These statistical moments are the mean μ, standard deviation σ, skewness *skew*, and kurtosis *kurt*. Let M_τ be the region containing the voxels of the ROI M with intensity value greater than a threshold τ. We define N_{M_τ} as the number of voxels in M_τ. The position of each voxel is determined by the 3D vector

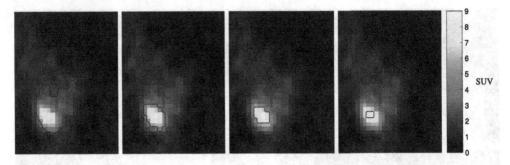

Figure 12.5 Resulting cropped ROI (in blue) when applying increasing metabolic thresholds τ on the initial ROI shown in Fig. 12.4.

ξ, and $f(\xi)$ corresponds to the intensity value at the position ξ. The four moments are defined as

$$\mu_{M_\tau} = \frac{1}{N_{M_\tau}} \sum_{\xi \in M_\tau} f(\xi), \tag{12.1}$$

$$\sigma_{M_\tau} = \sqrt{\frac{1}{N_{M_\tau}} \sum_{\xi \in M_\tau} \left(f(\xi) - \mu_{M_\tau}\right)^2}, \tag{12.2}$$

$$skew_{M_\tau} = \frac{1}{N_{M_\tau} \sigma_{M_\tau}^3} \sum_{\xi \in M_\tau} \left(f(\xi) - \mu_{M_\tau}\right)^3, \tag{12.3}$$

$$kurt_{M_\tau} = \frac{1}{N_{M_\tau} \sigma_{M_\tau}^4} \sum_{\xi \in M_\tau} \left(f(\xi) - \mu_{M_\tau}\right)^4. \tag{12.4}$$

The skewness measures the symmetry of the distribution. $skew = 0$ means that the distribution is symmetric, $skew < 0$ means that the distribution is more concentrated on the right of the mean μ (*i.e.*, high voxel values), and $skew > 0$ when the concentration is on the left of the μ. The kurtosis measures the tailedness of the distribution, *i.e.*, the concentration of the data around μ. The higher is the kurtosis, the more concentrated the data with a high peak at μ. The normal distribution has always $skew = 0$ because it is completely symmetric. Its kurtosis is always 3. Fig. 12.6 shows the distributions of the CT (HU) and PET (SUV) inside the unthresholded ROI M shown in Fig. 12.4. The normal distribution with identical μ and σ as the corresponding distribution is showed in black to illustrate the significance of skewness and kurtosis. The intensity statistics of M are specified in Table 12.1.

12.2.2.2 PET-specific intensity-based features

A set of PET-specific features based on the distribution of SUV intensity values has been identified in the literature as meaningful for assessing the metabolic properties of

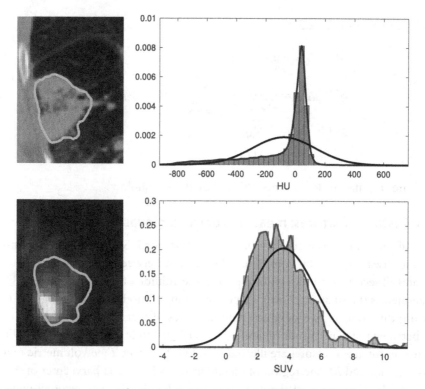

Figure 12.6 Distribution curves for each intensity measure in CT (top) and PET (bottom). The equivalent normal distributions with corresponding μ and σ are shown in black.

Table 12.1 Table with the values of the first four moments of the intensity distribution for the examples shown in Fig. 12.6

	μ_M	σ_M	$skew_M$	$kurt_M$
HU distribution	−78.0416	211.2127	−1.3821	9.4431
Normal distribution	−78.0416	211.2127	0	3
SUV distribution	3.5962	1.9587	1.5430	6.9098
Normal distribution	3.5962	1.9586	0	3

tumors [18,15]. These are (i) SUV_{max}, the maximum SUV value inside the ROI, (ii) SUV_{peak}, the peak SUV measuring the mean SUV value within a spherical neighborhood of 1.2 cm radius and centered at the position of SUV_{max}, (iii) MTV, the Metabolic Tumor Volume, and (iv) TLG, the Total Lesion Glycolysis combining both metabolic and volumetric information. Let S_{max} be the spherical neighborhood of SUV_{max}, and v_{vox} the volume of one single voxel. The measures are defined as (following the notation introduced in Section 12.2.2.1)

$$SUV_{\text{max}} = \max_{\xi \in M}(f(\xi)), \tag{12.5}$$

$$SUV_{\text{peak},\tau} = \frac{1}{N_{M_\tau \cap S_{\text{max}}}} \sum_{\xi \in M_\tau \cap S_{\text{max}}} f(\xi) = \mu_{M_\tau \cap S_{\text{max}}}, \tag{12.6}$$

$$MTV_\tau = \nu_{\text{vox}} \cdot N_{M_\tau}, \tag{12.7}$$

$$TLG_\tau = \mu_{M_\tau} \cdot MTV_\tau. \tag{12.8}$$

It worth noting that SUV_{max} is not affected by the threshold τ.

12.2.3 Distance features: measures of cancer invasiveness

Fried et al. introduced the concept of *disease solidity* for NSCLC in [19]. It consists of measuring disease spread by computing the relation between the volume of the main tumor and all secondary nodes with respect to the volume of their convex hull. The latter is defined as the smallest convex shape containing all nodules (including the tumor). Following this concept, we designed new measures of metastases spread based on distances between the main tumor (delimited in GTV_T) and the metastases (in GTV_N). The various distance measures are illustrated in Fig. 12.7. Let T be volumetric region of the main tumor and M_i the region of metastasis i. Let b_T be the barycenter of the main tumor, b_{M_i} the barycenter of the metastasis i, and b_M the barycenter of all metastases. The distance between T and M_i is defined as the Euclidean distance between their barycenters, *i.e.*,

$$d_{T,M_i} = ||b_T - b_{M_i}||_2. \tag{12.9}$$

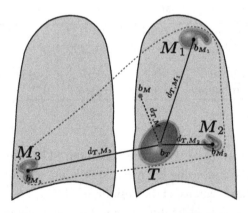

Figure 12.7 Distances features as measures of cancer invasiveness. Various distances d_{T,M_i} between the barycenter b_T of the primary tumor region T and the barycenter b_{M_i} of each metastasis region M_i are computed to measure the spread of disease.

The same formula is used to compute the distance between the main tumor and the barycenter of all metastases $d_{T,M}$. Considering this definition, and denoting the MTV in cubic millimeters (mm^3) of the metastasis M_i as MTV_i, seven different features are introduced as:

(i) the distance between the tumor and the barycenter of all metastases b_M

$$d_{T,M} = ||b_T - b_M||_2, \tag{12.10}$$

(ii) the sum of all tumor-metastasis distances

$$\sum_i d_{T,M_i} = \sum_i ||b_T - b_{M_i}||_2, \tag{12.11}$$

(iii) the sum of distances weighted by the respective MTV of the metastases

$$\sum_i MTV_i \cdot d_{T,M_i} = \sum_i MTV_i \cdot ||b_T - b_{M_i}||_2, \tag{12.12}$$

(iv) the metastasis remoteness

$$\max_i \left\{ d_{T,M_i} \right\} = \max_i \left\{ ||b_T - b_{M_i}||_2 \right\}, \tag{12.13}$$

(v) the metastasis remoteness weighted by the MTV of the corresponding metastasis

$$\max_i \left\{ MTV_i \cdot d_{T,M_i} \right\} = \max_i \left\{ MTV_i \cdot ||b_T - b_{M_i}||_2 \right\}, \tag{12.14}$$

(vi) the cumulative distance between each metastasis and the barycenter of all metastases

$$\sum_i d_{M_i,M} = \sum_i ||b_{M_i} - b_M||_2, \tag{12.15}$$

and (vii) the MTV-weighted version of the latter

$$\sum_i MTV_i \cdot d_{M_i,M} = \sum_i MTV_i \cdot ||b_{M_i} - b_M||_2. \tag{12.16}$$

12.2.4 Texture features

Another set of quantitative imaging features extracting measures of 3D texture, *i.e.*, the spatial transitions between the voxel values (see Section 1.2.1 of Chapter 1) are implemented. The three main groups of texture measures are 3D Gray-Level Co-occurrence Matrices (GLCM) [20], 3D Laplacians of Gaussians (LoG), and 3D Riesz wavelets [21]. Detailed descriptions and review can be found in Sections 3.3.1, 3.2.1, and 3.2.2 of Chapter 3, respectively. The latter were successfully used in various radiomics applications [22–24,14,25].

12.2.4.1 Isometric volumes

Inside the body, three-dimensional tissue architectures can be constituted of texture patterns characterized by transitions between voxel values along any directions of $\mathbb{R}^3$. To ensure the unbiased management of directions, the voxels of the images must be isometric, *i.e.*, the dimensions of each voxel $\Delta\xi_1$, $\Delta\xi_2$, $\Delta\xi_3$ must be equal (see Section 1.3.2 and Fig. 1.15 of Chapter 1). Moreover, the voxel dimensions must be normalized across patients to allow optimal interpatient comparisons of image scales and directions. As a preprocessing step before extracting any texture feature, CT, PET, and ROI volumes are resampled to have identical isometric voxel sizes. The latter are set to $0.75 \times 0.75 \times 0.75$ mm^3 as a trade-off between data size and image resolution. Resampling methods are based on cubic spline interpolation for the CT volume and on nearest neighbors for PET and ROI volumes.

12.2.4.2 3D gray level cooccurrence matrix (GLCM) features

GLCMs constitute a group of popular texture descriptors introduced by Haralick et al. in [20]. The 2D version of this descriptor is discussed in Section 3.3.1 of Chapter 3. 3D GLCMs were proposed for the description of 3D textures in [26,27]. The properties of 2D GLCMs presented in Table 3.8 of Chapter 3 are also valid for the 3D extension. 3D GLCMs are extending their 2D counterparts consisting in measuring the co-occurrences between two voxel values in a 3D neighborhood. Fig. 12.8 shows the 3D neighborhood of all possible spatial configurations between the central voxel (red) and another voxel (semitransparent) separated by a distance of 1 and 2 voxels. While the set of 2D directions $\boldsymbol{u}$ can be defined with one single angle θ in polar coordinates, $\boldsymbol{u}$ is parameterized by 2 angles (θ, ϕ) in spherical coordinates (see Section 2.2 of Chap-

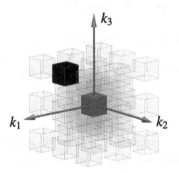

Figure 12.8 3D GLCM neighborhood of all possible spatial configurations between the central voxel (red) and another voxel (semitransparent) separated by a distance of 1 and 2 voxels. A displacement parameterized by the index vector $\Delta\boldsymbol{k} = (\Delta k_1 = 2, \Delta k_2 = 1, \Delta k_3 = 2)$ is exemplified between the black voxel and the center one (red). (For interpretation of the references to color in this figure legend, the reader is referred to the web version of this chapter.)

Table 12.2 The thirteen 3D directions u considered for computing GLCMs

Coordinates of $u \in \mathbb{R}^3$	θ	ϕ
$(0, 1, 0)$	$\frac{\pi}{2}$	0
$(1, 1, 0)$	$\frac{\pi}{4}$	0
$(1, 0, 0)$	0	0
$(1, -1, 0)$	$-\frac{\pi}{4}$	0
$(0, 1, 1)$	$\frac{\pi}{2}$	$\frac{\pi}{4}$
$(0, 0, 1)$	0	$\frac{\pi}{2}$
$(0, -1, 1)$	$-\frac{\pi}{2}$	$\frac{3\pi}{4}$
$(1, 1, 1)$	$\frac{\pi}{4}$	$\frac{\pi}{4}$
$(1, 0, 1)$	0	$\frac{\pi}{4}$
$(1, -1, 1)$	$-\frac{\pi}{4}$	$\frac{\pi}{4}$
$(1, 1, -1)$	$\frac{\pi}{4}$	$-\frac{\pi}{4}$
$(1, 0, -1)$	0	$-\frac{\pi}{4}$
$(1, -1, -1)$	$-\frac{\pi}{4}$	$-\frac{3\pi}{4}$

ter 2). The 13 equally-sampled directions used to approximate the full 3D neighborhood are detailed in Table 12.2 (see Fig. 1.14 of Chapter 1). The cumulative cooccurrences along every 13 directions allow building approximately locally rotation-invariant features in a similar fashion as 2D locally rotation-invariant GLCMs depicted in Fig. 3.10 of Chapter 3. This strategy has been commonly used in the literature [16]. 11 texture measurements are computed from the GLCMs to characterize 3D texture properties. These are: *contrast, correlation, energy, homogeneity, entropy, inverse different moment, sum average, sum entropy, sum variance, difference variance* and *difference entropy* (detailed formula can be found in [28]). The *contrast* and *energy* texture features correspond to our intuitive perception of texture in 2D.

12.2.4.3 3D Laplacian of Gaussians (LoG) features

One of the simplest texture descriptor based on convolution is the Laplacian of Gaussians (LoG). The LoG operator is defined for any D-dimensional image in Section 3.2.1 of Chapter 3. The formula of the operator function is specified in Eq. (3.5) and its main properties are summarized in Table 3.1. The parameter controlling the size of the LoG filters is σ and allows studying volumetric texture properties at multiple scales. This σ is referred in this section as σ_{LoG} to distinguish it from the standard deviation of intensity distributions introduced in Section 12.2.2.1. A set of multiscale and locally rotation-invariant (LoG are circularly/spherically symmetric) texture measurements is obtained by averaging the absolute values of the response maps of LoG operators in the ROI mask M for a series of increasing values of σ_{LoG}. Fig. 12.9 depicts the profile of the LoG filter for three different σ_{LoG} values. The response maps resulting from the convolutions

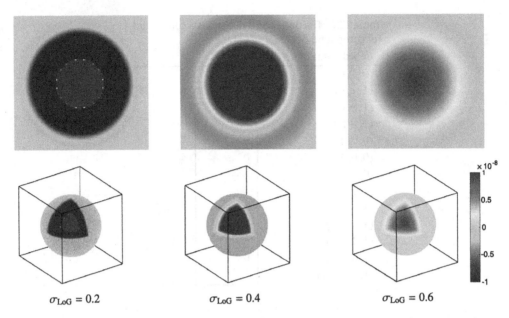

$\sigma_{\mathrm{LoG}} = 0.2$ $\sigma_{\mathrm{LoG}} = 0.4$ $\sigma_{\mathrm{LoG}} = 0.6$

Figure 12.9 3D LoG filters at various scales controlled by the parameter σ_{LoG}. The top row shows central sections of the filters and the bottom row shows spherical wedges.

of the three filters with two different images f_A and f_B are shown in Fig. 12.10. It can be observed that different values of σ_{LoG} can be used to characterize different texture scales.

As described in Fig. 1.12 of Chapter 1, it is possible to distinguish between the texture found on the margin of the ROI (M_{margin}) and the one found in the core region of the tumor ($M_{\mathrm{core}} \equiv GTV\_N$). To that end the LoG response maps are averaged in both regions separately (see Fig. 12.11). M_{margin} is obtained from the difference between M_{core} dilated and eroded with a spherical structural element (diameter = 2.25 mm). Therefore the tumor margin has an approximate thickness of 4.5 mm.

12.2.4.4 3D Riesz-wavelets features

Another convolutional feature group is extracted by the proposed pipeline and is based on 3D Riesz wavelets [21]. The latter are defined for any D-dimensional image and are detailed in Section 3.2.2 of Chapter 3. They consist in computing the real Riesz transform of a primal circularly/spherically symmetric wavelet decomposition, the latter being similar to LoG filters. The Lth-order real Riesz transform of an image $f(\boldsymbol{x})$ computes Lth-order all-pass derivatives of the latter. In other words, $L = 1$ computes first-order image derivatives and measures the slope of the spatial transition between the values of a set of aligned voxels, and the second-order transform ($L = 2$) measures the curvature of the transition. Consequently, they have an intuitive interpretation and have

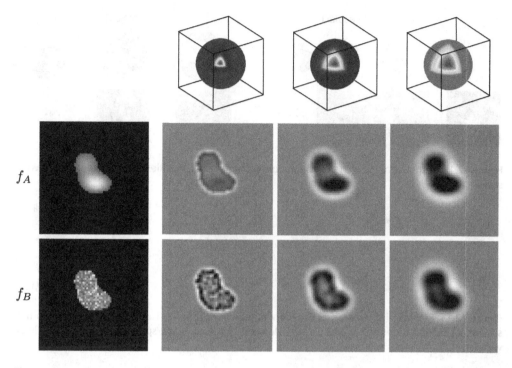

Figure 12.10 2D sections of the response maps of two different simulated tumors f_A and f_B for the LoG filters shown in Fig. 12.9.

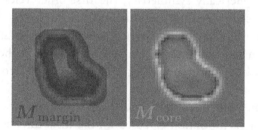

Figure 12.11 M_{margin} and M_{core} regions used for feature aggregation.

shown to provide relevant quantitative image measurements in the context of several medical applications [25,29,14,30–32]. We used the primal circularly/spherically symmetric wavelet presented in [24,33], which was found to have an optimal bandwidth. We limited a maximum Riesz order to 2 and a maximum of wavelet scales to 4. Riesz order 1 encodes the gradient of the image (first-order derivative), while order 2 also encodes the Hessian of the image (second-order derivative). The scale controls the size of the texture properties to be detected. Fig. 12.12 shows sliced response maps of first-order Riesz wavelets for scale 1 and 2 on a 2D slice of the synthetic tumor f_B (see Fig. 12.10).

scale 1   scale 2

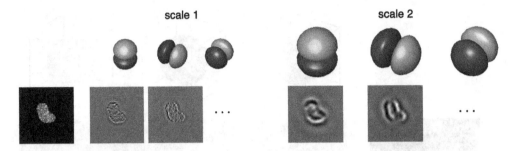

Figure 12.12 Response maps of the simulated tumor image f_B (left) for Riesz filterbanks of order 1 with scales 1 and 2.

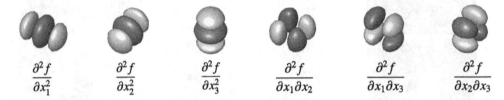

$$\frac{\partial^2 f}{\partial x_1^2} \qquad \frac{\partial^2 f}{\partial x_2^2} \qquad \frac{\partial^2 f}{\partial x_3^2} \qquad \frac{\partial^2 f}{\partial x_1 \partial x_2} \qquad \frac{\partial^2 f}{\partial x_1 \partial x_3} \qquad \frac{\partial^2 f}{\partial x_2 \partial x_3}$$

Figure 12.13 Profiles of the elements the Riesz filterbank of order 2. The latter corresponds to the six distinct second-order derivatives (*i.e.*, the Hessian) of the circularly/spherically symmetric primal wavelet.

While the LoG reacts equally to transitions between voxel values along any direction (LoGs are circularly/spherically symmetric), the first-order Riesz filterbank contains three different filters characterizing three orthogonal directions. Second-order Riesz filters characterize six different directions according to the Hessian (see Fig. 12.13). Sets of scalar texture measurements are obtained by computing the average energies of the response maps in both the tumor margin M_{margin} and the core M_{core} region (see Fig. 12.11).

In order to obtain both directional and locally rotation-invariant features, Riesz wavelets are aligned at each position ξ_0 based on the local direction u maximizing the energy of the gradient [34,21]. This alignment strategy is applicable to any Riesz order and is illustrated in Fig. 12.14 for order 1 (*i.e.*, the gradient). The benefits of locally aligning the filters are explained in Section 2.4.3 (moving frames) of Chapter 2 for 2D images. This local alignment is even more important in 3D as the amount of possible directions is much larger in volumetric neighborhoods. To demonstrate the importance of locally aligning the filters, we designed a synthetic example of possible local volumetric tissue architecture. The latter follows the examples provided in Section 2.4.1 of Chapter 2 showing two-dimensional natural and biomedical textures where the local organization of the image directions (LOID) is crucial for their unequivocal description (see Fig. 2.10). Our example extends the latter to 3D and contains three volumetric solid images with three tubular bars each (see Fig. 12.15). They simulate tubular struc-

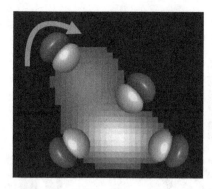

Figure 12.14 Illustration of the local alignment strategy of the Riesz filters on the boundary of a synthetic tumor. Riesz wavelets of order 1 are depicted.

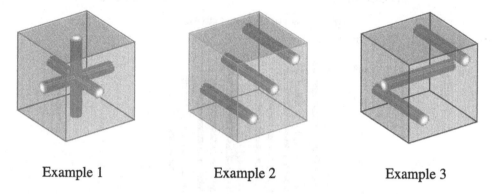

Example 1   Example 2   Example 3

Figure 12.15 Synthetic 3D images containing three tubular structures with varying spatial configurations. The latter simulate, *e.g.*, possible vascular structures.

tures crossing in the 3D space as it may occur with *e.g.*, intricate tissue vasculature. The importance of maintaining the directional information while obtaining locally rotation-invariant features is demonstrated in Fig. 12.16, where the discriminatory power of the various texture features, *i.e.*, 3D GLCMs (see Section 12.2.4.2), 3D LoGs (see Section 12.2.4.3), and aligned 3D Riesz wavelets are compared. GLCMs and LoGs yield almost identical feature values for the three synthetic images, while the aligned Riesz filters are able to discriminate each of them. For this example, seven values of σ_{LoG} were used for the LoG filters. Order 2 and one scale were used for the Riesz wavelets.

12.3 THE QUANTIMAGE ONLINE IMAGE ANALYSIS PLATFORM

The QuantImage online platform consists of a simple web-page where the physicians can upload batches of image series from multiple patients and download a CSV file containing the values of the features described in the previous sections. The underlying

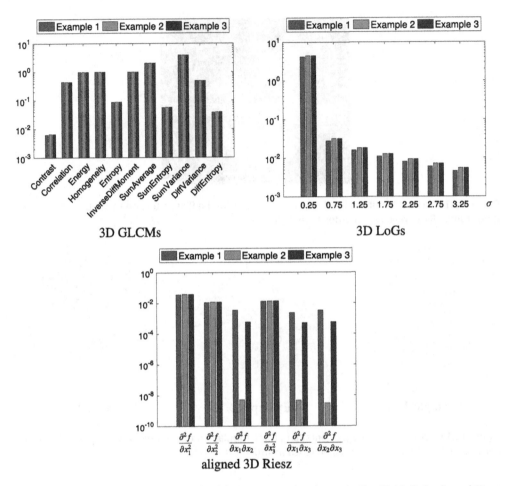

Figure 12.16 Feature values obtained for the three examples shown in Fig. 12.15. Only aligned Riesz wavelets are able to discriminate each of them.

system of the developed web-service follows the pipeline depicted in Fig. 12.3. It mainly consists of three steps from a user point of view: choosing system parameters, uploading DICOM files, and downloading the computed features. These three steps were carefully designed to be user-friendly and based on default parameter values, while users can also have the full control of the computed features and their parameterizations. The home page of QuantImage can be accessed at https://radiomics.hevs.ch and is depicted in Fig. 12.17.

12.3.1 Parameter setting

As a first step, the user can select the parameter values for the QuantImage computing pipeline. The latter are grouped in four consecutive blocks. The first block allows

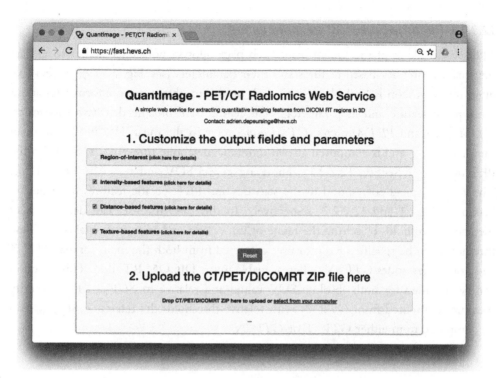

Figure 12.17 Home page of QuantImage, which can be accessed at https://radiomics.hevs.ch.

Figure 12.18 Options for performing axial-wise dilations of *GTV_N* ROIs provided in the DICOM RT files.

performing an initial dilation (in mm) of the ROIs (GTV_T and GTV_N). The ROI named GTV_T corresponds to the delineation of the main tumor (see Fig. 12.18), while GVT_N refers to secondary nodules. This optional dilation is performed in 2D for each axial slice of GTV_T and GTV_N. It is useful when the initial ROIs provided in the DICOM RT files do not cover the full tumor volumes. The subsequent metabolic thresholds τ will yield more precise metabolic subregions. The three following blocks refer to each feature group described in Sections 12.2.2, 12.2.3, 12.2.4 and are detailed in Sections 12.3.1.1, 12.3.1.2, 12.3.1.3, respectively.

12.3.1.1 Intensity-based features parameters

The second block of the QuantImage web page allows choosing whether to extract the various intensity-based features and their parameters (see Fig. 12.19). A clickable question mark icon is available next to each option to obtain more information about the type of feature and its parameters. The features are organized in two subgroups: *CT Measures* and *PET Measures*. *CT Measures* refer to the mean HU inside the main tumor GTV_T and is computed at two different metabolic thresholds τ_i based on the PET image (see Section 12.2.2.1). These are $\tau_1 = 3$ SUV and $\tau_2 = 42\%$ of SUV_{max}. The *PET Measures* correspond to the statistics and PET-specific features described in Sections 12.2.2.1 and 12.2.2.2, respectively. In this case, the user can define sets of metabolic thresholds by setting the range of values of τ and an incremental step. These thresholds can be specified for measures extracted from both the main tumor GTV_T and from other nodes GTV_N (when available in the DICOM RT files). This selection can be specified for both absolute SUV values and relative to SUV_{max}. For intensity measures computed from GTV_N with relative thresholds the reference SUV_{max} can be computed from either GTV_T or GTV_N.

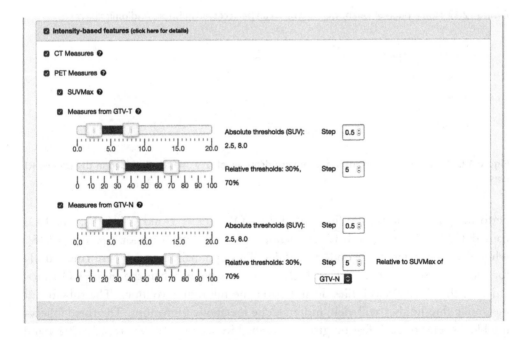

Figure 12.19 Parameter setting for intensity features.

Figure 12.20 Parameter setting for distance features.

12.3.1.2 Distance-based features parameters

The only parameter to set for distance based features is the metabolic threshold for all GTV regions (see Fig. 12.20). The features computed when this option is selected are described in Section 12.2.3.

12.3.1.3 Texture-based features parameters

The last block is dedicated to the parameter setting of the texture-based features described in Section 12.2.4 (3D GLCMs, 3D LoGs, and 3D Riesz). Only the most important parameters of the descriptors can be tuned to limit the complexity of use (see Fig. 12.21). The texture features are extracted from both CT and PET image series inside GTV_T. In the case of the GLCMs the user can choose whether to consider symmetric voxel pairs, the distance between voxels, and the number of gray levels (see Section 12.2.4.2). The features returned for this descriptor are the 11 measures mentioned previously. When selecting 3D LoGs the user must select the range of scales parameterized by σ_{LoG} in between 0.25 and 3.25 with a fixed step of 0.5. As explained in Section 12.2.4.3, these features are computed from both the core and the margin of the ROI. The last section of this block concerns the parameter setting for Riesz wavelets, *i.e.*, the order of the image derivatives and the number of scales. The Riesz energy features are computed from both the core and the margin of the ROI.

12.3.2 File upload

Once the parameters are defined by the user the next step is to upload the DICOM files of the batch of patients to be analyzed on the server. This file transfer includes encryption to protect patient data. The user must provide one unique zip file containing all CT, PET, and DICOM RT files of all patients. The files must be all DICOM files (CT and PET), and DICOM RT (ROI). The file structure inside the zip file is not important, because the system is able to parse the inner structure of the zip file by reading the DICOM headers and to regroup files based on patients and image series. The DICOM RT file must at least include one ROI named GTV_T corresponding to the main tumor. It may contain another optional ROI named GTV_N with other nodes of interest (*e.g.*, metastasis regions). If no GTV_N ROI is provided the corresponding features

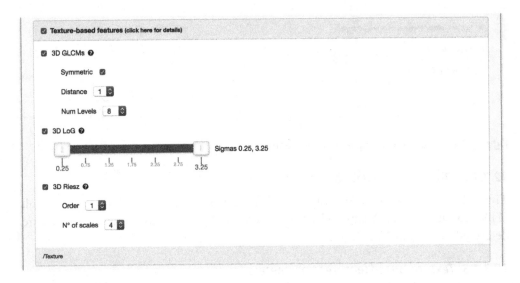

Figure 12.21 Parameter setting for 3D texture features.

will not be computed and will have *NaN* (not a number) values in the returned CSV data structure. After creating the zip file containing the above-mentioned files the user can upload it in the block entitled "2. Upload the CT/PET/DCOMRT ZIP file here" shown in Fig. 12.17. The upload can be done either by browsing the file system or with a simple drag-and-drop. After clicking the button upload a progress bar will show the progress of file upload, followed by the progress of file processing and feature extraction.

12.3.3 Output CSV data structure

The output of the web-service is a CSV file containing the feature values (see Fig. 12.22). The first two rows of the file contain the values of the parameters chosen by the user, allowing reproducibility of the experiments. The next nonempty row of the file contains a header with the names of the computed features. Each following row corresponds to each single patient included in the uploaded zip file. Tables 12.4, 12.5, 12.6, 12.7, and 12.8 in the Appendix detail the various feature names used and their signification. In order to shorten the list of names, we used the following naming conventions in the feature-code field:

- a list of strings in between { } means that one and only one string can be present in the final name,
- a letter in between [ ] means that this letter corresponds to a parameter and it will be encoded by its substituted value,
- a parameter called *charDir* was included for Riesz features that will be replaced by the composition of the letters *X*, *Y*, and *Z*, representing the direction of the Riesz

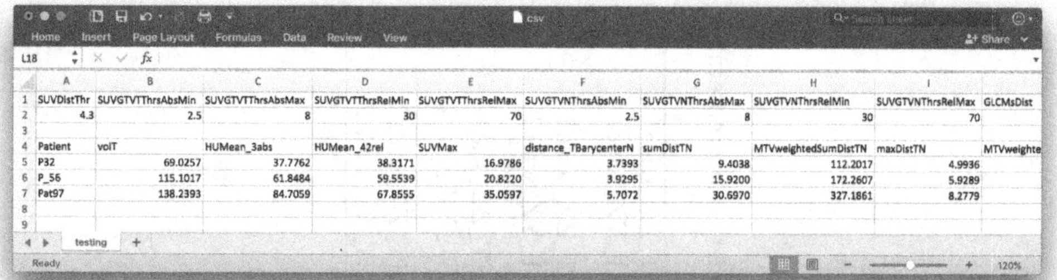

	A	B	C	D	E	F	G	H	I	
1	SUVDistThr	SUVGTVTThrsAbsMin	SUVGTVTThrsAbsMax	SUVGTVTThrsRelMin	SUVGTVTThrsRelMax	SUVGTVNThrsAbsMin	SUVGTVNThrsAbsMax	SUVGTVNThrsRelMin	SUVGTVNThrsRelMax	GLCMsDist
2	4.3	2.5	8	30	70	2.5	8	30	70	
3										
4	Patient	volT	HUMean_3abs	HUMean_42rel	SUVMax	distance_TBarycenterN	sumDistTN	MTVweightedSumDistTN	maxDistTN	MTVweighte
5	P32	69.0257	37.7762	38.3171	16.9786	3.7393	9.4038	112.2017	4.9936	
6	P_56	115.1017	61.8484	59.5539	20.8220	3.9295	15.9200	172.2607	5.9289	
7	Pat97	138.2393	84.7059	67.8555	35.0597	5.7072	30.6970	327.1861	8.2779	
8										
9										

Figure 12.22 Screenshot of the resulting CSV data structure returned by the system. In this particular case, three patients with identifiers "P32," "P_56," and "Pat96" were processed (see lines 5–7). The parameter values used for feature extraction are displayed in lines 1 and 2.

filter following the partial directional image derivatives (*e.g.*, XX corresponds to $\partial^2/\partial x_1^2$, XY corresponds to $\partial^2/\partial x_1 \partial x_2$).

As an example, let us consider the following feature-code:

*tex_InnerRieszN[n]_*{CT,PET}*_[*charDir*]_scale[s].*

Let then suppose that the user selected a Riesz order of 2, and 2 wavelet scales. Then this feature-code will be denoted in the CSV file with all the forms listed in Table 12.3.

12.4 DISCUSSIONS AND CONCLUSIONS

We presented an online radiomics tool called QuantImage. The latter is based on a user-friendly web-service allowing the extraction of current state-of-the-art three-dimensional quantitative imaging features from PET and CT images. To the best of our knowledge, this is the first radiomics-specific tool that can be simply accessed through a secure web platform. The latter enables the extraction of quantitative imaging parameters from large cohorts of patients without the need of neither software nor hardware installation. The online tool can therefore be accessed directly from a clinical environment. We believe that this effort is timely to further validate the relevance of radiomics imaging biomarkers with large-scale clinical studies in the context of well-defined and well-controlled oncological contexts. The features extracted by the system include intensity-, distance-, and texture-based descriptors. While all feature groups are preconfigured with default parameter values, the user has the possibility to further choose which features to extract and to tune their most important parameters. The most relevant aspects of these features are explained in this chapter, which can be used as a reference manual of the QuantImage system. The output of the system is in the standard CSV data format, offering the possibility to manually analyze the feature values with data plots, as well as their inclusion in more complex systems including advanced uni

Table 12.3 Variations of the feature-code *tex_InnerRieszN[n]_{CT,PET}_[charDir]_scale[s]* when using Riesz wavelets of order 2 and 2 consecutive scales

tex_InnerRieszN2_CT_XX_scale1	tex_InnerRieszN2_CT_YY_scale1	tex_InnerRieszN2_CT_ZZ_scale1
tex_InnerRieszN2_CT_XY_scale1	tex_InnerRieszN2_CT_XZ_scale1	tex_InnerRieszN2_CT_YZ_scale1
tex_InnerRieszN2_CT_XX_scale2	tex_InnerRieszN2_CT_YY_scale2	tex_InnerRieszN2_CT_ZZ_scale2
tex_InnerRieszN2_CT_XY_scale2	tex_InnerRieszN2_CT_XZ_scale2	tex_InnerRieszN2_CT_YZ_scale2
tex_InnerRieszN2_PET_XX_scale1	tex_InnerRieszN2_PET_YY_scale1	tex_InnerRieszN2_PET_ZZ_scale1
tex_InnerRieszN2_PET_XY_scale1	tex_InnerRieszN2_PET_XZ_scale1	tex_InnerRieszN2_PET_YZ_scale1
tex_InnerRieszN2_PET_XX_scale2	tex_InnerRieszN2_PET_YY_scale2	tex_InnerRieszN2_PET_ZZ_scale2
tex_InnerRieszN2_PET_XY_scale2	tex_InnerRieszN2_PET_XZ_scale2	tex_InnerRieszN2_PET_YZ_scale2

Table 12.4 Feature-codes used for general information in the CSV file

Feature-code	Description
Patient	Patient ID found in DICOM header.
volT	Volume of GTV_T.

Table 12.5 Feature-codes for intensity features used in the CSV file (part I): features extracted in ROI GTV_T

Feature-code	Description
int_HUMean_3abs	Mean HU of the main tumor voxels with SUV >3.
int_HUMean_42rel	Mean HU of the main tumor voxels with SUV >42 % of SUVMax.
int_SUVMax	Maximum SUV inside the main tumor.
int_SUVMean_T_[τ]abs	Mean SUV of the main tumor voxels with SUV $> \tau$.
int_SUVVariance_T_[τ]abs	SUV variance of the main tumor voxels with SUV $> \tau$.
int_SUVSkewness_T_[τ]abs	SUV skewness of the main tumor voxels with SUV $> \tau$.
int_SUVKurtosis_T_[τ]abs	SUV kurtosis of the main tumor voxels with SUV $> \tau$.
int_SUVPeak_T_[τ]abs	Mean SUV of the main tumor voxels with SUV $> \tau$ in a spherical neighborhood of a radius of 8 voxels centered at the voxel with maximum SUV.
int_MTV_T_[τ]abs	Volume (in mm) of the ROI containing the main tumor voxels with SUV $> \tau$.
int_TLG_T_[τ]abs	Total lesion glycosis of the ROI containing the main tumor voxels with SUV $> \tau$.
int_SUVMean_T_[τ]rel	Mean SUV of the main tumor voxels with SUV $> \tau$ % of SUVMax.
int_SUVVariance_T_[τ]rel	SUV variance of the main tumor voxels with SUV $> \tau$ % of SUVMax.
int_SUVSkewness_T_[τ]rel	SUV skewness of the main tumor voxels with SUV $> \tau$ % of SUVMax.
int_SUVKurtosis_T_[τ]rel	SUV kurtosis of the main tumor voxels with SUV $> \tau$ % of SUVMax.
int_SUVPeak_T_[τ]rel	Mean SUV of the main tumor voxels with SUV $> \tau$ % of SUVMax in a spherical neighborhood of a radius of 8 voxels centered at the voxel with maximum SUV.
int_MTV_T_[τ]rel	Volume (in mm) of the ROI containing the main tumor voxels with SUV $> \tau$ % of SUVMax.
int_TLG_T_[τ]rel	Total lesion glycosis of the ROI containing the main tumor voxels with SUV $> \tau$ % of SUVMax.

Table 12.6 Feature-codes for intensity features used in the CSV file (part II): features extracted in ROI *GTV_N*

Feature-code	Description
int_SUVMean_N_[τ]abs	Mean SUV of the metastases voxels with SUV > τ.
int_SUVVariance_N_[τ]abs	SUV variance of the metastases voxels with SUV > τ.
int_SUVSkewness_N_[τ]abs	SUV skewness of the metastases voxels with SUV > τ.
int_SUVKurtosis_N_[τ]abs	SUV kurtosis of the metastases voxels with SUV > τ.
int_MTV_N_[τ]abs	Volume (in mm) of the ROI containing the metastases voxels with SUV > τ.
int_TLG_N_[τ]abs	Total lesion glycosis of the ROI containing the metastases voxels with SUV > τ.
int_SUVMean_N_[τ]rel	Mean SUV of the metastases voxels with SUV > τ % of SUVMax.
int_SUVVariance_N_[τ]rel	SUV variance of the metastases voxels with SUV > τ % of SUVMax.
int_SUVSkewness_N_[τ]rel	SUV skewness of the metastases voxels with SUV > τ % of SUVMax.
int_SUVKurtosis_N_[τ]rel	SUV kurtosis of the metastases voxels with SUV > τ % of SUVMax.
int_MTV_N_[τ]rel	Volume (in mm) of the ROI containing the metastases voxels with SUV > τ % of SUVMax.
int_TLG_N_[τ]rel	Total lesion glycosis of the ROI containing the metastases voxels with SUV > τ % of SUVMax.

Table 12.7 Feature-codes for distance features used in the CSV file

Feature-code	Description
dst_TBarycenterN	Distance between barycenter of the main tumor and the barycenter of all metastases (see Eq. (12.10)).
dst_sumDistTN	Cumulative distance between the barycenter of each metastasis and the barycenter of the main tumor (see Eq. (12.11)).
dst_MTVweighted-SumDistTN	MTV-weighted cumulative distance between the barycenter of each metastasis and the barycenter of the main tumor (see Eq. (12.12)).
dst_maxDistTN	Maximum metastasis remoteness (see Eq. (12.13)).
dst_MTVweighted-MaxDistTN	MTV-weighted maximum metastasis remoteness (see Eq. (12.14)).
dst_sumDistNBarycenterN	Cumulative distance between the barycenter of each metastasis and the barycenter of all metastases (see Eq. (12.15)).
dst_MTVweightedSumDist-NBarycenterN	MTV-weighted cumulative distance between the barycenter of each metastasis and the barycenter of all metastases (see Eq. (12.16)).

Table 12.8 Feature-codes for texture features used in the CSV file

Feature-code	Description
tex_GLCMsContrast_{CT,PET}_[d]_[n]	Contrast measure of the GLCMs computed with distance d and number of gray levels n for CT or PET.
tex_GLCMsCorrelation_{CT,PET}_[d]_[n]	Correlation measure of the GLCMs computed with distance d and number of gray levels n for CT or PET.
tex_GLCMsEnergy_{CT,PET}_[d]_[n]	Energy measure of the GLCMs computed with distance d and number of gray levels n for CT or PET.
tex_GLCMsHomogeneity_{CT,PET}_[d]_[n]	Homogeneity measure of the GLCMs computed with distance d and number of gray levels n for CT or PET.
tex_GLCMsEntropy_{CT,PET}_[d]_[n]	Entropy measure of the GLCMs computed with distance d and number of gray levels n for CT or PET.
tex_GLCMsInverseDiffMoment_{CT,PET}_[d]_[n]	Inverse difference moment measure of the GLCMs computed with distance d and number of gray levels n for CT or PET.
tex_GLCMsSumAverage_{CT,PET}_[d]_[n]	Sum average measure of the GLCMs computed with distance d and number of gray levels n for CT or PET.
tex_GLCMsSumEntropy_{CT,PET}_[d]_[n]	Sum entropy measure of the GLCMs computed with distance d and number of gray levels n for CT or PET.
tex_GLCMsSumVariance_{CT,PET}_[d]_[n]	Sum variance measure of the GLCMs computed with distance d and number of gray levels n for CT or PET.
tex_GLCMsDiffVariance_{CT,PET}_[d]_[n]	Difference variance measure of the GLCMs computed with distance d and number of gray levels n for CT or PET.
tex_GLCMsDiffEntropy_{CT,PET}_[d]_[n]	Difference entropy measure of the GLCMs computed with distance d and number of gray levels n for CT or PET.
tex_InnerLoG_{CT,PET}_sigma[σ]	LoG in the core of the main tumor with sigma σ for CT or PET.
tex_MarginLoG_{CT,PET}_sigma[σ]	LoG in the margin of the main tumor with sigma σ for CT or PET.
tex_InnerRieszN[n]_{CT, PET}_[$charDir$]_scale[s]	Riesz energy of the Riesz filter with direction $charDir$ of order n and scale s in the core of the main tumor for CT or PET.
tex_MarginRieszN[n]_{CT, PET}_[$charDir$]_scale[s]	Riesz energy of the Riesz filter with direction $charDir$ of order n and scale s in the margin of the main tumor for CT or PET.

and multivariate statistical methods or machine learning techniques such data clustering, linear discriminant analysis, LASSO regression, or support vector machines to name a few. Whereas all intensity- and texture-based quantitative imaging features were previously validated in various applicative contexts [25,14,35–37,35,38–40,22,41,11,42–52]. The presented system was initially validated in the context of head and neck [15,53,54] tumors at the Lausanne University Hospital (CHUV). Future work includes the implementation of covariance-based aggregation functions as an alternative to the average for LoGs and Riesz wavelets [25], as well as three-dimensional shape features for the characterization of the tumoral contour.

ACKNOWLEDGMENTS

This work was partly supported by the Swiss National Science Foundation with grant agreements 320030–146804 and PZ00P2_154891, and the CIBM.

APPENDIX 12.A

This section details the various feature names and their signification as listed in the CSV data structure returned by the system. General information, intensity (parts I and II), distance and texture feature-codes are detailed in Tables 12.4, 12.5, 12.6, 12.7, and 12.8, respectively.

REFERENCES

[1] R.A. Gatenby, O. Grove, R.J. Gillies, Quantitative imaging in cancer evolution and ecology, Radiology 269 (1) (2013) 8–14.

[2] R. Mirnezami, J. Nicholson, A. Darzi, Preparing for precision medicine, N. Engl. J. Med. 366 (6) (2012) 489–491.

[3] M. Gerlinger, A.J. Rowan, S. Horswell, J. Larkin, D. Endesfelder, E. Gronroos, P. Martinez, N. Matthews, A. Stewart, P. Tarpey, I. Varela, B. Phillimore, S. Begum, N.Q. McDonald, A. Butler, D. Jones, K. Raine, C. Latimer, C.R. Santos, M. Nohadani, A.C. Eklund, B. Spencer-Dene, G. Clark, L. Pickering, G. Stamp, M. Gore, Z. Szallasi, J. Downward, P.A. Futreal, C. Swanton, Intratumor heterogeneity and branched evolution revealed by multiregion sequencing, N. Engl. J. Med. 366 (10) (2012) 883–892.

[4] V. Kumar, Y. Gu, S. Basu, A. Berglund, S.A. Eschrich, M.B. Schabath, K. Forster, H.J.W.L. Aerts, A. Dekker, D. Fenstermacher, D.B. Goldgof, L.O. Hall, P. Lambin, Y. Balagurunathan, R.A. Gatenby, R.J. Gillies, Radiomics: the process and the challenges, Magn. Reson. Imaging 30 (9) (2012) 1234–1248.

[5] C.C. Jaffe, Imaging and genomics: is there a synergy?, Radiology 264 (2) (2012) 329–331.

[6] R.T.H. Leijenaar, G. Nalbantov, S. Carvalho, W.J.C. van Elmpt, E.G.C. Troost, R. Boellaard, H.J.W.L. Aerts, R.J. Gillies, P. Lambin, The effect of SUV discretization in quantitative FDG-PET radiomics: the need for standardized methodology in tumor texture analysis, Sci. Rep. 5 (August 2015) 11075.

[7] H.J.W.L. Aerts, E.R. Velazquez, R.T.H. Leijenaar, C. Parmar, P. Grossmann, S. Carvalho, J. Bussink, R. Monshouwer, B. Haibe-Kains, D. Rietveld, F. Hoebers, M.M. Rietbergen, C.R. Leemans,

A. Dekker, J. Quackenbush, R.J. Gillies, P. Lambin, Decoding tumour phenotype by noninvasive imaging using a quantitative radiomics approach, Nat. Commun. 5 (June 2014).

[8] O. Gevaert, J. Xu, C.D. Hoang, A.N. Leung, Y. Xu, A. Quon, D.L. Rubin, S. Napel, S.K. Plevritis, Non-small cell lung cancer: identifying prognostic imaging biomarkers by leveraging public gene expression microarray data — methods and preliminary results, Radiology 264 (2) (2012) 387–396.

[9] O. Gevaert, L.A. Mitchell, A.S. Achrol, J. Xu, S. Echegaray, G.K. Steinberg, S.H. Cheshier, S. Napel, G. Zaharchuk, S.K. Plevritis, Glioblastoma multiforme: exploratory radiogenomic analysis by using quantitative image features, Radiology 273 (1) (2014) 168–174.

[10] S. Napel, M. Giger, Special section guest editorial: radiomics and imaging genomics: quantitative imaging for precision medicine, J. Med. Imag. 2 (4) (2015) 41001.

[11] B. Ganeshan, V. Goh, H.C. Mandeville, Q.S. Ng, P.J. Hoskin, K.A. Miles, Non-small cell lung cancer: histopathologic correlates for texture parameters at CT, Radiology 266 (1) (2013) 326–336.

[12] M. Ravanelli, D. Farina, M. Morassi, E. Roca, G. Cavalleri, G. Tassi, R. Maroldi, Texture analysis of advanced non-small cell lung cancer (NSCLC) on contrast-enhanced computed tomography: prediction of the response to the first-line chemotherapy, Eur. Radiol. 23 (12) (2013) 3450–3455.

[13] S.A. Mattonen, D.A. Palma, C.J.A. Haasbeek, S. Senan, A.D. Ward, Early prediction of tumor recurrence based on CT texture changes after stereotactic ablative radiotherapy (SABR) for lung cancer, Med. Phys. 41 (3) (2014) 1–14.

[14] A. Depeursinge, M. Yanagawa, A.N. Leung, D.L. Rubin, Predicting adenocarcinoma recurrence using computational texture models of nodule components in lung CT, Med. Phys. 42 (2015) 2054–2063.

[15] J. Castelli, B. De Bari, A. Depeursinge, A. Simon, A. Devilliers, G. Roman Jimenez, J. Prior, R. De Crevoisier, J. Bourhis, Overview of the predictive value of quantitative 18 FDG PET in head and neck cancer treated with chemoradiotherapy, Crit. Rev. Oncol./Hematol. 108 (2016) 40–51.

[16] A. Depeursinge, A. Foncubierta-Rodríguez, D. Van De Ville, H. Müller, Three-dimensional solid texture analysis and retrieval in biomedical imaging: review and opportunities, Med. Image Anal. 18 (1) (2014) 176–196.

[17] Y.-H.D. Fang, C.-Y. Lin, M.-J. Shih, H.-M. Wang, T.-Y. Ho, C.-T. Liao, T.-C. Yen, Development and evaluation of an open-source software package "CGITA" for quantifying tumor heterogeneity with molecular images, BioMed Res. Int. 2014 (2014) 248505, http://dx.doi.org/10.1155/2014/248505.

[18] F. Orlhac, M. Soussan, J.-A. Maisonobe, C.A. Garcia, B. Vanderlinden, I. Buvat, Tumor texture analysis in ^{18}F-FDG PET: relationships between texture parameters, histogram indices, standardized uptake values, metabolic volumes, and total lesion glycolysis, J. Nucl. Med. 55 (3) (2014) 414–422.

[19] D.V. Fried, O. Mawlawi, L. Zhang, X. Fave, S. Zhou, G. Ibbott, Z. Liao, L.E. Court, Stage III non-small cell lung cancer: prognostic value of FDG PET quantitative imaging features combined with clinical prognostic factors, Radiology 278 (1) (2016) 214–222.

[20] R.M. Haralick, K. Shanmugam, I. Dinstein, Textural features for image classification, IEEE Trans. Syst. Man Cybern. 3 (6) (November 1973) 610–621.

[21] N. Chenouard, M. Unser, 3D steerable wavelets in practice, IEEE Trans. Image Process. 21 (11) (2012) 4522–4533.

[22] B. Ganeshan, K. Skogen, I. Pressney, D. Coutroubis, K. Miles, Tumour heterogeneity in oesophageal cancer assessed by CT texture analysis: preliminary evidence of an association with tumour metabolism, stage, and survival, Clinical Radiology 67 (2) (2012) 157–164.

[23] D.V. Fried, O. Mawlawi, L. Zhang, X. Fave, S. Zhou, G. Ibbott, Z. Liao, L.E. Court, Potential use of ^{18}F-fluorodeoxyglucose positron emission tomography-based quantitative imaging features for guiding dose escalation in stage III non-small cell lung cancer, Int. J. Radiat. Oncol. Biol. Phys. 94 (2) (2016) 368–376.

[24] A. Depeursinge, P. Pad, A.C. Chin, A.N. Leung, D.L. Rubin, H. Müller, M. Unser, Optimized steerable wavelets for texture analysis of lung tissue in 3-D CT: classification of usual interstitial pneumonia, in: IEEE 12th International Symposium on Biomedical Imaging, ISBI 2015, IEEE, apr 2015, pp. 403–406.

[25] P. Cirujeda, Y. Dicente Cid, H. Müller, D. Rubin, T.A. Aguilera, B.W. Loo Jr., M. Diehn, X. Binefa, A. Depeursinge, A 3-D Riesz-covariance texture model for prediction of nodule recurrence in lung CT, IEEE Trans. Med. Imaging 26 (4) (Dec. 2016) 2620–2630, http://dx.doi.org/10.1109/TMI. 2016.2591921.

[26] L.K. Soh, C. Tsatsoulis, Texture analysis of SAR sea ice imagery using gray level co-occurrence matrices, IEEE Trans. Geosci. Remote Sens. 37 (2) (mar 1999) 780–795.

[27] D.A. Clausi, An analysis of co-occurrence texture statistics as a function of grey level quantization, Can. J. Remote Sens. 28 (1) (2002) 45–62.

[28] R.M. Haralick, Statistical and structural approaches to texture, Proc. IEEE 67 (5) (may 1979) 786–804.

[29] A. Depeursinge, C. Kurtz, C.F. Beaulieu, S. Napel, D.L. Rubin, Predicting visual semantic descriptive terms from radiological image data: preliminary results with liver lesions in CT, IEEE Trans. Med. Imaging 33 (8) (August 2014) 1–8.

[30] J. Barker, A. Hoogi, A. Depeursinge, D.L. Rubin, Automated classification of brain tumor type in whole-slide digital pathology images using local representative tiles, Med. Image Anal. 30 (2016) 60–71.

[31] A. Depeursinge, A. Foncubierta-Rodríguez, D. Van De Ville, H. Müller, Multiscale lung texture signature learning using the Riesz transform, in: Medical Image Computing and Computer-Assisted Intervention MICCAI 2012, in: Lecture Notes in Computer Science, vol. 7512, Springer, Berlin/Heidelberg, October 2012, pp. 517–524.

[32] S. Otálora, A. Cruz Roa, J. Arevalo, M. Atzori, A. Madabhushi, A. Judkins, F. González, H. Müller, A. Depeursinge, Combining unsupervised feature learning and Riesz wavelets for histopathology image representation: application to identifying anaplastic medulloblastoma, in: N. Navab, J. Hornegger, W.M. Wells, A. Frangi (Eds.), Medical Image Computing and Computer-Assisted Intervention — MICCAI 2015, in: Lecture Notes in Computer Science, vol. 9349, Springer International Publishing, oct 2015, pp. 581–588.

[33] J.P. Ward, P. Pad, M. Unser, Optimal isotropic wavelets for localized tight frame representations, IEEE Signal Process. Lett. 22 (11) (nov 2015) 1918–1921.

[34] Y. Dicente Cid, H. Müller, A. Platon, P.-A. Poletti, A. Depeursinge, 3D solid texture classification using locally-oriented wavelet transforms, IEEE Trans. Image Process. 26 (4) (April 2017) 1899–1910, http://dx.doi.org/10.1109/TIP.2017.2665041.

[35] O.S. Al-Kadi, D. Van De Ville, A. Depeursinge, Multidimensional texture analysis for improved prediction of ultrasound liver tumor response to chemotherapy treatment, in: Medical Image Computing and Computer-Assisted Interventions (MICCAI), vol. 9900, Springer International Publishing, 2016, pp. 619–626.

[36] N. Michoux, L. Bollondi, A. Depeursinge, L. Fellah, I. Leconte, H. Müller, Is tumor heterogeneity quantified by 3D texture analysis of MRI able to predict non-response to NAC in breast cancer?, in: European Society for Magnetic Resonance in Medicine and Biology, ESMRMB, 2016.

[37] N. Michoux, L. Bollondi, A. Depeursinge, A. Geissbuhler, L. Fellah, H. Müller, I. Leconte, Predicting non-response to NAC in patients with breast cancer using 3D texture analysis, in: European Congress of Radiology (ECR2015), ECR2015, Vienna, Autria, mar 2015.

[38] G. Sonn, L.D. Hahn, R. Fan, J. Barker, A. Depeursinge, D.L. Rubin, Quantitative image texture analysis predicts malignancy on multiparametric prostate MRI, in: 91st Annual Meeting of the Western Section of American Urological Association, WSAUA, October 2015.

[39] A. Madabhushi, M.D. Feldman, D.N. Metaxas, D. Chute, J.E. Tomaszewski, A novel stochastic combination of 3D texture features for automated segmentation of prostatic adenocarcinoma from high resolution MRI, in: Medical Image Computing and Computer-Assisted Intervention — MICCAI 2003, 6th International Conference, November 2003, pp. 581–591.

[40] B. Ganeshan, S. Abaleke, R.C.D. Young, C.R. Chatwin, K.A. Miles, Texture analysis of non-small cell lung cancer on unenhanced computed tomography: initial evidence for a relationship with tumour glucose metabolism and stage, Cancer Imaging 10 (jul 2010) 137–143.

[41] B. Ganeshan, E. Panayiotou, K. Burnand, S. Dizdarevic, K.A. Miles, Tumour heterogeneity in non-small cell lung carcinoma assessed by CT texture analysis: a potential marker of survival, Eur. Radiol. 22 (4) (2012) 796–802.

[42] D. Mahmoud-Ghoneim, G. Toussaint, J.-M. Constans, J.D. de Certaines, Three dimensional texture analysis in MRI: a preliminary evaluation in gliomas, Magn. Reson. Imaging 21 (9) (November 2003) 983–987.

[43] S.B. Antel, D.L. Collins, N. Bernasconi, F. Andermann, R. Shinghal, R.E. Kearney, D.L. Arnold, A. Bernasconi, Automated detection of focal cortical dysplasia lesions using computational models of their MRI characteristics and texture analysis, NeuroImage 19 (4) (August 2003) 1748–1759.

[44] A. Huisman, L.S. Ploeger, H.F.J. Dullens, T.N. Jonges, J.A.M. Belien, G.A. Meijer, N. Poulin, W.E. Grizzle, P.J. van Diest, Discrimination between benign and malignant prostate tissue using chromatin texture analysis in 3D by confocal laser scanning microscopy, Prostate 67 (3) (February 2007) 248–254.

[45] W. Chen, M.L. Giger, H. Li, U. Bick, G.M. Newstead, Volumetric texture analysis of breast lesions on contrast-enhanced magnetic resonance images, Magn. Reson. Med. 58 (3) (sep 2007) 562–571.

[46] T.Y. Kim, H.K. Choi, Computerized renal cell carcinoma nuclear grading using 3D textural features, in: IEEE International Conference on Communications Workshops, 2009. ICC Workshops 2009, jun 2009, pp. 1–5.

[47] D. Kontos, P.R. Bakic, A.-K. Carton, A.B. Troxel, E.F. Conant, A.D.A. Maidment, Parenchymal texture analysis in digital breast tomosynthesis for breast cancer risk estimation: a preliminary study, Acad. Radiol. 16 (3) (2009) 283–298.

[48] P. Georgiadis, D. Cavouras, I. Kalatzis, D. Glotsos, E. Athanasiadis, S. Kostopoulos, K. Sifaki, M. Malamas, G. Nikiforidis, E. Solomou, Enhancing the discrimination accuracy between metastases, gliomas and meningiomas on brain MRI by volumetric textural features and ensemble pattern recognition methods, Magn. Reson. Imaging 27 (1) (January 2009) 120–130.

[49] T.Y. Kim, H.J. Choi, H. Hwang, H.K. Choi, Three-dimensional texture analysis of renal cell carcinoma cell nuclei for computerized automatic grading, J. Med. Syst. 34 (4) (aug 2010) 709–716.

[50] G. De Nunzio, G. Pastore, M. Donativi, A. Castellano, A. Falini, A CAD system for cerebral glioma based on texture features in DT-MR images, Nucl. Instrum. Methods Phys. Res., Sect. A, Accel. Spectrom. Detect. Assoc. Equip. 648 (Suppl (0)) (2011) 100–102.

[51] S. Basu, L.O. Hall, D.B. Goldgof, Y. Gu, V. Kumar, J. Choi, R.J. Gillies, R.A. Gatenby, Developing a classifier model for lung tumors in CT-scan images, in: IEEE International Conference on Systems, Man, and Cybernetics, October 2011, pp. 1306–1312.

[52] S.A. Christe, B.V. Kumari, A. Kandaswamy, Experimental study for 3D statistical property based intracranial brain tumor classification, J. Sci. Ind. Res. 71 (1) (2012) 36–44.

[53] J. Castelli, A. Depeursinge, V. Ndoh, J. Prior, E. Ozsahin, A. Devillers, H. Bouchaab, E. Chajon, R. De Crevoisier, N. Scher, F. Jegoux, B. Laguerre, B. De Bari, J. Bourhis, Development and validation of a PET-based nomogram predictive for survival of oropharyngeal cancer patients, Eur. J. Cancer 75 (2017) 222–230.

[54] J. Castelli, A. Depeursinge, B. De Bari, A. Devillers, R. De Crevoisier, J. Bourhis, J.O. Prior, Metabolic tumor volume and total lesion glycolysis in oropharyngeal cancer treated with definitive radiotherapy: which threshold is the best predictor of local control?, Clin. Nucl. Med. 42 (6) (2017) e281–e285, http://dx.doi.org/10.1097/RLU.0000000000001614.

CHAPTER 13

Web-Based Tools for Exploring the Potential of Quantitative Imaging Biomarkers in Radiology
Intensity and Texture Analysis on the ePAD Platform

Roger Schaer*, Yashin Dicente Cid*,†, Emel Alkim§, Sheryl John§, Daniel L. Rubin§, Adrien Depeursinge*,‡

*University of Applied Sciences Western Switzerland (HES-SO), Institute of Information Systems, Sierre, Switzerland
†University of Geneva, Computer Vision and Multimedia Laboratory, Geneva, Switzerland
‡École Polytechnique Fédérale de Lausanne (EPFL), Biomedical Imaging Group, Lausanne, Switzerland
§Stanford University, Department of Radiology and Medicine (Biomedical Informatics), Stanford, CA, USA

Abstract

Recent papers in the field of radiomics showed strong evidence that novel image biomarkers based on structural tissue properties have the potential to complement and even surpass invasive and costly biopsy-based molecular assays in certain clinical contexts. To date, very few translations of these research results to clinical practive have been carried out. In addition, a majority of the identified imaging biomarkers are perceived as black boxes by end-users, hindering their acceptance in clinical and research environments. We present a suite of plugins called Quantitative Feature Explore (QFExplore) for the open-access cloud-based ePAD platform enabling the exploration and validation of new imaging biomarkers in a clinical environment. The latter include the extraction, visualization and comparison of intensity- and texture-based quantitative imaging features, regional division of regions of interest to reveal tissue diversity, as well as the construction, use and sharing of user-personalized statistical machine learning models. No software installation is required and the platform can be accessed through any web browser. The relevance of the developed tools is demonstrated in the context of various clinical use-cases. The software is available online.

Keywords

Quantitative imaging biomarkers, Radiomics, Texture analysis, Medical imaging, Radiology, Web technologies, Machine learning

13.1 INTRODUCTION

Modern multidimensional imaging in radiology yields much more information than the naked eye can appreciate. As a result, errors and variations in interpretations are currently the weakest aspect of clinical imaging. Computerized quantitative image analysis may provide solutions for ensuring the quality of medical image interpretation by yielding exhaustive, comprehensive, and reproducible analysis of imaging features, which

Biomedical Texture Analysis
DOI: 10.1016/B978-0-12-812133-7.00013-2

spawned the new fields of *radiomics* and *imaging genomics* [1] (see Chapter 8). It has the potential to complement, and possibly even surpass invasive biopsy-based molecular assays with the ability to capture intralesional heterogeneity in a noninvasive way [2]. Two large initiatives have recently been created in North America to coordinate and promote research and development of quantitative imaging methods for cancer: the Quantitative Imaging Network (QIN)[1] from the National Cancer Institute (NCI), and the Quantitative Imaging Biomarkers Alliance (QIBA)[2] from the Radiological Society of North America (RSNA). Although the research community in radiomics provided strong evidence that the novel imaging biomarkers may have a very high potential clinical value, very few translations of these biomarkers into clinical trials or medical practice were carried out so far. In 2012, the United States (US) Food and Drugs Administration (FDA) issued a comprehensive report[3] containing guidance for industry on how to describe and evaluate computer-assisted detection devices applied to radiology data in order to obtain FDA approval.

Quantitative imaging features can be difficult to interpret, as they can appear abstract to a medical doctor, since their meaning in the clinical context may not be directly obvious. It is very timely to develop front-end software enabling the exploration and validation of the imaging biomarkers in a clinical environment. Tools are required to allow experimentation with, *e.g.*, various features, parameter settings. Such tools can lead to a better understanding of relatively intangible numerical values. It may help in understanding which values are affected by a certain pathology, how one value relates to another, how certain values generalize across a cohort of patients, or how a set of variables can differentiate between an unhealthy patient and a control case. During the past 10 years, an increasing number of commercial, freeware, and open-source quantitative imaging software for radiology data have been developed. Some of them are for general purpose, or either organ-, imaging modality-, or disease-specific. Noteworthy examples are listed in Table 13.1. Along with Original Equipment Manufacturer (OEM) image analysis software, some of the examples listed in Table 13.1 reported high acceptance and usefulness in the clinical environment. The market for these tools appears to be growing at a higher rate as compared to the equipment market.[4] Most of current software available implementing quantitative biomarkers is based on image intensity and shape only. Moreover, the existing software applications leveraging structural tissue properties (*i.e.*, image texture) are most often not easy to use for non-medical experts and do not have application-specific clinical workflows implemented (*e.g.*, Slicer, Analyze, MeVisLab). In addition, they generally provide only basic texture features (*i.e.*, Haralick, isotropic

[1] http://imaging.cancer.gov/programsandresources/specializedinitiatives/qin, as of December 5, 2016.
[2] https://www.rsna.org/qiba/, as of December 5, 2016.
[3] http://www.fda.gov/RegulatoryInformation/Guidances/ucm187249.htm, as of December 5, 2016.
[4] http://www.marketsandmarkets.com/Market-Reports/medical-image-analysis-software-market-846.html, as of December 5, 2016.

Table 13.1 Overview of existing quantitative imaging software

Imaging software	Medical structure	Imaging modality	Quantitative biomarkers	License	Type
AMIDE[a]	multiple	multiple	intensity	freeware	standalone
Analyze[b]	multiple	multiple	intensity, shape, texture	commercial	standalone
Apollo MIT[c]	multiple	CT, DCE-MRI	intensity	commercial	multiple
CGITA[d] [3]	multiple	PET, CT, MRI	intensity, texture	open-source	standalone
Dynamika[e]	multiple	(DCE-) MRI	intensity	multiple	cloud
ePAD	multiple	multiple	intensity	open-source	cloud
LifeX[f]	multiple	PET, CT, MRI	intensity, texture	freeware	standalone
MaZda[g]	multiple	multiple	intensity, texture	open-source	standalone
MeVis[h], MeVisLab[i]	multiple	multiple	intensity, shape, texture	multiple	multiple
Medis[j]	cardiovascular	multiple	intensity, shape	commercial	standalone
MIPAV[k]	multiple	multiple	intensity, shape	open-source	standalone
mint Lesion[l]	multiple	multiple	intensity	commercial	multiple
Pie Medical Imaging[m]	cardiovascular	multiple	intensity, shape	commercial	standalone
Slicer[n]	multiple	multiple	intensity, shape, texture	open-source	standalone
TexRAD[o]	multiple	CT, MRI	texture	commercial	cloud
VIDA diagnostics[p]	lungs	CT	intensity, shape	commercial	cloud

[a] http://amide.sourceforge.net/index.html, as of December 5, 2016.
[b] http://analyzedirect.com, as of December 5, 2016.
[c] Apollo Medical Imaging Technology, http://www.apollomit.com/mrm.htm, as of December 5, 2016.
[d] https://code.google.com/archive/p/cgita/, as of December 5, 2016.
[e] http://www.imageanalysis.org.uk/demo-dynamika as of December 5, 2016.
[f] http://www.lifexsoft.org, as of December 5, 2016.
[g] http://www.eletel.p.lodz.pl/programy/mazda/, as of December 5, 2016.
[h] http://www.mevis.de, as of December 5, 2016.
[i] http://www.mevislab.de, as of December 2016.
[j] http://www.medis.nl, as of 5 December 5, 2016.
[k] http://mipav.cit.nih.gov, as of December 5, 2016.
[l] https://mint-medical.com/products-solutions/, as of December 5, 2016.
[m] http://www.piemedicalimaging.com, as of December 5, 2016.
[n] http://www.slicer.org, as of 5 December 5, 2016.
[o] http://texrad.com, as of 5 December 5, 2016.
[p] http://vidadiagnostics.com, as of December 5, 2016.

filtering), which have shown limited abilities to fully leverage multiscale and directional texture patterns when compared to recent approaches in the field [4–6] (see Chapter 3).

The electronic Patient Annotation Device (ePAD) imaging platform[5] [7–9] is a freely available quantitative imaging informatics platform, developed by the Rubin Lab at Stanford University. Its main objectives include providing a rich web client application that runs on any modern browser and does not require software installation on the client side, catalyzing the adoption and use of quantitative imaging data in clinical practice, and providing a flexible architecture enabling third-party developers to extend the capabilities of ePAD and adapt it to more specific needs and personalized use-cases. Thanks to its plugin architecture, ePAD can be used to support a wide range of imaging-based projects. The platform has been widely used by radiologists at various academic centers in the United States and showed to provide state-of-the-art solutions for visualizing, collecting, annotating, and sharing data between clinicians from multiple institutions.

Using ePAD as an entry-point for medical doctors is appealing, as the system is accessible and usable from any computer and mimics the behavior of a standard radiology image workstation closely. It provides the opportunity to easily set up and share an advanced collaborative environment for clinical and research purposes. It is therefore a convenient tool to let doctors interact and familiarize themselves with general-purpose imaging tools and features, allowing them to gain intuition from intensity or texture measurements that can at first seem very abstract, but which can carry a lot of information and meaning in a given context. We also aim to provide medical doctors with tools that can be used to create personalized machine learning models for a given study, giving full control to the doctor of what should or should not be included in a given model for classification. This is in contrast with recent developments in the fields of *big data* and *deep learning*, where effective models can be achieved, but understanding the inner decision rules of the algorithms can be difficult to grasp. Moreover, deep learning methods require large amounts of data for building the models, which significantly complicates the investigation of novel promising imaging biomarkers. Using simple models that can be efficiently trained with small amounts of data (*e.g.*, Support Vector Machines (SVM)) allows for quick evaluation of the relevance of a given quantitative imaging measure in a user-defined applicative context.

Our main goal is to provide tools allowing medical doctors and researchers to explore and understand how to use quantitative imaging biomarkers in a specific medical application context (and especially texture biomarkers, which may be conceptually difficult to comprehend but can provide important cues for diagnosis of various pathologies). We developed a suite of plugins called Quantitative Feature Explore (QFExplore) for the ePAD ecosystem and set up our own server architecture for performing complex computations and saving data for further analysis. The software is available online.[6]

---

[5] http://epad.stanford.edu, as of December 5, 2016.
[6] https://epad.stanford.edu/plugins, as of July 3, 2017.

13.2 METHODS

This section describes the architecture of the ePAD platform and the developed plugins. The implemented quantitative imaging features are described. The system set up and deployment, as well as the tools and libraries used are detailed.

13.2.1 Overview of ePAD

This section presents a general overview of the ePAD platform with a focus on the web interface and its associated features. The ePAD web interface can be accessed from any modern browser and any location. The access to the platform is secured by a login and password, ensuring that only authorized users can visualize potentially sensitive data. Once the user has logged in, the main interface is shown. Fig. 13.1 depicts an overview of the interface, which is composed of the following sections:

1. **Navigation menu** – This menu provides access to the different views available in ePAD, such as the hierarchical structure of patients, their studies, images series and annotations under the menu entry *Search* (see Fig. 13.1). Other views include a global list of all annotations (menu entry *Annotate*) and the Digital Imaging and Communications in Medicine (DICOM) image viewer (menu entry *Display*). An *Edit* menu entry provides common actions for any item in the platform (*e.g.*, *Delete*, *Copy & Paste*, *Undo*).

2. **Command toolbar** – This contextual toolbar provides different options depending on the current view. It allows uploading files in the *Search* or *Annotate* view or navigating and manipulating DICOM images within the system's image viewer, for instance.

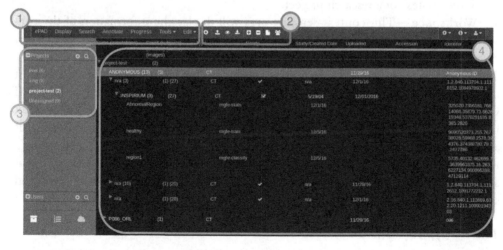

Figure 13.1 General overview of the ePAD web interface, highlighting the main areas: 1. Navigation menu, 2. Command toolbar, 3. Project list, and 4. Workspace.

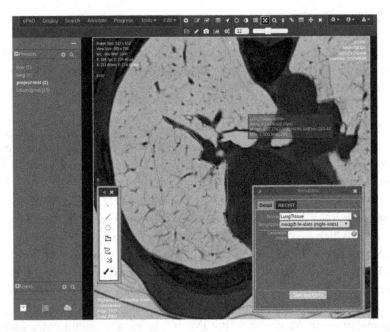

Figure 13.2 Screenshot of the ePAD image viewer, showcasing several of the features such as drawing annotations, inverting the image, zooming and panning in the image as well as selecting a template for running a plugin.

3. **Project list** – Users can assign patients, plugins and other resources to various projects, allowing to group similar cases together for example. This panel allows the user to switch from one project to another, and to set up a collaborative environment in the context of a research project.

4. **Workspace** – This context-sensitive area will display the information of the currently selected view (*e.g.*, image viewer, patient list). It constitutes the central area where most interactions with the user take place.

The *Display* view containing the image viewer and annotation tools is where most user interactions will be performed. It allows visualizing and navigating in DICOM volumes, adjusting the window width and level, zooming in on specific areas of an image, Multi-Planar Rendering (MPR) visualizations, to name a few possibilities. It is also the main entry point for saving annotations and executing plugins. When saving a region, a window will be shown allowing the user to enter information related to that annotation and selecting a template which is linked to a specific plugin in the back-end system. Fig. 13.2 shows an overview of the features provided in the viewer.

13.2.1.1 Typical workflow in ePAD

This section presents a typical workflow for a medical doctor using the ePAD platform in clinical routine:

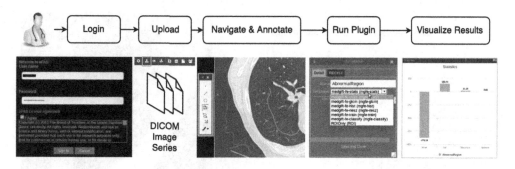

Figure 13.3 Overview of a typical image series visualization and analysis workflow in ePAD.

1. The user logs into the platform from his/her web browser.
2. DICOM image series can be uploaded to the platform by choosing local files, or entire studies can be pushed from the Picture Archiving and Communication System (PACS).
3. The image series is opened inside the viewer, where several features are available, such as:
   - volumetric navigation,
   - adjust the window level settings for specific image types (*e.g.*, lung, brain, abdomen),
   - invert pixel values, zooming, taking screen captures,
   - display metadata about the series as an image overlay,
   - annotate images with 2D Region Of Interest (ROI) using several drawing tools.
4. Once an annotation has been drawn, the user can optionally call a plugin. A selection window allows choosing the type of information to provide about the annotation and which analysis to perform on the selected ROI.
5. The chosen plugin is executed and the user is notified about the status of the plugin's state (started, running, complete, failed).
6. Depending on the plugin, various event types can be triggered in the browser once a plugin has completed. For example, a chart displaying the computed features can be shown.

Fig. 13.3 shows a schematic overview of the workflow with screenshots of the ePAD interface.

13.2.2 Quantitative imaging features

Various types of quantitative imaging features are implemented in the proposed ePAD plugins. This section details the basics of these image measurements and their relevance to practical use-cases. Two distinct feature groups are considered in the following subsections: intensity- and texture-based attributes.

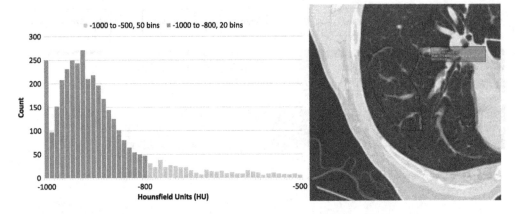

Figure 13.4 Two overlapping histograms of the same ROI with different upper bounds. The wide plot (gray) shows the distribution of values where the minimum and maximum edges of the histogram are set to −1000 and −500 Hounsfield Units (HUs), with a bin size of 10 (*i.e.*, 50 bins are generated). The narrower plot (blue) shows the distribution of values where the minimum and maximum edges are set to −1000 and −800 HU, with a bin size of 10 (*i.e.*, 20 bins are generated).

13.2.2.1 Intensity-based features

Intensity-based features are based on the statistical distribution of the pixel/voxel values (see Section 1.2.1 of Chapter 1). They are both fundamental and interpretable. In the context of the ePAD plugin development, we implemented two general-purpose intensity feature subtypes: *Statistics* and *Histogram*.

The *Statistics* are meant to summarize the intensity distribution of gray-levels within the chosen ROI. The first four statistical moments are computed to characterize the distributions: (i) mean, (ii) standard deviation (Std), (iii) kurtosis, and (iv) skewness. Exact definitions and significance are detailed in Section 12.2.2.1 of Chapter 12.

The **Histogram** shows the discretized distribution of gray-level values in the selected ROI and can be customized in two ways: setting the minimum and maximum values (*i.e.*, bounds) of the histogram, and defining the resolution of the histogram (*i.e.*, number of bins). This allows the user to focus on a given range of gray levels and to choose the level of detail of the chart. Fig. 13.4 shows an example of two histograms of the same ROI with different upper bounds in high-resolution chest Computed Tomography (CT).

13.2.2.2 Texture-based features

Texture features are encoding the spatial transitions between the pixel/voxel values, which is independent from the distribution of the latter (see Fig. 1.1 of Chapter 1). Texture measurements are therefore complementary to their intensity counterpart. They provide exclusive cues for assessing, *e.g.*, diagnosis, prognosis, treatment response in a

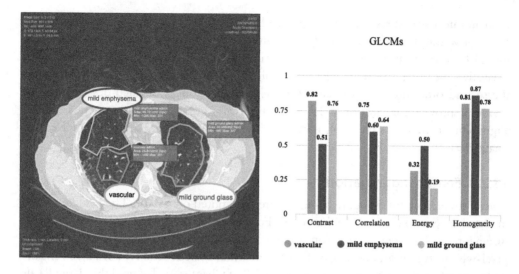

Figure 13.5 *GLCMs* features extracted from three full ROIs, showing global values for each of them.

variety of diseases and imaging modalities [10,4]. There is a large variety of methods for extracting quantitative texture measurements from medical images, and the most popular categories are extensively reviewed in Chapter 3. Two groups of texture measurements were integrated into the feature extraction plugins developed for ePAD.

The first one is based on Gray-Level Cooccurrence Matrices (GLCMs), which estimates the statistics of gray-level cooccurrences between pixel pairs separated by a user-defined distance and direction. A detailed description and review is presented in Section 3.3 of Chapter 3. Thanks to their simplicity and interpretability, GLCMs are the most widely used group of texture measurements. In particular, GLCM statistics such as *contrast* or *energy* correspond to our intuitive perception of textured image patterns [11] (see Fig. 13.5). However, a major limitation of this feature group is their lack of ability to characterize transition between pixels that are far apart. It also requires drastic reduction and merging of the gray-level values, which can potentially result in a loss of information. Important parameters are the *number of gray levels* considered for the reduction (8, 16, 32) as well as the *distance* between pixels for computing the cooccurrences (*e.g.*, 1, 2, 3). The GLCM measures are made invariant to image rotation by averaging the statistics over the four main image directions (0, 45, 90, and 135 degrees, see Section 3.3.1 of Chapter 3).

The second texture group is based on Riesz wavelets. The latter are a subtype of convolutional approaches and can quantify image derivatives of any order and at multiple scales. Moreover, the image derivatives are maximized for each position independently (*i.e.*, the filters are "aligned" along dominant local orientations), allowing characterization of the local organization of image direction with invariance to the local orientation of anatomical structures (*e.g.*, bronchovascular structures). The image derivatives have an intuitive interpretation and the Riesz features have shown to provide valuable imaging measurements in various medical applications [12–17]. For instance, the first-order derivative measures the slope of the spatial transition between the pixel values, and the second-order measures the curvature of the transition. A detailed description and review is provided in Section 3.2.2 of Chapter 3. The important parameters are the alignment of the filters (yes/no), number of scales (*e.g.*, $1, \ldots, 5$) and the order of the image derivatives (*e.g.*, 0, 1, 2, 3).

13.2.3 Feature aggregation

Several strategies exist for aggregating features computed on a ROI. They are described in detail in Section 1.3.1 of Chapter 1. For instance, the intensity or texture measurement can be averaged or counted over a user-defined ROI. However, averaging pixel-wise feature values over large ROIs containing mixed tissue type results in meaningless imaging measurements (see Section 2.3.2 of Chapter 2). To avoid this, large ROIs can be automatically split into smaller patches to reveal the diversity of tissue types (*e.g.*, tumor habitats [18]).

To that end, two aggregating strategies are implemented for each feature extraction plugin. First, pixel-wise features can be aggregated over the entire ROI, yielding a single feature vector for the entire ROI (see Fig. 13.5). The second method divides ROIs into a series of small circular patches on which pixel-wise features are aggregated. This allows characterizing local tissue subtypes of the region. Feature-specific inter and intraregional variability can be visualized and investigated with scatter plots of the first two principal components. Principal Component Analysis (PCA) is computed over the collection of patches from the multiple regions. Fig. 13.6 illustrates how this method can be used to compare multiple regions.

13.2.4 Classification

Apart from extracting visual features and displaying them to the user, two more ePAD plugins were developed for creating user-personalized machine learning models. A first plugin allows selecting a series of training ROIs and to build a SVM model. A second plugin allows the submission of one or several ROIs for classification. SVMs are able to learn maximum margin separating hyperplanes from a sparse number of training instances called support vectors. This learning rule showed to generalize well on unseen instances. It is therefore well suited to quickly explore the potential of a feature subset based on a small number of training ROIs. The Waikato Environment for Knowledge Analysis (WEKA) data mining software was used for the implementation of the SVM classifier [19]. The pairwise coupling strategy was used to output class-wise probabilities [20]. A two-phase approach was taken for the creation and use of SVM models, consisting of the following steps (see Fig. 13.7 for a detailed flowchart of the steps involved

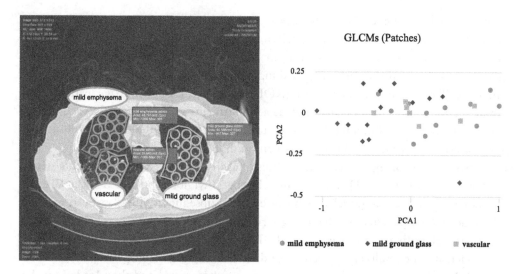

Figure 13.6 *GLCMs* features extracted from multiple ROIs split into circular patches. The feature vectors are transformed into a two-dimensional PCA space, where each patch is represented by a point in the scatter plot.

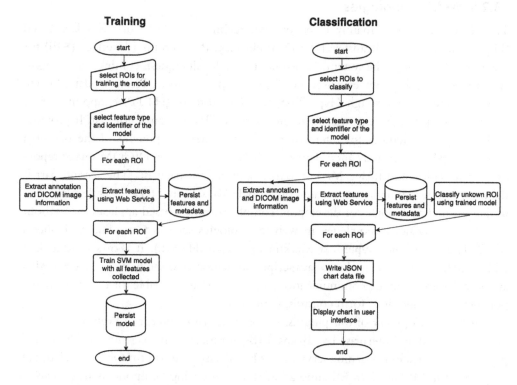

Figure 13.7 Flowchart of the training and classification plugins' processes.

in the process). The beginning of the process is similar in both cases: the user selects one or more regions to train or classify, specifying a feature group (*i.e.*, *Statistics*, *Histogram*, *GLCMs* or *Riesz*), as well as an identifier for the created model (*e.g.*, "healthyVSabnormal_Riesz"). Each ROI is then processed independently, and the initial steps include (i) extracting all the necessary information (ROI coordinates, DICOM image, patient data) from the annotation and associated metadata, and (ii) extracting features for each region (see Fig. 13.12). The computed feature vectors are persisted in the database. From that point onward, the processes diverge between training and classification:

1. During the **training** process, the features of each ROI are collected in a data structure, normalized in [0, 1], and then used by the SVM library included in the WEKA toolkit to generate a model which can be used subsequently for classification. The model is then persisted as a file using serialization.

2. For the **classification** process, each ROI is classified using the previously generated model, and the results are collected in a data structure. A JavaScript Object Notation (JSON) file containing all the classification results is saved and then loaded by the client's web browser to display the chart.

13.2.5 Web technologies

The ePAD platform is mainly based on two technologies. The first is a Java-based REpresentational State Transfer (REST) Application Programming Interface (API) for the implementation of the system's back-end: it handles all requests coming from plugins and clients using the front-end user interface. It exposes Create, Read, Update, Delete (CRUD) operations through HyperText Transfer Protocol (HTTP) endpoints for all aspects of the platform (*e.g.*, users, patients, images, files, plugins, projects). It provides tools and utility methods for interacting with the other components of the platform such as the My Structured Query Language (MySQL) database, DICOM image repository, Annotation and Image Markup (AIM) annotations database. The second part is the front-end user interface. It is based on Google Web Toolkit (GWT), a development toolkit created by Google for developing complex browser-based applications without the need of advanced knowledge in web technologies such as Cascading StyleSheets (CSS), JavaScript, and HyperText Markup Language (HTML). It allows writing Java code which is then translated into JavaScript and deployed to the browser. Several APIs are provided for handling communication, and widgets are available for the visual components (*e.g.*, panels, lists, form controls, menu bars).

For the development of our plugins, we used the Java Enterprise Edition (Java EE) platform, including Enterprise Java Beans (EJBs) for lifecycle management of the image processing and machine learning services, the Java Persistence API for Object-Relational Mapping (ORM), and Java EE annotations for configuring components (*e.g.*, *stateless* beans for web services or *singleton* beans for keeping one single shared reference to

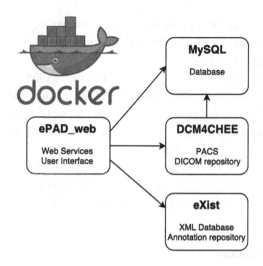

Figure 13.8 Docker container architecture of the base ePAD distribution.

heavy MATLAB objects). The Highcharts[7] library was used for generating attractive and dynamic charts. Highcharts is a powerful JavaScript library for generating a variety of graphs and charts in 2D and 3D.

13.2.6 System architecture

This section focuses on the technical architecture of ePAD and related plugin ecosystem.

13.2.6.1 Docker architecture

The ePAD platform is based on Docker, which enables the packaging and deployment of applications inside software containers [21]. It allows running isolated processes in light-weight, self-contained, and single-purpose containers. The main advantages are the ease of deploying containers on any host running any operating system with a consistent behavior, as well as the ability to avoid conflicts between different system libraries and tools that may be required for running each process. Fig. 13.8 illustrates how the system is designed. It is composed of the following docker containers:

- **ePAD_web** – This container runs the Java web server. It hosts both the user interface and the back-end of the ePAD platform. It includes a web service tier with a REST API, a GWT-based interface and utilities for communicating between the two tiers.

---

[7] http://www.highcharts.com/, as of 5 December 2016.

- **MySQL** – This container runs an instance of the MySQL database system and is responsible for storing the organizational structures of ePAD (*e.g.*, users, projects, image series, plugins, files).
- **DCM4CHEE** – This container hosts the DICOM for Che Enterprise (DCM4CHEE) service, a Java EE-based DICOM archive. All uploaded DICOM image series and segmentations are stored in this repository. A web interface allows searching for patients and images. It is also linked to the **ePAD_web** container for adding uploaded images to the repository.
- **eXist** – This container runs eXist, a native eXtensible Markup Language (XML) database used for storing image annotations in the AIM format [22].

Combining and linking the four containers mentioned above constitutes the complete ePAD platform.

13.2.6.2 Plugin architecture

There exist two main approaches for server-side ePAD plugin development:
- **Local plugins** – These plugins are executed directly on the Java application server running the ePAD back-end. The main advantage of these plugins is data locality, but drawbacks include the dependency on the server's runtime environment (installed libraries such as the MATLAB Compiler Runtime), as well as the risk of altering ePAD's responsiveness when computationally heavy plugins are called.
- **Remote plugins** – These plugins are executed on another server, and communication with ePAD is performed through HTTP calls. Advantages include the separation of concerns, running computations on a more powerful distant computer and being independent of the ePAD server's runtime environment. Drawbacks include the necessity for transferring data across the network, which can increase the latency of the plugin.

The second approach was chosen, as the benefits of running computations on an independent server with more power and being able to reuse the developed web services outside the scope of ePAD were very important for us. Each plugin consists of three parts: a Java plugin class that is integrated into ePAD, a web service class hosted on a remote application server, and a MATLAB routine that contains the image processing code. The MATLAB code is compiled into a Java library and added to the remote application server's dependencies. Using this technique, only the royalty-free MATLAB Runtime Compiler, which requires neither to purchase nor to use MATLAB licenses for execution, is needed on the application server. Fig. 13.9 shows the architecture and interaction between the ePAD platform and the developed Java EE web service layer:

1. A plugin execution is started from within the ePAD Docker container, and the corresponding plugin class is called within the same container.
2. The container communicates via standard HTTP requests and responses with a remote Java EE server.

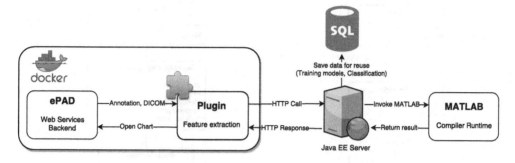

Figure 13.9 Overview of the ePAD plugin architecture.

3. The Java EE server runs the plugin code (*e.g.*, a MATLAB routine) and performs additional tasks such as persisting data in the database.
4. Plugin results are returned to the ePAD Docker container, where the user will be notified of the status of the plugin's execution (success or failure).

13.2.6.3 Java EE for plugin development

The Java EE application server is used for multiple purposes. It serves as the main resource for heavy computations required by certain plugins. The physical machine hosting the server has the following performance characteristics:

- 2 x Intel(R) Xeon(R) Central Processing Unit (CPU) E7-4820 @ 2.00 GigaHertz (GHz) processors (64 cores in total),
- 128 GigaByte (GB) main memory,
- Gigabit Ethernet network connection,
- 550 GB Solid State Drive storage.

The Java EE server architecture allows persisting all processed images, annotations, patients, along with computed features and generated results in a relational database for further analysis, reuse for training machine learning algorithms, extraction of statistics, etc. Fig. 13.10 shows a model of the various business objects involved in the persistence of image annotations and all linked metadata. It consists of the following main classes:

- Person – abstract class defining the basic properties of a person, *i.e.*, an identifier (for the database) and a name.
- User & Patient – implementations of Person with additional fields (user ID and role for User, and personal information such as the birth date and sex for Patient).
- ImageReferenceEntity – description and content of a DICOM image.
- ImageAnnotation – properties and coordinates of ROIs.
- QuantitativeImagingFeatures – representation of the computed visual features extracted by a quantitative image analysis plugin.

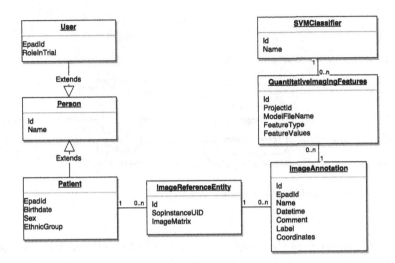

Figure 13.10 Business objects of the Java EE plugin platform.

- <u>SVMClassifier</u> – instances of SVM models built from one or more computed feature vectors stored in QuantitativeImagingFeatures.

The persistence of business objects is fully managed through the Java Persistence API (JPA). Java EE contains several security mechanisms, allowing to ensure data privacy and security in all layers of the system (applicative, network transmission, storage). All components of the application (web services, business objects, image processing classes) are easily managed through the use of EJBs. Their execution environment (*i.e.*, EJB containers) handles the role, scope and lifecycle of every component. Web Services and business classes used for short-lived operations are defined as *stateless* session beans. JPA entities are used to annotate business object classes for persistence. A *singleton* bean is used for managing the MATLAB factory, which initializes a unique reference to the heavy objects used for executing the MATLAB routines. Its lifecycle is as long as the application. The persistence bean contains a persistence context, which allows establishing a connection to the MySQL server and executing queries through the *EntityManager* interface. Fig. 13.11 shows the various EJBs and their links to other managed components of the system.

13.3 USE-CASES

This section describes the various plugins that were developed for the ePAD platform in the context of general-purpose use-cases. These plugins are designed to accelerate the exploration of quantitative imaging parameters' value in diagnosis, prognosis or treatment planning tasks. The use-cases are divided into three major subsections.

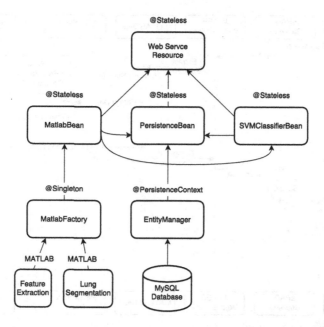

Figure 13.11 Java EE EJBs. The stateless Web Service Resource (top) is used by the ePAD plugins using a web service client class. It holds references to the three main EJBs (MATLAB Bean, Persistence Bean and SVM Classifier Bean), which are obtained through dependency injection. The MATLAB Bean handles the call to the MATLAB routines using a singleton factory object. The Persistence Bean handles saving and reading the business objects described in Fig. 13.10 to the MySQL database. The SVM ClassifierBean handles the creation and reuse of SVM models using WEKA.

Section 13.3.1 describes various examples of (i) the extraction of intensity and texture measurements aggregated from whole user-defined ROIs, and (ii) the display of the latter in bar plots. Section 13.3.2 details the use of patch-based analysis to reveal the diversity of tissue components contained in user-defined ROIs with scatter plots. Section 13.3.3 illustrates the construction of personalized statistical models from a collection of user-defined ROIs and associated labels. The obtained models can be used to classify new ROIs with unknown labels.

13.3.1 Analysis of whole ROIs

The goal of this plugin is to allow medical doctors to annotate a specific region in a medical image and evaluate the value of associated quantitative imaging features. Fig. 13.12 details the process of executing the plugin and the interactions between the various layers of the application (ePAD front-end and back-end, Java EE server and EJBs, as well as the MySQL database). The process for one ROI is:

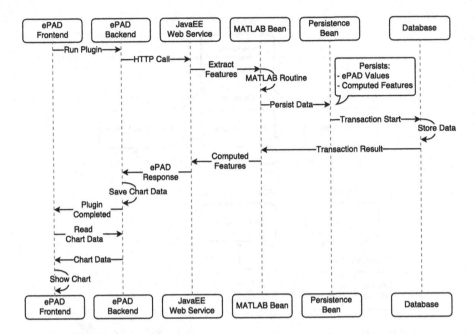

Figure 13.12 Sequence diagram showing the execution of the feature extraction plugins.

1. The user creates an image annotation and selects a feature extraction plugin to run based on various feature types. He/she clicks on a button to initiate the plugin execution.
2. The ePAD back-end receives this call, containing all the required information about the image, selected region, patient, etc. The input is transformed into an object sent (via an HTTP request) to the web service layer of the remote Java EE server.
3. The web service receives the query, parses the input parameters and calls the MATLAB Bean to execute the corresponding feature extraction routine.
4. Once feature extraction is complete, the received input data and the generated output results are sent to the Persistence Bean.
5. The Persistence Bean begins a transaction for saving all the supplied objects, connecting to the MySQL database and executing the appropriate queries to create or update rows. If any step of this procedure fails, the entire transaction is rolled back to ensure data integrity.
6. Once the objects have been persisted, the results are sent back to the web service layer, which will respond to the ePAD back-end with the computed features.
7. The ePAD back-end saves the computed features along with chart formatting parameters in a format that the Highcharts library can process, as a JSON file that is saved within the file hierarchy of ePAD. An event is then generated to inform the user that the plugin's execution has completed.

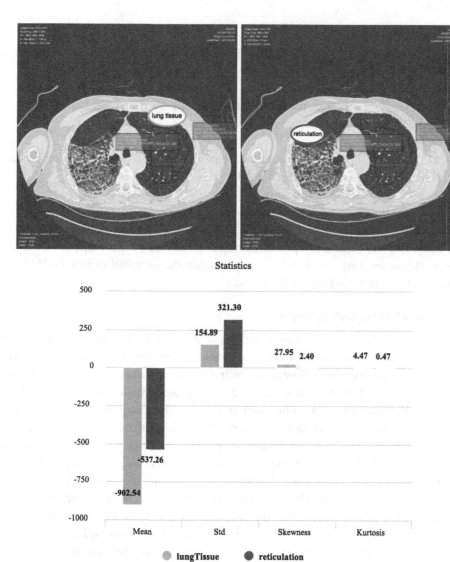

Figure 13.13 Basic statistical feature extraction on two ROIs.

8. The front-end receives the event and runs a query to receive the newly created JSON file from the back-end and renders the chart with the computed values and given configuration (see Fig. 13.13).

Each feature type may contain one or more parameters that can be specified at runtime. The parameters of the plugin vary according to the selected feature category. Below is a list describing the currently available parameters:

- All feature types
  - Patches – Boolean parameter to determine whether the ROI is analyzed as a whole or split up into smaller patches (see Section 13.3.2).
- *Riesz*
  - RieszOrder – order of the Riesz transform (see Section 13.2.2.2),
  - NumScales – number of wavelet scales.
- *Histogram*
  - HistMin, HistMax – minimum and maximum bounds for the intensity values of the histogram (see Section 13.2.2.1),
  - HistNumBins – number of bins.
- *GLCMs*, *Statistics* – no additional parameters (see Sections 13.2.2.2 and 13.2.2.1).

By testing different feature types and parameter values, the user can gain intuition about the meaning and discriminatory power of the raw features in the context of a given pathology. Below are two concrete use-cases to show the potential of using ePAD and the developed plugin for whole ROI analysis.

13.3.1.1 Liver lesion quantification

In this use-case, the goal is to compare regions of the liver in terms of texture features (*Riesz* and *GLCMs* were used for this use-case). A first run uses the Riesz feature extraction plugin, using the following default parameter values (RieszOrder = 0, NumScales = 5, Patches = false). Fig. 13.14 (bottom left) shows the results of the plugin execution on four ROIs. The chart shows that the energies of scales 2, 3, and 5 (*i.e.*, *feat2*, *feat3*, *feat5*, note that the *y*-axis uses a logarithmic scale) is higher for the lesioned regions than those of healthy tissue (the scales vary from fine to coarse, see Section 3.2.2 of Chapter 3). Scale 1 is mostly characterizing small scale CT noise. Scale 4 is more represented in normal tissue when compared to lesions. A second run was performed on the same ROIs, using *GLCM* features with the following parameter value: Patches = false. The results of the plugin execution are also shown in Fig. 13.14 (bottom right). In this case, there is no clear trend separating healthy and lesioned tissue. The values of the computed features vary between regions of the same type, and have similar values as regions of the other type. We can conclude that *GLCMs* may not be optimal for specifically characterizing liver tissue in CT.

13.3.1.2 Liver metastasis quantification

In this use-case, the objective is to quantify the liver tissue density in three ROIs in CT. Two hypodense ROIs contain metastasized tissue and the third one contains a majority of normal liver parenchyma. The intensity-based features (*Statistics* and *Histogram*) were used for this use-case. A first run uses the basic *Statistics* feature extraction plugin, using the following parameter value: Patches = false. In Fig. 13.15 (top right chart) the mean shows a clear distinction between healthy and metastasized liver tissue. Using

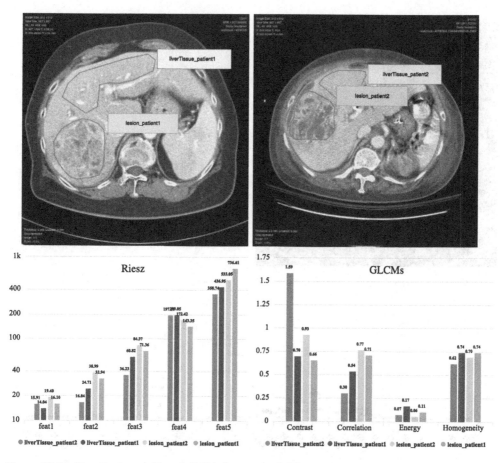

Figure 13.14 *Riesz* (bottom left) and *GLCMs* (bottom right) feature extraction on four ROIs located in the liver (top).

simple statistical analysis therefore seems appropriate in this specific situation, as it is effective and easily interpretable. A second run was performed on the same ROIs using the *Histogram* feature extraction with: HistMin = 0, HistMax = 200, HistNumBins = 30, Patches = false. Results are shown in Fig. 13.15 (bottom chart). Note that the histograms are not normalized across ROIs. The hypodensity of metastases is well characterized by the intensity histograms, which enables fine-grained visualization of the tissue contained in each region. It is valuable to assess similarities in terms of density between various tissue types.

13.3.2 Patch-based ROI analysis

The plugin described in 13.3.1 can also be used in *patch* mode, where the selected ROI is split into several smaller patches. It can be used to reveal the diversity of one or several

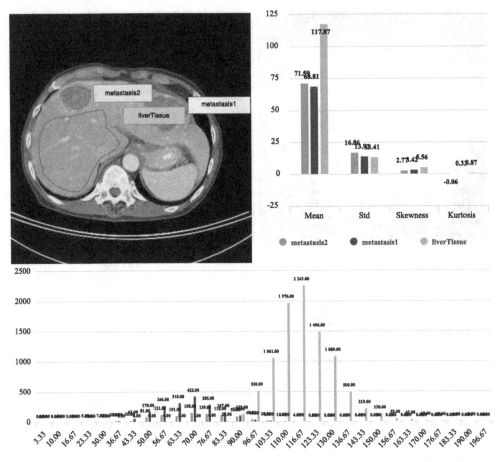

Figure 13.15 *Statistics* (top right) and *Histogram* (bottom, unnormalized) feature extraction on three ROIs located in the liver (top left).

regions. The user can visualize the spread of patches when represented in the first two PCA dimensions of a chosen feature space. Each selected ROI is split into several smaller patches (patch radius $r = 28$ pixels). The features are computed for each patch and the results aggregated in a data structure. When all features have been computed, the PCA decomposition is computed and the first two components explaining most variance are used as the dimensions of a scatter plot where each point represents one patch instance. Fig. 13.16 shows the result of extracting *GLCMs* features from two large ROIs divided into patches. The unhealthy lung parenchyma (gray) appears to have more variation along the first PCA component (PCA1) when compared to the healthy lung tissue (blue). Similarly, the healthy lung tissue has more variation along the PCA2 component than the unhealthy lung parenchyma tissue.

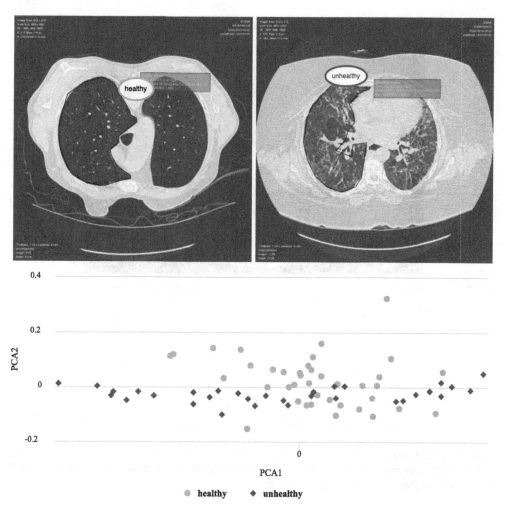

Figure 13.16 Scatter plot (bottom) of a patch-based *GLCMs* feature extraction on two ROIs (top). The first two PCA dimensions are represented.

13.3.2.1 *Patch-based liver lesion quantification*

In this use-case, the goal is similar to Section 13.3.1.1, but the global features are substituted for locally computed features based on patches. The basic *Statistics* features were used for this use-case, with Patches = true (see Fig. 13.17). It can be observed that the patches in the healthy ROI are occupying a well-defined region of the PCA space, whereas the ROI containing a heterogeneous lesion is spread out (some patches also contain healthy tissue, which also clearly visible with the naked eye).

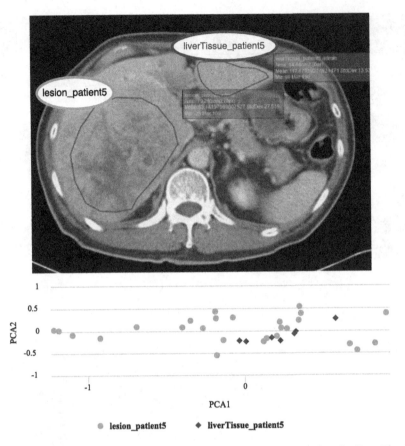

Figure 13.17 Patch-based *Statistics* feature extraction on two ROIs located in the liver. The first two PCA dimensions are represented. The diversity in terms of tissue density of the liver lesion is revealed by scattered patches instances in the PCA space.

13.3.3 Training a statistical modal and classifying ROIs

Visualizing features and comparing them can be useful, but manual visual inspection and analysis is not necessarily the easiest or most efficient way of differentiating ROIs. A more automated and reproducible approach is to enable the training of user-personalized statistical models based on annotations with known labels (*e.g.*, normal versus abnormal, good versus poor treatment response). With a sufficient number of training examples (this depends on the complexity of the recognition task), SVMs can be further used to classify ROIs of unknown classes with probability output. The goal is to allow the user to fully control which ROI instances and quantitative imaging features to use for the generation of the predictive model. This makes the classification process easier to understand and trust (the decisions taken by the machine learning algorithm

Figure 13.18 ROI selection for statistical model training.

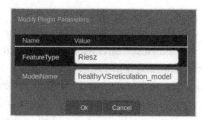

Figure 13.19 Plugin parameter definition for statistical model training. Both the visual feature type and a user-defined model name are required.

are backed up with concrete examples, rather than abstract concepts gathered from huge datasets with limited quality control as it is the case for big data and deep learning).

The first step of this process consists in selecting several ROIs of known class for training a new model. The labels of each class present in the generated model are currently based on the name given to each ROI (*e.g.*, "normal," "abnormal"). Fig. 13.18 shows the ePAD interface used for selecting the training ROIs. The next step is to determine the feature type for the creation of the model, as well as a user-defined model name for the created model (allowing a user to create and share multiple models for different purposes). This is done through the input of two plugin parameters by the user in the ePAD web interface, as shown in Fig. 13.19. The features are then extracted from each region, and the resulting instances are normalized and used to train the model. A notification is provided to the user when the model training is complete, which can subsequently be used for classification of new ROIs with unknown label.

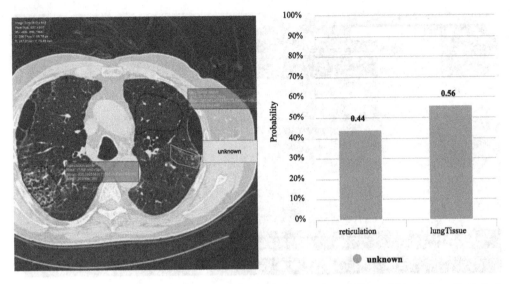

Figure 13.20 Classification of a ROI with unknown label (two-class SVM model based on Statistical features).

For classifying a ROI, the user selects it in the *Annotations* list and runs the classification plugin based on the classifier specified by the feature type and the model name previously used for the training plugin. The model is loaded by the server and the classification is performed using the normalized features extracted from the unseen ROI. The result is sent back to the client's browser, where a bar plot shows the probability (in percent) for each class (see Fig. 13.20). Using these two plugins (training and classification), the user can easily create and share statistical models, test and refine them by providing new examples increasing the size of the ground truth, improving the classification accuracy of the generated model over time.

13.3.3.1 Lung tissue classification

In this use-case, the goal is to train a model that can distinguish between healthy lung parenchyma and reticulation. The model is trained using 12 ROIs (6 normal, 6 reticulation) taken from 2 different patients. Fig. 13.21 shows 4 of the 12 training ROIs. Once the model is trained, the classification is tested with 3 ROIs from an unseen patient: 1 normal, 1 reticulation, and 1 unknown. Distinct models based on the same collection of training ROIs but based on different feature types (*i.e.*, *Riesz*, *GLCMs*, *Statistics*, and *Histogram*) are compared in Fig. 13.22. In each case, the `Patches` parameter was set to false. For all feature types but *Histogram*, the two control ROIs were correctly classified. *Histogram* feature type wrongly classified healthy tissue as reticulation by a small margin. The latter can be explained by the inadequate choices of the *Histogram* parameters, the latter characterizing the intensity between $[0, 100]$ HU whereas both healthy lung and

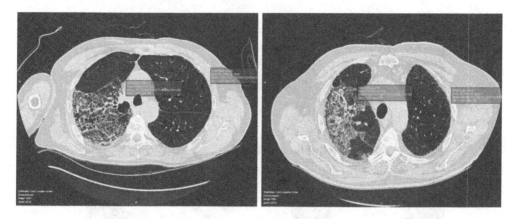

Figure 13.21 Four among the twelve ROIs of known class (normal versus reticulation) from two different patients used for training.

reticulation mainly have intensity values below 0 HU. The largest classification confidence for the control ROIs are obtained by the *GLCM* features, with a probability of 89% for the reticulated tissue and 92% for the healthy lung tissue. The unknown ROI is classified as normal lung tissue when using either *Riesz*, *GLCMs* or *Statistics* features, with a probability of 53%, 64%, and 56%, respectively.

13.3.4 Helper tools for segmentation

In addition to the specific use-cases mentioned previously, another plugin was developed to investigate and showcase the numerous features provided by the ePAD platform. It allows automatically segmenting lungs in a DICOM volume, using the method described in [23]. The detected lung regions are highlighted using colored overlay in the image viewer of ePAD (see Fig. 13.23).

This type of plugin along with the image overlay functionality showcase the possibility to automate the annotation process, yielding reproducible ROIs that can be further used by the quantitative image analysis plugins. The execution process of this plugin differs significantly from the previously mentioned ones. In this case, the analysis is performed not only on a single DICOM image, but on the entire volume. In addition, it does not require a user-defined ROI for initialization and is entirely automated. ePAD provides utilities for downloading all images of the current DICOM series and storing segmentation information in the DICOM archive using the standardized DICOM Segmentation Object (DSO) format. It also handles displaying this information in the front-end interface. Similar plugins may be used in combination with quantitative image analysis methods (*e.g.*, intensity, shape or texture) to highlight areas in an image representing a distinct tissue type using different colors, enabling the quantification and

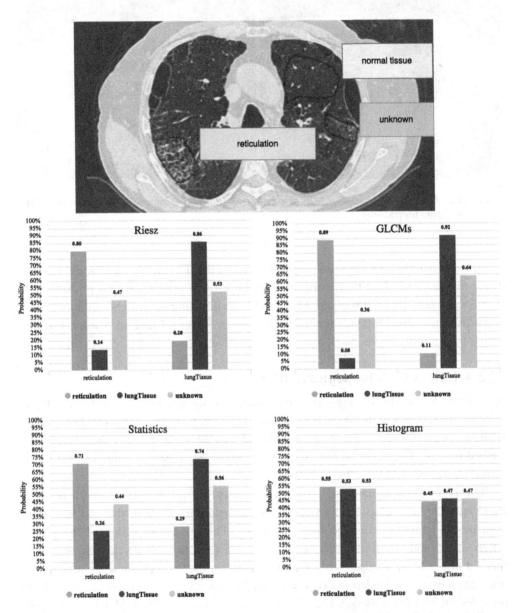

Figure 13.22 Lung tissue classification using *Riesz* features (middle left, `RieszOrder = 0`, `Num-Scales = 5`), *GLCMs* features (middle right), *Statistics* features (bottom left) and *Histogram* features (bottom right, `HistMin = 0`, `HistMax = 100`, `HistNumBins = 20`, the bin values are used as features).

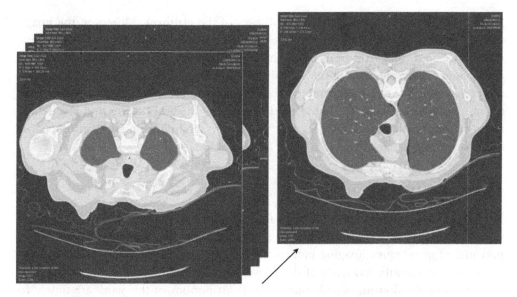

Figure 13.23 Two images of automatically segmented lungs in a DICOM volume. The detected lung regions are highlighted with green image overlays in the front-end of ePAD.

simplified visual inspection of images using visual cues of pathological findings within an affected area.

13.4 DISCUSSION

This section is dedicated to the analysis of the obtained results and experience working with the ePAD platform. The ePAD environment is a fully-featured imaging, annotation, and analysis platform that is already in use by several medical doctors and researchers. This provides the exciting opportunity to finally include end-users in the development cycle of quantitative image analysis in radiology, as all developed plugins and tools are within immediate reach of the clinical context. The methods developed in this work represent efforts to reduce the black-box effect of current quantitative imaging biomarkers. The plugins allow quickly evaluating the significance and discriminatory power of intensity or texture imaging biomarkers in an applicative context that the end user fully controls. The possibility to offload the main computational aspects of the plugins to a remote server designed to our own specifications added a layer of complexity (*e.g.*, managing the transfer of data between the two servers), but was ultimately a judicious decision, because it allowed freedom for the design of the plugins and the associated frameworks (*e.g.*, use of Java EE for component lifecycle management and persistence, use of a version of MATLAB that differs from the version for which ePAD is configured). The use of standardized protocols and formats such as HTTP and

JSON allowed interoperability between all tiers of the platform. Moreover, the image analysis services deployed on the remote server can be accessed by any other client with or without the ePAD front-end, which allows eased reusability of the developed components. The relevance of developed plugins has been demonstrated with several use-cases. Using the developed tools, we can bring quantitative imaging biomarkers and texture analysis closer to the clinical world and allow medical doctors and researchers to experiment with various feature types, parameters, and perform comparisons across image series and patients with ease.

13.5 CONCLUSIONS

We developed a suite of web-based tools for medical doctors and researchers called the Quantitative Feature Explore (QFExplore) and allowing the rapid exploration of the potential of quantitative imaging biomarkers in a medical applicative context. Recent and extensive scientific evidence of the relevance of such biomarkers was provided by researchers in the domain of radiomics. The contributions of this work are timely to ease the adoption of quantitative imaging biomarkers in the medical domain. Focusing on a small set of use-cases and features, the developed tools enable the creation of personalized methods and models that can be adapted to various specific medical contexts. They are a proof-of-concept showing that it is possible to bring complex texture analysis methods such as Riesz or GLCMs to the user in an approachable manner, using experimentation as the key to understanding. Extracting quantitative features from several ROIs, comparing between images and patients can provide intuition as to whether a given feature type can be helpful in a given clinical setting, and can be used to identify optimal features for a medical image analysis task. We recognize limitations of this work, including the current impossibility to deal with three-dimensional ROIs and texture features, which are presented in Chapters 11 and 12. In the future, we aim to enhance the current plugins and create new plugins, performing more complex tasks while remaining user-friendly. The possibility of segmenting an image and highlighting different texture components is very promising, for example. We also plan to integrate new types of features such as circular harmonic wavelets [24] and to improve the training/classification processes with the addition of several machine learning methods and inner cross-validations for automatically choosing hyperparameters. There is still much work to be done, and we want to develop tools that will make the frequent use of quantitative imaging biomarkers analysis in hospitals a reality. The software is available online.

ACKNOWLEDGMENTS

This work was supported by the Swiss National Science Foundation (under grant PZ00P2_154891), the Center for Biomedical Imaging (CIBM), and grants from the National Cancer Institute, National Institutes of Health, U01CA142555, 1U01CA190214, and 1U01CA187947.

REFERENCES

[1] V. Kumar, Y. Gu, S. Basu, A. Berglund, S.A. Eschrich, M.B. Schabath, K. Forster, H.J.W.L. Aerts, A. Dekker, D. Fenstermacher, D.B. Goldgof, L.O. Hall, P. Lambin, Y. Balagurunathan, R.A. Gatenby, R.J. Gillies, Radiomics: the process and the challenges, Magn. Reson. Imaging 30 (9) (2012) 1234–1248.

[2] M. Gerlinger, A.J. Rowan, S. Horswell, J. Larkin, D. Endesfelder, E. Gronroos, P. Martinez, N. Matthews, A. Stewart, P. Tarpey, I. Varela, B. Phillimore, S. Begum, N.Q. McDonald, A. Butler, D. Jones, K. Raine, C. Latimer, C.R. Santos, M. Nohadani, A.C. Eklund, B. Spencer-Dene, G. Clark, L. Pickering, G. Stamp, M. Gore, Z. Szallasi, J. Downward, P.A. Futreal, C. Swanton, Intratumor heterogeneity and branched evolution revealed by multiregion sequencing, N. Engl. J. Med. 366 (10) (2012) 883–892.

[3] Y.-H.D. Fang, C.-Y. Lin, M.-J. Shih, H.-M. Wang, T.-Y. Ho, C.-T. Liao, T.-C. Yen, Development and evaluation of an open-source software package CGITA for quantifying tumor heterogeneity with molecular images, BioMed Res. Int. 2014 (March 2014).

[4] A. Depeursinge, A. Foncubierta-Rodríguez, D. Van De Ville, H. Müller, Three-dimensional solid texture analysis and retrieval in biomedical imaging: review and opportunities, Med. Image Anal. 18 (1) (2014) 176–196.

[5] A. Depeursinge, A. Foncubierta-Rodríguez, D. Van De Ville, H. Müller, Rotation-covariant texture learning using steerable Riesz wavelets, IEEE Trans. Image Process. 23 (2) (February 2014) 898–908.

[6] O. Ronneberger, P. Fischer, T. Brox, U-net: convolutional networks for biomedical image segmentation, in: N. Navab, J. Hornegger, W.M. Wells, A.F. Frangi (Eds.), Medical Image Computing and Computer-Assisted Intervention — MICCAI 2015, in: Lecture Notes in Computer Science, vol. 9351, Springer International Publishing, 2015, pp. 234–241.

[7] D.L. Rubin, D. Willrett, M.J. O'Connor, C. Hage, C. Kurtz, D.A. Moreira, Automated tracking of quantitative assessments of tumor burden in clinical trials, Transl. Oncol. 7 (1) (January 2014) 23–35.

[8] A. Snyder, D. Willrett, D.A. Moreira, K.J.A. Serique, P. Mongkolwat, V. Semeshko, D.L. Rubin, ePAD: a cross-platform semantic image annotation tool for quantitative imaging, ResearchGate 1 (January 2011) 1.

[9] D.A. Moreira, C. Hage, E.F. Luque, D. Willrett, D.L. Rubin, 3D markup of radiological images in ePAD, a web-based image annotation tool, in: Proceedings of the 2015 IEEE 28th International Symposium on Computer-Based Medical Systems, CBMS '15, IEEE Computer Society, Washington, DC, USA, 2015, pp. 97–102.

[10] R.M. Summers, Texture analysis in radiology: does the emperor have no clothes?, Abdom. Radiol. 42 (2) (2017) 342–345, http://dx.doi.org/10.1007/s00261-016-0950-1.

[11] R.M. Haralick, Statistical and structural approaches to texture, Proc. IEEE 67 (5) (May 1979) 786–804.

[12] P. Cirujeda, Y. Dicente Cid, H. Müller, D. Rubin, T.A. Aguilera, B.W. Loo Jr., M. Diehn, X. Binefa, A. Depeursinge, A 3-D Riesz-covariance texture model for prediction of nodule recurrence in lung CT, IEEE Trans. Med. Imaging 35 (12) (2016) 2620–2630, http://dx.doi.org/10.1109/TMI.2016.2591921.

[13] A. Depeursinge, C. Kurtz, C.F. Beaulieu, S. Napel, D.L. Rubin, Predicting visual semantic descriptive terms from radiological image data: preliminary results with liver lesions in CT, IEEE Trans. Med. Imaging 33 (8) (August 2014) 1–8.

[14] A. Depeursinge, M. Yanagawa, A.N. Leung, D.L. Rubin, Predicting adenocarcinoma recurrence using computational texture models of nodule components in lung CT, Med. Phys. 42 (2015) 2054–2063.

[15] J. Barker, A. Hoogi, A. Depeursinge, D.L. Rubin, Automated classification of brain tumor type in whole-slide digital pathology images using local representative tiles, Med. Image Anal. 30 (2016) 60–71.

[16] A. Depeursinge, A. Foncubierta-Rodríguez, D. Van De Ville, H. Müller, Multiscale lung texture signature learning using the Riesz transform, in: Medical Image Computing and Computer–Assisted

Intervention MICCAI 2012, in: Lecture Notes in Computer Science, vol. 7512, Springer, Berlin/Heidelberg, October 2012, pp. 517–524.

[17] S. Otálora, A. Cruz Roa, J. Arevalo, M. Atzori, A. Madabhushi, A. Judkins, F. González, H. Müller, A. Depeursinge, Combining unsupervised feature learning and Riesz wavelets for histopathology image representation: application to identifying anaplastic medulloblastoma, in: N. Navab, J. Hornegger, W.M. Wells, A. Frangi (Eds.), Medical Image Computing and Computer-Assisted Intervention — MICCAI 2015, in: Lecture Notes in Computer Science, vol. 9349, Springer International Publishing, October 2015, pp. 581–588.

[18] R.A. Gatenby, O. Grove, R.J. Gillies, Quantitative imaging in cancer evolution and ecology, Radiology 269 (1) (2013) 8–14.

[19] M. Hall, E. Frank, G. Holmes, B. Pfahringer, P. Reutemann, I.H. Witten, The WEKA data mining software: an update, ACM SIGKDD Explor. Newsl. 11 (1) (2009) 10–18.

[20] T.-F. Wu, C.-J. Lin, R.C. Weng, Probability estimates for multi-class classification by pairwise coupling, J. Mach. Learn. Res. 5 (August 2004) 975–1005.

[21] D. Merkel, Docker: lightweight Linux containers for consistent development and deployment, Linux J. 2014 (239) (March 2014).

[22] P. Mongkolwat, D.S. Channin, V. Kleper, D.L. Rubin, Informatics in radiology: an open-source and open-access cancer biomedical informatics grid annotation and image markup template builder, Radiographics 32 (4) (2012) 1223–1232.

[23] Y. Dicente Cid, O.A. Jiménez-del Toro, A. Depeursinge, H. Müller, Efficient and fully automatic segmentation of the lungs in CT volumes, in: Proceedings of the VISCERAL Challenge at ISBI, in: CEUR Workshop Proceedings, vol. 1390, April 2015.

[24] A. Depeursinge, Z. Püspöki, J.-P. Ward, M. Unser, Steerable wavelet machines (SWM): learning moving frames for texture classification, IEEE Trans. Image Process. 26 (4) (2017) 1626–1636.

INDEX

A

Absolute values, 15, 59–62, 68–70, 73–75, 201, 359

Aggregation function design, 3, 30–32, 48

Aggregation functions, 12, 14, 16, 17, 22, 23, 30, 36, 41, 49, 60, 62, 71, 80, 82, 89, 90, 163

Analysis of histopathology images, 288, 289, 299

ANNs, *see* Artificial neural network

Application programming interface, 334, 390, 391

Artificial neural network, 97, 98, 103, 254

B

Band-pass, 59, 60, 62, 64, 66, 69, 74, 78, 81, 83, 84, 86, 88

Base classifiers, 169, 171, 172, 177–179, 183, 185

Biomedical image analysis, 1, 20, 23, 147, 281

Biomedical image processing, 322

Biomedical image texture analysis, 248

Biomedical images, 1, 12, 19, 20, 23, 38, 46, 96, 115, 122, 248, 334

Biomedical imaging, 2, 23, 50, 55, 77, 109, 126, 215, 239, 408

Biomedical texture information, 1–3, 29

Biomedical textures, 1, 2, 11, 16, 20, 23, 40, 41, 55, 89, 362

Breast cancer histological images, 296, 299

Breast lesions, 228, 255, 256, 259

C

CBIR, *see* Content-based image retrieval

Channels, 74–77, 102, 104, 306, 307, 324, 325, 328, 343, 344

Class label, 172, 175, 176, 178, 235, 251, 338

Classes, 46, 98, 104, 110, 111, 113, 115, 116, 121–123, 163, 164, 169–172, 174, 175, 177–186, 208, 209, 251, 331, 332, 403, 404
  correct, 171, 174, 175
  reference, 336, 338, 340, 341
  wrong, 171, 174, 175, 185

Classification, 75, 76, 96, 106–108, 119, 120, 134, 169–172, 185, 186, 253, 254, 260, 261, 264, 265, 288, 292–294, 302–306, 335–337, 403, 404
  biomedical texture, 73, 95

Classification accuracy, 124, 265, 288, 340, 341, 404

Classification recall, 182–184

Classifiers, 164, 168–170, 172, 184, 185, 212, 253, 262, 265, 301, 319, 330–332, 334, 337, 341, 342, 404

CNNs, *see* Convolutional neural networks

Color image, 321, 324, 330

Computed tomography, 3, 8, 131, 157, 194, 224, 232, 248, 255, 321, 350, 352–355, 367, 373, 381

Computer-aided detection, 247, 248, 252, 253

Computer-aided diagnosis, 247, 248, 253, 255, 256, 261, 262, 265, 269, 273

Computer-aided diagnosis framework, 254, 257, 258

Computer-aided diagnosis system, 253, 254, 256, 259

Content, 224, 225, 233, 393

Content-based image retrieval, 167, 265, 267, 293

Convolution layers, last, 106, 108, 109, 111, 114, 115

Convolutional layers, 74, 75, 250

Convolutional neural networks, 4, 12, 55, 56, 71, 78, 79, 90, 95, 96, 101–109, 115, 116, 167, 202, 262–264, 271–273, 292, 304, 305

Coverage of image directions, 60, 62, 64, 66, 69, 74, 78, 81, 83, 84, 86, 88

Coverage of image scales, 60, 62, 64, 66, 69, 74, 78, 81, 83, 84, 86, 88

CT, *see* Computed tomography

CT images, 155, 248, 252, 259, 260, 262, 264, 268, 350, 352

CT scans, 232, 238, 249, 251, 255, 259–261, 266, 352

D

Data vectors, 333, 338, 340, 341

Datasets, 105, 111, 113–115, 118, 120, 121, 123, 125, 177, 179, 180, 182, 184, 236, 263, 294–297, 307

DCNNs, *see* Deep convolutional neural network

Deep convolutional neural network, 55, 71, 250, 251, 262–264, 273

411

Printed in the United States
by Baker & Taylor Publisher Services